Games, Norms and Reasons

SYNTHESE LIBRARY

STUDIES IN EPISTEMOLOGY, LOGIC, METHODOLOGY, AND PHILOSOPHY OF SCIENCE

VOLUME 353

For further volumes:
http://www.springer.com/series/6607

Games, Norms and Reasons

Logic at the Crossroads

Edited by

Johan van Benthem
ILLC, University of Amsterdam, The Netherlands and Stanford University, USA

Amitabha Gupta
Indian Institute of Technology Bombay, India

and

Eric Pacuit
Tilburg University, Tilburg Institute for Logic and Philosophy of Science, The Netherlands

Springer

Editors
Prof. Johan van Benthem
University of Amsterdam
Institute for Logic
Language and Computation (ILLC)
Science Park, P.O. Box 94242
1090 GE Amsterdam
The Netherlands
johan@science.uva.nl

Prof. Amitabha Gupta
Adi Shankaracharya Marg
503 Whispering Woods
Powai Vihar, Bldg. 3
700076 Powai, Mumbai
India
agcg503@gmail.com

Asst. Prof. Eric Pacuit
Tilburg University
Tilburg Institute for Logic
and Philosophy of Science
Warandelaan 2
5037 AB Tilburg
The Netherlands
e.j.pacuit@uvt.nl

ISBN 978-94-007-3577-4 ISBN 978-94-007-0714-6 (eBook)
Set ISBN 978-94-007-0920-1
DOI 10.1007/978-94-007-0714-6
Springer Dordrecht Heidelberg London New York

Printed on acid-free paper

Springer is part of Springer Science+Business Media (www.springer.com)

Preface

This book continues a series called "Logic at the Crossroads" of which the first volume appeared in 2007.[1] The series title reflects a view of our discipline as a vantage point for seeing new developments across academic fields, as well as a catalyst in making them happen. In particular, the editors feel that the deep insights from the classical phase of mathematical logic can form a harmonious mixture with a new, more ambitious research agenda of understanding and enhancing human reasoning and intelligent interaction in their full extent, whence our title "Games, Norms and Reasons".

This broad view of logic animated the Second Indian Conference on "Logic and its Relationship with other Disciplines" held at IIT Bombay, Mumbai 2007, the origin of most of the papers in this volume, while a few additional ones have been solicited by the editors to round out coverage. But the Mumbai conference also added a very concrete focus, by honouring one person who embodies the general trend described, namely Professor *Rohit Jivanlal Parikh*. Rohit stands at the crossroads of many academic disciplines, and of two major human cultures, and why and how this is so, is admirably explained in the Congratulatory Citation that was read at a special event at IIT Bombay celebrating Rohit's 70th birthday. This document has been reprinted here, after the following brief editorial description of contents.

The papers in this volume do two things. First, they paint a lively portrait of modern logic as it is expanding today with many new themes and concerns. But at the same time, they include a gallery of friends and colleagues of Rohit's who eagerly responded to our call for honouring him. The chapters to follow highlight major dimensions of his work and influence, in particular, his "Social Software" program for studying the procedural behaviour of interactive human agents with beliefs and preferences.

Horacio Arló Costa and Arthur Paul Pedersen start off with a key interest of Parikh's, viz. bridging between realistic models of human behaviour and logic based

[1] *Logic at the Crossroads: An Interdisciplinary View*, Amitabha Gupta, Rohit Parikh and Johan van Benthem, editors, Allied Publishers, New Delhi. Recently reissued by Springer Publishers under the new title *Proof, Computation and Agency: Logic at the Crossroads*.

ones. They analyze Gigerenzer's "Take the Best" algorithm for successful decisions by bounded-rational agents, and show how it can be understood and taken further using formal models of choice and preference. Sergei Artemov continues with an account of "justification logics", which merge ideas from proof theory and provability logic with epistemic logic to arrive at rich and more realistic accounts of evidence in human reasoning that link up with formal epistemology. Next, Akeel Bilgrami discusses a key aspect of what holds communication together, viz. meaningful discourse, and argues that linguistic meanings are not normative in any deep sense – while still fully acknowledging Wittgenstein's insights about the normativity of human intentions. Melvin Fitting follows up on his long-standing formal studies of logics of proofs, and gives a new proof of a "realization theorem" linking the epistemic logic S5 to its justification counterpart. Next, Sujata Ghosh and Fernando Velázquez-Quesada discuss current models of belief and belief change based on networks, and relate these to the more qualitative ones found in epistemic and doxastic logic, including an account of successive dynamic steps that change a current belief network. Patrick Girard analyzes another crucial dynamic social scenario, viz. merge of individual preferences into a group-based one over a set of outcomes. He shows how preference merging in structured groups receives a perspicuous semantic and axiomatic formulation using the tools of modal and dynamic preference logic. Amy Greenwald, Amir Jafari and Casey Marks develop new links between the fields of game theory and computational learning theory, both crucial ingredients of intelligent interaction – based on a new notion of "no-Φ-regret". Vincent Hendricks returns to Parikh's interests in the fit between formal models and our natural ways of understanding knowledge and information. He gives an insightful discussion of the intuitive distinction between "first person" and "third person" perspectives, linking this to a wide range of issues from philosophy to economics. Dexter Kozen and Ganesh Ramanarayanan present another innovative social use of classical ideas from proof theory, developing a universal Horn equational logic with dynamic operations of "publish", "cite" and "forget" which describe the actual construction and maintenance of mathematical knowledge. Lawrence Moss returns to the use of natural language in human interaction, and shows how reasoning in natural language has interesting syllogistic subsystems allowing for negations that form a hitherto hardly explored level in between the classical Syllogistic and full-fledged first-order predicate logic. Jeff Paris and Alena Vencovská consider another major aspect of our natural reasoning, namely, its use of probabilistic features, whether quantitative, or qualitative with tags like "probably". Linking up with De Finetti's foundations of probability, they develop a sophisticated new version of Carnap's inductive logic for monadic predicates which handles the right inference relations between agents' degrees of belief. They also generalize this approach to binary and higher-order predicates, and develop its mathematical theory further, using a new principle of "Spectrum Exchangeability". Finally, R. Ramanujam and S.P. Suresh look at the central role of knowledge in reasoning about security protocols. The semantics of conventional epistemic logics throws up challenges, since cryptographic primitives in communication lead to problems of unboundedness and undecidability. The authors propose an alternative epistemic logic which is expressive enough while its protocol verification procedure is decidable.

Taken together, these papers show how logic can extend its scope without any sacrifice in rigour or focus. Moreover, their quality, and their cast of authors, speaks for itself about Rohit Parikh's intellectual breadth and influence. It has been a pleasure to arrange this tribute, and beyond logistics, the editors have learned a lot in the process.

We end with a citation which was read during the 2007 Mumbai conference. The present volume has taken some time to appear, due to circumstances beyond our control, but the editors feel that the sentiments expressed then are as fresh and appropriate as ever.

Stanford — *Johan van Benthem*
Bombay — *Amitabha Gupta*
Tilburg — *Eric Pacuit*

Dedication

On the occasion of his seventieth birthday, we, the participants and organizers of the "Second Indian Conference on Logic and its Relationship with Other Disciplines", acclaim and honour Professor Rohit Jivanlal Parikh and celebrate his many extraordinary achievements and outstanding contributions in his pursuit of knowledge.

Born on November 20, 1936, he was educated at Mumbai, and then at Harvard University, where he obtained degrees with highest honors in Physics (A.B.) and Mathematics (Ph.D.). Parikh's romance with Logic began with a course by Quine. Later on he was supervised by Burton Dreben and Hartley Rogers, writing his dissertation on transfinite progressions, influenced by and extending work of Kreisel. Subsequently, Rohit Parikh taught at several prestigious institutions, including Stanford, Boston and Punjab University. He has had a long and distinguished professional career that continues to be productive and influential. A full survey would be lengthy indeed, but the following may be highlighted:

- formal language theory, especially inherently ambiguous languages and semilinear sets,
- mathematical logic, specifically lengths of proofs and feasible arithmetic,
- logic in computer science, specifically dynamic logic and logics of knowledge,
- philosophical logic, especially vagueness, nonmonotonic reasoning, and belief revision,
- social software, including the logic of games.

In a research career spanning five decades, these contributions were not only seminal, but shaped the direction of a field, sometimes even creating it. The work ranges broadly, from mathematics, to computer science, economics, and philosophy. No wonder then that his current affiliation is: "Distinguished Professor of Computer

Science, Mathematics and Philosophy, Graduate Center, and Dept. of Computer Science, Brooklyn College, City University New York."

At the start of his career when working with Noam Chomsky as a research assistant, Parikh published a fundamental paper in automata theory. This subject essentially started with a paper of Kleene in 1956, and Parikh's paper appeared in 1966, so he was there almost at the beginning. Today Parikh's theorem and the Parikh map are standard tools. The Parikh map sends each word over an n-letter alphabet to an n-dimensional vector whose components give the number of occurrences of the letters in the word. His theorem says, in part, that the image under this map of any context-free language is always a semi-linear set, a finite union of linear sets. It follows that the emptiness problem for context-free languages is decidable. As with many of Parikh's contributions, this was not an end-point, but a starting point for others. A quick internet search turns up many current papers using the automata-theoretic tools that Parikh introduced.

It is a common observation today that some things computable in principle are not so in practice. Here Parikh made important early contributions. To study the strength of exponentiation relative to addition and multiplication, one needs a weak background theory. In 1971 Parikh introduced bounded arithmetic, which is both weak and strong enough for the purpose. A vast stream of research in the foundations of proof and computation has made essential use of this theory. One fascinating result from Parikh's 1971 paper is his demonstration that there is a theorem of Peano Arithmetic for which any proof is substantially longer than a proof that the formula has a proof. Another concerns "feasible" integers, and is related to the Sorites paradox. Suppose we add to Peano Arithmetic conditions asserting that 0 is a feasible number, adding 1 to a feasible number yields another feasible number, and there is an explicitly designated huge number that is not feasible. Of course this is an inconsistent theory, but Parikh showed that any result provable in this theory without the use of the notion of feasibility in inductions, and whose proof is of feasible length, will in fact be a theorem of Peano Arithmetic itself. In particular, provable contradictions must have infeasible proof lengths–the feasible part of the theory is safe to use.

Moving to logics of computation, propositional dynamic logic is a famous logic of action, designed to establish facts about computer program behavior. Completeness of a proposed axiomatization was a difficult problem, with more than one incorrect solution proposed at the time. Parikh (and independently, Gabbay) gave the first correct completeness proof, then he gave a second along different lines in a joint paper with Dexter Kozen. In addition to its impact on dynamic logic itself, this work has had a substantial spill-over effect, since Parikh's methods turned out applicable to similar logics as well.

In recent years, much of Parikh's work has fallen under the general program of "social software." Here the idea is to investigate the reasoning used by people in social interactions. Many apparently separate topics come together under this rubric. Our reasoning often involves vague terms. Parikh has created and analyzed algorithms that work even with vagueness present (in one paper this is whimsically applied to sorting socks into pairs, where color matching can be only approximate).

Also, everyday reasoning is often non-monotonic—we tend to make plausible default assumptions, but may have to retract them when further information comes in, and Parikh has worked on this, as well as well as the related area of belief revision. Next, common sense reasoning may be approximate, but the approximation may be improved at the cost of further work. Parikh has, with his colleagues, created an interesting family of logics that take expenditure of effort into account. Finally, how society values things, from election candidates to stocks and norms of social behavior, is studied mathematically in terms of game theory. In recent years, Parikh has made interesting contributions to game theory, and has trained students who have continued this work.

It is not easy to say, in simple terms, what the common thread to an oevre of Parikh's scope and depth might be, over many years and across several areas. But perhaps it is this. Our actual reasoning, as opposed to the ideal represented by the tradition beginning with Principia Mathematica, is subject to limitations of time, space, uncertainty of real-world facts, vagueness of concepts, and the complexity of interaction between people. Parikh has made a series of formal contributions that, together, say that what we do in practice is not without form. There is logic and structure there, though it may be complex and still imperfectly understood. Parikh has let the light of reason shine on some of the terra incognita of human thought and understanding. Thus, he is an investigator in the Wittgenstein tradition, rather than in that of Russell.

Today, his students, colleagues and friends wish to express their admiration for all that Rohit Parikh has done and the many contributions he has made. He has been a source of inspiration for all of us.

Also practically, the idea of creating a logic base at IIT Bombay is as much his as ours. He has been instrumental in urging his Indian colleagues to put in place an ambitious plan of not merely organizing Conferences and Schools, but building something that will last for years, eventually becoming the foundation of a strong Indian School in Logic. He says: "There are already strong Indian logicians outside India. What we need now is a strong group inside India to train brilliant Indian minds and to take our rightful place in this rich intellectual community in the world."

This citation is presented to Professor Rohit Jivanlal Parikh in recognition of his seminal contributions and outstanding achievements in research, teaching, institution building, and especially, for his dedication and wisdom, admired by his colleagues and students, and respected by all.

The Participants of ICLA 2
IIT Bombay, Mumbai, January 9 - 11, 2007

Contents

List of Contributors

Horacio Arló Costa
Department of Philosophy, Carnegie Mellon University, Pittsburgh, PA 15213, USA, e-mail: hcosta@andrew.cmu.edu

Sergei Artemov
Graduate Center CUNY, New York City, NY 10016, USA, e-mail: sartemov@gc.cuny.edu

Akeel Bilgrami
Department of Philosophy, Columbia University, New York, NY 10027, USA, e-mail: ab41@columbia.edu

Melvin Fitting
Department of Mathematics and Computer Science, Lehman College (CUNY), Bronx, NY 10468-1589, USA, e-mail: melvin.fitting@lehman.cuny.edu

Sujata Ghosh
Institute of Artificial Intelligence, University of Groningen, Groningen, The Netherlands, e-mail: sujata@ai.rug.nl

Patrick Girard
Department of Philosophy, University of Auckland, Auckland, New Zealand, e-mail: p.girard@auckland.ac.nz

Amy Greenwald
Department of Computer Science, Brown University, Providence, RI 02912, USA, e-mail: amy@cs.brown.edu

Vincent F. Hendricks
Department of Philosophy, University of Copenhagen, DK-2300 Copenhagen S, Denmark and Department of Philosophy, Columbia University, New York, NY 10027, USA, e-mail: vincent@hum.ku.dk

Amir Jafari
Mathematics Department, Duke University, Durham, NC 27708, USA,
e-mail: amir@math.duke.edu

Dexter Kozen
Department of Computer Science, Cornell University, Ithaca, NY 14853-7501,
USA, e-mail: kozen@cs.cornell.edu

Johann A. Makowsky
Department of Computer Science, Technion–Israel Institute of Technology, Haifa,
Israel, e-mail: janos@cs.technion.ac.il

Larry Moss
Department of Mathematics, Indiana University, Bloomington, IN 47405, USA,
e-mail: lsm@cs.indiana.edu

Casey Marks
Five Hut Consulting, Providence, RI 02906, USA, e-mail: casey@fivehut.com

Jeff B. Paris
School of Mathematics, University of Manchester, Manchester M13 9PL, UK,
e-mail: jeff.paris@manchester.ac.uk

Arthur Paul Pedersen
Department of Philosophy, Carnegie Mellon University, Pittsburgh, PA 15213,
USA, e-mail: apaulpedersen@andrew.cmu.edu

R. Ramanujam
Chennai Mathematical Institute, Chennai, India, e-mail: jam@imsc.res.in

Ganesh Ramanarayanan
Department of Computer Science, Cornell University, Ithaca, NY 14853-7501,
USA, e-mail: graman@cs.cornell.edu

S.P. Suresh
Chennai Mathematical Institute, Chennai, India, e-mail: spsuresh@cmi.ac.in

Fernando R. Velázquez-Quesada
Institute for Logic, Language and Computation, University of Amsterdam,
Amsterdam, The Netherlands, e-mail: F.R.VelazquezQuesada@uva.nl

Alena Vencovská
School of Mathematics, University of Manchester, Manchester M13 9PL, UK,
e-mail: alena.vencovska@manchester.ac.uk

Chapter 1
Bounded Rationality: Models for Some Fast and Frugal Heuristics

Horacio Arló Costa and Arthur Paul Pedersen

1.1 Introduction

Herb Simon pioneered the study of *bounded* models of rationality. Simon famously argued that decision makers typically *satisfice* rather than *optimize*. According to Simon, a decision maker normally chooses an alternative that meets or exceeds specified criteria, even when this alternative is not guaranteed to be unique or in any sense optimal. For example, Simon argued that an organism – instead of scanning all the possible alternatives, computing each probability of every outcome of each alternative, calculating the utility of each alternative, and thereupon selecting the optimal option with respect to expected utility – typically chooses the first option that satisfies its "aspiration level."

Simon insisted that real minds are *adapted* to real-world environments. Simon writes, "Human rational behavior is shaped by a scissors whose two blades are the structure of task environments and the computational capabilities of the actor" [19, p. 7]. Yet by paying undivided attention to the heuristics used by real agents, a very influential line of work on models of bounded rationality privileges one of these two blades over the other. This is the tradition initiated by the heuristics-and-biases program championed by Kahneman and Tversky (see the classic [12] and the more recent [11]). According to followers of this tradition, laws of probability, statistics, and logic constitute normative laws of human reasoning; descriptively, nevertheless, human reasoners follow heuristics that are systematically biased and error-prone. But this is hardly Simon's idea, who insisted on the accuracy of adapted and bounded methods.

Gerd Gigerenzer and his research group in Germany have followed Simon's view about bounded rationality more closely than many other researchers also deriving

Horacio Arló Costa
Department of Philosophy, Carnegie Mellon University, Pittsburgh, PA 15213, USA,
e-mail: hcosta@andrew.cmu.edu
Arthur Paul Pedersen
Department of Philosophy, Carnegie Mellon University, Pittsburgh, PA 15213, USA,
e-mail: apaulpedersen@andrew.cmu.edu

J. van Benthem et al. (eds.), *Games, Norms and Reasons*, Synthese Library 353,
DOI 10.1007/978-94-007-0714-6_1, © Springer Science+Business Media B.V. 2011

inspiration from Simon. The following gives an idea of Gigerenzer's main interests and goals:

> We wish to do more than just oppose the Laplacian demon view. We strive to come up with something positive that could replace this unrealistic view of the mind. What are these simple, intelligent heuristics capable of making near-optimal inferences? How fast and how accurate are they?...[W]e propose a class of heuristics that exhibit bounded rationality in both of Simon's senses [this is a reference to Simon's scissors metaphor]. These 'fast and frugal' heuristics operate with simple psychological principles that satisfy the constraints of limited time, knowledge and computational might, rather than those of classical rationality. At the same time they are designed to be fast and frugal without a significant loss of inferential accuracy because they can exploit the structure of the environments.

A way to reconcile this view with the normative status of laws of logic and statistics is to argue that the cognitive role of the fast and frugal heuristics is to approximate the standards of these laws: When the methods in question are sufficiently well adapted to the environment, we will have a happy convergence on recommendations of these normative laws. This is, for example, Dawkins' view of the "rules of thumb" converging on optimality provided by natural selection [5]. Indeed, this is a nice way of interpreting the role of the fast and frugal heuristics that Gigerenzer discusses. But Gigerenzer has quite emphatically denied that this is his view. In fact, Gigerenzer has argued that it is rational to use his fast and frugal heuristics even when they conflict with the dictates of the traditional conception of rationality.

In this article we focus on one fast and frugal heuristic that Gigerenzer and his colleagues advertise called Take The Best. There are at least two varieties of the heuristic, defined in [7–9]. So more precisely, there are two heuristics, and we will investigate these heuristics in this article. One goal of this article is a study of the choice-theoretical properties of these heuristics. In order to do so, both varieties of Take The Best have to be extended beyond two-alternative-choice tasks, presently the primary application of Gigerenzer's fast and frugal heuristics. We provide here a first approximation to this problem. Our strategy is to extend choice beyond two-alternative-choice tasks by maximizing the binary preference relations induced by both varieties of Take The Best. This permits the study of abstract conditions of choice that completely characterize both varieties. The methodology employed extends techniques that are commonly used in the theory of choice functions.

The characterization results presented in this article are complemented by a different model offered in a companion article [3]. In this article we consider an extension that is psychologically more plausible (in the sense that it only appeals to satisficing methods).

We will see that one version of Take The Best violates some basic rationality constraints like transitivity and that important axiomatic properties of choice functions are violated as well. We will also see that the other variant of the heuristic is fairly well behaved and compatible with some salient norms of rationality and axiomatic properties of choice. Accordingly, we claim that speed and frugality can be obtained without renouncing basic norms of rationality. Gigerenzer seems to argue in various places against this view, suggesting that speed and frugality cannot be reconciled with the traditional norms of rationality. Perhaps his claim can be reconstructed as asserting that *psychologically plausible* fast and frugal heuristics cannot be reconciled with norms of rationality. But even in this case there is a general theory of choice functions that can be developed for these fast and frugal heuristics. Of course, the interpretation of the choice functions used in the characterization of (both varieties of) Take The Best is not the usual one. The idea is to develop an empirical (descriptive) theory of choice functions. The main interpretation of axiomatic properties of choice is not that of imposing normative constraints on choice, but as describing patterns of inference permitted by the underlying heuristic, which is taken as an epistemological primitive.

1.2 Take The Best

The best manner of making clear Gigerenzer's position is by means of an example. Consider the heuristic Gigerenzer, Goldstein, and their collaborators call Take The Best [8]. The main idea behind this heuristic is that for a given domain of objects X, an agent decides between two objects with respect to a set of *cues* ordered by *cue validity*. We can make this idea more precise as follows.

Definition 1.1. Let X be a collection of objects.

(i) A *cue* for X is a function $Cue : X \to \{-,?,+\}$,
(ii) A *recognition heuristic* for X is a function $Rec : X \to \{-,+\}$. If $Rec(a) = +$, we say that a is *recognized.*
(iii) Let Y be the collection of all subsets of X of cardinality 2. A *randomizer* for X is a function $Ran : Y \to X$ such that $Ran(A) \in A$ for each $A \in Y$. If $Ran(\{a,b\}) = a$, we say that a is *chosen at random.*

Definition 1.2. Let $(Cue_i)_{i<n}$ be a finite collection of cues for X, where Cue_0 is a recognition heuristic. We call the pair $(X,(Cue_i)_{i<n})$ a *cue validity ordering* over X (and say that $(Cue_i)_{i<n}$ is *ordered by cue validity*) if for every $a \in X$, if $Cue_0(a) = -$, then for each i with $0 < i < n$, $Cue_i(a) = ?$.

Definition 1.3. If $\mathfrak{C}$ is a cue validity ordering over a set X and *Ran* is a randomizer for X, we call the pair $(\mathfrak{C}, Ran)$ a *Take The Best frame* over X.

We denote Cue_0 by *Rec*. Intuitively, $(Cue_i)_{i<n}$ is ordered according to the reliability of the cues, a recognizer is nothing more than a cue with the demand that its range is $\{-,+\}$, and a randomizer is employed to make a guess when no cue provides sufficient information to make a choice between a pair of objects, i.e., when no cue discriminates or is decisive between a pair of objects.

The following algorithm for binary choices with respect to a Take The Best frame over X comprises the Take The Best heuristic.

Step 0: *Recognition Heuristic.* If only one of two possible objects is recognized, then choose the recognized object. If neither of the two objects is recognized, then choose randomly between them. If both of the objects are recognized, then proceed to Step 1.

Step 1: *Search Rule*. Choose the cue with the highest validity that has not yet been tried for this choice task. Look up the cue values of the two objects.

Step 2: *Stopping Rule*. If one object has a positive cue value and the other does not, then stop the search and go to Step 3. Otherwise go back to Step 1 and search for another one. If no further cue is found then guess.

Step 3: *Decision Rule* (*one-reason decision making*). Predict that the object with the positive cue value has the higher value on the criterion.

Step 0 demands that if $Rec(a) \neq Rec(b)$, then one should choose the object from $\{a,b\}$ to which *Rec* assigns a positive value. But if $Rec(a) = Rec(b) = -$, then one should choose the object from $Ran(\{a,b\})$. In the remaining case when $Rec(a) = Rec(b) = +$, one should proceed to Step 1. Observe that in the case that $Rec(a) = Rec(b) = -$, it follows that for each i with $0 < i < n$, $Cue_i(a) = Cue_i(b) = ?$.

Hence, Steps 0–3 together require one to search for the least $i < n$ for which $Cue_i(a) = +$ and $Cue_i(b) \in \{-,?\}$ or $Cue_i(b) = +$ and $Cue_i(a) \in \{-,?\}$, whereby the object from $\{a,b\}$ to which Cue_i assigns a positive value is chosen. However, if there is no such i, then one should choose the object from $Ran(\{a,b\})$.

Observe that search operates only on a fraction of the total knowledge in memory and is stopped when a cue discriminates between a pair of objects. Thus, Take The Best is similar to Simon's satisficing algorithm insofar as it terminates search when it finds the first cue that discriminates between a pair of objects and thereupon makes a choice. Also observe that the algorithm does not integrate information, reducing all aspects of information to a single dimension; rather, it employs one-reason decision making, thereby conflicting with the classical economic view of human behavior.

Gigerenzer and his collaborators considerstwo-alternative-choice tasks in various contexts where inferences on a quantitative dimension must be made from memory under the constraints of limited time and knowledge. Gigerenzer offers various examples of two-alternative-choice tasks in [7]. As an illustration, consider the following prompt:

Which city has a larger population?

(a) Hamburg
(b) Cologne

Now suppose that an agent neither knows nor can deduce the answer to this question and so must make an inductive inference based on related real-world knowledge. Thus, the problem is how such an inference can be made. According to the theory of *probabilistic mental models* that Gigerenzer and his colleagues adopt, the first step is to search knowledge about a reference class, e.g., "Cities in Germany." The knowledge itself consists of probability cues and their corresponding values for the objects in the reference class. For example, a candidate for a cue may be a function of whether a city has a professional soccer team in the major league. According to this cue, if one city has a team in the major league and the other does not, the city with the professional soccer team in the major league is likely – but perhaps not certain – to have a larger population than the other city. Furthermore, according to Gigerenzer and Goldstein's Take The Best heuristic, if this cue is the first cue to discriminate between the two cities, the city with the professional soccer team is inferred to have a larger population than the other city. Other examples of cues for the population demographics task are functions of answers to the following questions: (*Intercity Train*) "Is the city on the Intercity line?"; (*University*) "Is the city home to a University?"; (*Industrial Belt*) "Is the city in the Industrial Belt?"

Gigerenzer and his collaborators utilize probabilistic mental models which are sensible to the fact that an agent typically does not know all the information upon which it could base an inference. These probabilistic mental models reflect at least two aspects of such limited knowledge: On the one hand, an agent can have incomplete knowledge of the objects in the relevant reference class; on the other hand, the person might have limited knowledge of values of cues.

We have seen that each cue has an associated cue validity. The validity of a cue specifies the predictive reliability or predictive power of the cue. More precisely, the *ecological validity* of a cue is the relative frequency (with respect to a given reference class) that the cue correctly predicts the target variable (e.g., population). Returning to the aforementioned example, in 87 percent of the pairs in which one city has a professional soccer team and the other does not, the city with the professional soccer team has a larger population. Thus, 0.87 specifies the ecological validity of the soccer team cue (see [7, p. 176]). In the model adopted by Gigerenzer and his collaborators, it is required that there are only finitely many cues and that no two cues have the same ecological validity.

In Gigerenzer's original article [7, pp. 125–163], Take The Best has a different stopping rule. In this earlier version, search terminates only when one object has a positive value and the other has a negative value:

> **Step 2′**: *Stopping Rule.* If one object has a positive cue value and the other has a negative value, then stop the search and go to Step 3. Otherwise go back to Step 1 and search for another one. If no further cue is found then guess.

According to Gigerenzer, the Take The Best heuristic presented above follows empirical evidence that agents tend to use the fast, simpler stopping rule expressed in Step 2 [7, p. 175]. Nonetheless, we will consider both varieties of the heuristic in this article. We will thereby call the original heuristic (with Step 2′ instead of Step 2) the Original Take The Best heuristic.

Definition 1.4. Let $(X, (Cue_i)_{i<n}, Ran)$ be a Take The Best frame over X. For each $i < n$, define a binary relation $\succ$ over X by setting for all $a, b \in X$,

$$a \succ_i b \text{ if and only if } Cue_i(a) = + \text{ and } Cue_i(b) = -.$$

Similarly, for each $i < n$, define a binary relation $>$ over X by setting for all $a, b \in X$,

$$a >_i b \text{ if and only if } Cue_i(a) = + \text{ and } Cue_i(b) \in \{?, -\}.$$

For each $i < n$, we say that Cue_i is *decisive* for $a, b \in X$ if $a \succ_i b$ or $b \succ_i a$. We also call Cue_i *discriminating* for $a, b \in X$ if $a >_i b$ or $b >_i a$.

We are now in a position to define the notion of a Take The Best model with respect to a Take The Best frame.

Definition 1.5. Let $(X, (Cue_i)_{i<n}, Ran)$ be a Take The Best frame over X. Define a binary relation $\succ$ over X by setting for every $a, b \in X$,

$$a \succ b \text{ if and only if } a \succ_i b \text{ for the least } i < n \text{ for which } Cue_i \text{ is decisive for } a, b, \\ \text{or there is no decisive cue for } a, b \text{ and } Ran(\{a, b\}) = a.$$

Similarly, define a binary relation $>$ over X by setting for $a, b \in X$,

$$a > b \text{ if and only if } a >_i b \text{ for the least } i < n \text{ for which } Cue_i \text{ is discriminating for } a, b, \\ \text{or there is no discriminating cue for } a, b \text{ and } Ran(\{a, b\}) = a.$$

We thereby call the quadruple $(X, (Cue_i)_{i<n}, Ran, \succ)$ an *Original Take The Best model.* Similarly, we call the quadruple $(X, (Cue_i)_{i<n}, Ran, >)$ a *Take The Best model.*

Definition 1.6. Let $(X, (Cue_i)_{i<n}, Ran, \succ)$ be an Original Take The Best model, and let $(X, (Cue_i)_{i<n}, Ran, >)$ be a Take The Best model.

(i) We say that $\succ$ is *decisive* if there is a decisive cue for every distinct $a, b \in X$.
(ii) We say that $>$ is *discriminating* if there is a discriminating cue for every distinct $a, b \in X$.
(iii) We call $(X, (Cue_i)_{i<n}, Ran, \succ)$ *decisive* (asymmetric, connected, transitive, modular, etc.) if $\succ$ is decisive (asymmetric, connected, transitive, modular, etc.).
(iv) We call $(X, (Cue_i)_{i<n}, Ran, >)$ *discriminating* (asymmetric, connected, transitive, modular, etc.) if $>$ is discriminating (asymmetric, connected, transitive, modular, etc.).

Observe that by definition both $\succ$ and $>$ are asymmetric (and so irreflexive) and connected.

As Gigerenzer points out, both the Original Take The Best heuristic and the Take The Best heuristic are hardly standard statistical tools for inductive inference: they do not use all available information and are non-linear. Furthermore, the binary relations for models of both heuristics may violate transitivity if they fail to be discriminating or decisive. But an Original Take The Best model may violate transitivity even if it is decisive. Let's see an example of the latter feature.

Example 1.1. Consider a collection X of three objects, a, b and c, a recognition heuristic *Rec*, and three cues, Cue_1, Cue_2, and Cue_3 (any randomizer will do), as presented in the table below:

	a	b	c
Rec	+	+	+
Cue_1	+	−	?
Cue_2	−	+	−
Cue_3	−	−	+

Clearly, $a \succ b$ in view of the decisive cue Cue_1 for a and b. Also, $b \succ c$ in light of the decisive cue Cue_2 for a and b. But according to the third cue, $c \succ a$.

□

Not only does this constitute a violation of transitivity, but it also induces a cycle of preference. So an Original Take The Best model may violate various salient norms of rationality, even if there is no guessing involved, i.e., even if it is decisive.

The first thing that one can say in defense of Take The Best is that it seems descriptively adequate.[1] This is interesting *per se*, but this does not indicate anything regarding other potential cognitive virtues of the heuristic. It is at this point that Gigerenzer turns to the ecological aspect of the argument for bounded rationality. Gigerenzer contends that if fast and frugal heuristics are well-tuned ecologically, then they should not fail outright. So in [7], Gigerenzer proposes a competition among Take The Best and other inferential methods which are more costly but which integrate information. These rival methods (like linear regression) obey norms of rationality that Take The Best could violate. The idea of the competition is that the contest will go to the strategy that scores the highest proportion of correct inferences (accuracy) using the smallest number of cues (frugality).

And in [7] Gigerenzer presents surprising results showing that Take The Best wins the contest. All results are computed for simulations taking into account empirical constraints that are determined by the results of behavioral experiments.

As Bermúdez points out in [4], Gigerenzer claims to have replaced criteria of rationality based upon logic and probability theory with a heuristic-based criteria of real-world performance. There is an innocuous way of interpreting such a claim via a "long run" argument. One might claim that it is beneficial to stick to certain heuristics given the long run benefits it provides, even if in the short run it might lead you astray. But this does not seem to be the argument that Gigerenzer sets forth for his heuristics. Apparently he wants to argue that something more controversial than this flows from the result of the competition between different inferential methods (after all, the long run argument still presupposes the normativity of laws of logic and statistics). The fast and frugal heuristics can trump normative theories when adapted to the ecological environment.

Simon was centrally interested in discovering the inferential strategies that humans use in the real world. His interest was not to provide a new theory of rationality but to engage in psychology. It is unclear the extent to which Gigerenzer might be interested in replacing standard norms of rationality with fast and frugal strategies of inference that, when adapted, are also accurate and successful. Undoubtedly, the strategies in question remain interesting under a purely descriptive point of view. In this article we propose to begin a study of the theory of choice that applies to these methods. In a companion article we intend to apply this theory of choice to study the structure of bounded methods of belief change.

[1] See [7, pp. 174–175] for an account of the empirical evidence supporting Take The Best.

1.3 Rational Choice: The Received View

One of the central tools in the theory of rational choice is the notion of *choice function*. Here we present this notion, largely following the classical presentations of Amartya Sen [17] and Kotaro Suzumura [20].

Definition 1.7. Let X be set, and let $\mathcal{S}$ be a collection of non-empty subsets of X. We call the pair $(X, \mathcal{S})$ a *choice space*. A *choice function* on $(X, \mathcal{S})$ is a function $C : \mathcal{S} \to \mathcal{P}(X)$ such that $C(S) \subseteq S$ for every $S \in \mathcal{S}$.

For a choice function C on a choice space $(X, \mathcal{S})$, we call $C(S)$ the *choice set* for S. Intuitively, a choice function C selects the "best" elements of each S, and $C(S)$ represents the "choosable" elements of S.

When the context is clear, we will often talk of a choice function C on a choice space $(X, \mathcal{S})$ without reference to the choice space. Thus, we may simply speak of a choice function C. For our purposes it is convenient to restrict our attention to choice functions each of which has a domain that is the family of all nonempty finite subsets of a collection of objects. We wish to emphasize, however, that many of the results presented in the remainder of this article are preserved when less restrictive conditions are imposed on the domains of choice functions.

The following condition demarcates a special class of choice functions. Often in the literature it is assumed that choice functions satisfy this condition.

Regularity. For each $S \in \mathcal{S}$, $C(S) \neq \varnothing$.

We call a choice function satisfying Regularity *regular*.

The idea that a choice function C picks the "best" elements of each $S \in \mathcal{S}$ has been made more precise by assuming that there is some binary relation over the elements of X according to which $C(S)$ distinguishes the best elements of each $S \in \mathcal{S}$. A choice function with this property is called a *rational* choice function. Two formalizations of the idea of a rational choice function have been widely utilized in the literature. We discuss these formalizations here.

The first formalization of a rational choice function demands that a choice function C selects the optimal (greatest) elements from each set $S \in \mathcal{S}$. Here we need some notation. The set of greatest elements of a set S with respect to a binary relation $\geq$ is defined as follows:

$$G(S, \geq) := \{x \in S : x \geq y \text{ for all } y \in S\}$$

Using this notation, we now offer a definition.

Definition 1.8. A binary relation $\geq$ on a universal set X *G-rationalizes* (or is a *G-rationalization* of) a choice function C on a choice space $(X, \mathcal{S})$ if for every $S \in \mathcal{S}$, $C(S) = G(S, \geq)$.

We also say that a choice function C is *G-rational* if there is a binary relation $\geq$ that G-rationalizes C. This formalization of a rational choice function is what Sen and others often call an *optimizing* notion of rationality.

The second formalization of a rational choice function captures a less stringent notion of rationality, demanding only that a choice function selects the *maximal* elements from each set $S \in \mathcal{S}$. Again, we require some notation. The set of maximal elements of a set S with respect to a binary relation $>$ is defined as follows:

$$M(S,>) := \{x \in S : y > x \text{ for no } y \in S\}$$

With this notation at hand, we again offer a definition.

Definition 1.9. A binary relation $>$ on a universal set X *M-rationalizes* (or is a *M-rationalization* of) a choice function C on a choice space $(X, \mathcal{S})$ if for every $S \in \mathcal{S}$, $C(S) = M(S,>)$.

As with G-rationality, we say that a choice function C is *M-rational* if there is a binary relation $>$ that M-rationalizes C. Sen and others call this a *maximizing* notion of rationality. We will assume this notion of rationality in this article.[2]

In the following, let $\mathcal{P}_{fin}(X)$ denote the family of all finite subsets of a collection of objects X. The next proposition establishes that any reasonable binary relation that M-rationalizes a choice function on a choice space $(X, \mathcal{P}_{fin}(X))$ is unique.

Proposition 1.1. *Let C be a choice function on a choice space $(X, \mathcal{P}_{fin}(X))$. Then if $>$ is a irreflexive binary relation on X that M-rationalizes C, $>$ uniquely M-rationalizes C.*

Proof. Let $>$ be a irreflexive binary relation on X that M-rationalizes C, and let $>'$ be a binary relation on X that also M-rationalizes C. Now since $>$ is irreflexive, $C(\{x\}) = \{x\}$ for every $x \in X$, whereby it follows that $>'$ is irreflexive. Now for *reductio ad absurdum*, suppose $> \neq >'$. Without loss of generality, we may assume that for some $x, y \in X$, $x > y$ but $x \not>' y$. Then $y \notin M(\{x,y\},>)$, but $y \in M(\{x,y\},>')$, yielding a contradiction.

Remark 1.1. Observe that the irreflexivity of $>$ is a necessary condition for Proposition 1.1. To illustrate, consider a choice space $(X, \mathcal{P}_{fin}(X))$, where $X := \{x,y\}$, and choice function C on $(X, \mathcal{P}_{fin}(X))$ defined by setting $C(\{x\}) := \varnothing$, $C(\{y\}) = \{y\}$, and $C(X) := \{y\}$. We can define two binary relations $>$ and $>'$ by setting $>:= \{(x,x)\}$ and $>' := \{(x,x),(y,x)\}$. Clearly neither $>$ nor $>'$ is irreflexive, both $>$ and $>'$ M-rationalize C, yet $> \neq >'$.

We call a choice function C *acyclic* (asymmetric, connected, transitive, modular, etc.) *M-rational* if C is M-rationalized by an acyclic (asymmetric, connected, transitive, modular, etc.) binary relation. Using this terminology we can state the following lemma with ease.

Lemma 1.1 ([20], p. 35). *A choice function C on a choice space $(X, \mathcal{P}_{fin}(X))$ is acyclic M-rational if and only if it is regular M-rational.*

---

[2] It is well known that if we require that $\geq$ be understood as the asymmetric part of $>$, then *an M-rational choice function is G-rational, but not vice-versa in general.*

Proof. For the direction from left to right, suppose C is acyclic M-rational. We must show that C is regular. For *reductio ad absurdum*, assume that C is not regular, and let $S \in \mathcal{P}_{fin}(X)$ be such that $C(S) = \varnothing$. Choose $x_0 \in S$. Since S is finite, there are $x_1, \ldots, x_{n-1} \in S$ such that $x_0 > x_{n-1} > \cdots > x_1 > x_0$, contradicting that C is acyclic M-rational.

For the direction from right to left, suppose C is regular M-rational, and let $>$ be an M-rationalization of C. For *reductio ad absurdum*, assume that $>$ is not acyclic. Then there are $x_0, \ldots, x_{n-1} \in X$ such that $x_0 > x_{n-1} > \cdots > x_1 > x_0$. But then $C(\{x_0, \ldots, x_{n-1}\}) = M(\{x_0, \ldots, x_{n-1}\}, >) = \varnothing$, contradicting that C is regular.

There has been some work on the project of characterizing the notion of rationality axiomatically using so-called *coherence constraints*. One salient coherence constraint is the condition Sen calls *Property* α [17, p. 313], also known as *Chernoff's axiom* [20, p. 31].

Property α. For all $S, T \in \mathcal{S}$, if $S \subseteq T$, then $S \cap C(T) \subseteq C(S)$.

There are two lines of argument for the characterization of rationality, one proposed by Sen in [17] (and reconsidered in [18]) and another proposed by Kotaro Suzumura in [20]. Both use the notion of *base preference* ([17, p. 308, 18, p. 64, 20, p. 28]). We modify their respective arguments in terms of maximization.

Definition 1.10 (Base Preference). Let C be a choice function on a choice space $(X, \mathcal{P}_{fin}(X))$. We define *(strict) base preference* by setting

$$>^C := \{(x,y) \in X \times X : x \in C(\{x,y\}) \text{ and } y \notin C(\{x,y\})\}$$

Observe that $>^C$ must be asymmetric and so irreflexive.

We now present Suzumura's line of argument. As a first step we can add a new coherence constraint [20, p. 32] which we have reformulated in terms of maximization:

Generalized Condorcet Property. For all $S \in \mathcal{S}$, $M(S, >^C) \subseteq C(S)$

The first general result that Suzumura presents is the following (modified for our purposes):

Theorem 1.1 ([20], p. 35).
A choice function C on a choice space $(X, \mathcal{P}_{fin}(X))$ is acyclic M-rational if and only if it is regular and satisfies Property α and the Generalized Condorcet Property.

Proof. By Lemma 1.1, it suffices to show that a regular choice function C on a space $(X, \mathcal{P}_{fin}(X))$ is M-rational if and only if it satisfies Property α and the Generalized Condorcet Property.

($\Rightarrow$) Suppose C is M-rational, and let $>$ be a M-rationalization of C.

Property α. Let $S, T \in \mathcal{P}_{fin}(X)$. If $S \subseteq T$ and $x \in S \cap C(T)$, then $y > x$ for no $y \in T$, whence $y > x$ for no $y \in S$, so $y \in C(S)$.

Generalized Condorcet Property. Let $S \in \mathcal{P}_{fin}(X)$, and suppose $x \in M(S, >^C)$. Then because C is regular, $x \in C(\{x,y\})$ for all $y \in S$, whereby $y > x$ for no $y \in S$ and so $x \in C(S)$.

($\Leftarrow$) Suppose C satisfies Property α and the Generalized Condorcet Property. We must show that $>^C$ M-rationalizes C. By the Generalized Condorcet Property, we must only verify that $C(S) \subseteq M(S, >^C)$ for all $S \in \mathcal{P}_{fin}(X)$. So let $S \in \mathcal{P}_{fin}(X)$, and suppose $x \in C(S)$. Then by Property α, for every $y \in S$, $x \in C(\{x,y\})$ and so $y \not>^C x$, whereby $x \in M(S, >^C)$, as desired.

The proof relies heavily on the assumption that choice functions are to pick *nonempty* subsets of the sets from which they make selections. So Suzumura's axiomatic characterization of rationality depends in both cases on the use of *regular* choice functions.

We now present the second line of argument for the characterization of rationality due to Sen [17]. The argument also proceeds in terms of regular choice functions and (as indicated above) uses the notion of base preference. We need an additional axiom in order to present the argument:

Property γ. For every nonempty $I \subseteq \mathcal{S}$ such that $\bigcup_{S \in I} S \in \mathcal{S}$,

$$\bigcap_{S \in I} C(S) \subseteq C(\bigcup_{S \in I} S)$$

With this we can state the following theorem:

Theorem 1.2. *A choice function C on a choice space $(X, \mathcal{P}_{fin}(X))$ is acyclic M-rational if and only if it is regular and satisfies Property α and Property γ.*

Proof. By Lemma 1.1, it suffices to show that a regular choice function C on a space $(X, \mathcal{P}_{fin}(X))$ is M-rational if and only if it satisfies Property α and Property γ.

($\Rightarrow$) Suppose C is M-rational, and let $>$ be a M-rationalization of C. In light of the proof of Theorem 1.1, we only show that C satisfies Property γ. Let $I \subseteq \mathcal{P}_{fin}(X)$ be such that $\bigcup_{S \in I} S \in \mathcal{P}_{fin}(X)$, and suppose $x \in \bigcap_{S \in I} C(S)$. Then for each $S \in I$, $y > x$ for no $y \in S$, so $x \in C(\bigcup_{S \in I} S)$.

($\Leftarrow$) Suppose C satisfies Property α and Property γ. We must show that $>^C$ M-rationalizes C. Again, in light of the proof of Theorem 1.1, we only show that $M(S, >^C) \subseteq C(S)$ for all $S \in \mathcal{P}_{fin}(X)$. So let $S \in \mathcal{P}_{fin}(X)$, and suppose $x \in M(S, >^C)$. Then $y >^C x$ for no $y \in S$ and therefore by Regularity $x \in \bigcap_{y \in S} C(\{x,y\})$, so by Property γ, $x \in C(S)$.

Corollary 1.1 (cf.[20], p. 28). *A regular choice function C is M-rational if and only if $>^C$ uniquely M-rationalizes C.*

Proof. The direction from right to left is trivial. For the other direction, observe that by Theorem 1.2, if C is M-rational, then C satisfies Property α and Property γ, so by the proof of Theorem 1.2, $>^C$ M-rationalizes C, whence by Proposition 1.1 $>^C$ uniquely M-rationalizes C.

We will see below that Theorem 1.2 can be generalized to a larger class of choice functions. Indeed, we will see that this result holds for choice functions that *fail* to be regular.

Many important results involving the role of Chernoff's axiom presuppose that the choice functions used are regular. As Sen points out, Fishburn, Blair, and Suzumura seem to think that Property α guarantees that the base relation is acyclic. But it is easy to see that this is incorrect, for it is Regularity that corresponds to acyclicity of the base relation, and Property α is independent of Regularity.

1.4 Choice Functions for Take The Best

The first step in the definition of a choice function for Gigerenzer and his colleagues' heuristics is to articulate a notion of binary preference for a Take The Best frame. We offered two such notions of preference in Definition 1.5. Thus, we are now in a position to define choice functions for heuristics.

Definition 1.11. Let $\mathfrak{H} = (X, (Cue_i)_{i<n}, Ran, \succ)$ be an Original Take The Best model. We define a choice function $C_{\mathfrak{H}}$ on $(X, \mathcal{P}_{fin}(X))$ by setting for every $S \in \mathcal{P}_{fin}(X)$, $C_{\mathfrak{H}}(S) := M(S, \succ)$. We thereby call $C_{\mathfrak{H}}$ an *Original Take The Best choice function*, or an *OTB choice function* for short, and we say that $C_{\mathfrak{H}}$ is based on $\mathfrak{H}$.

Definition 1.12. Let $\mathfrak{H} = (X, (Cue_i)_{i<n}, Ran, >)$ be a Take The Best model. We define a choice function $C_{\mathfrak{H}}$ on $(X, \mathcal{P}_{fin}(X))$ by setting for every $S \in \mathcal{P}_{fin}(X)$, $C_{\mathfrak{H}}(S) := M(S, >)$. We call $C_{\mathfrak{H}}$ a *Take The Best choice function*, or just a *TB choice function* for short, and we say that $C_{\mathfrak{H}}$ is based on $\mathfrak{H}$.

To illustrate the use of the choice functions defined above, let us return for a moment to the population demographics task described earlier. It is clear that we can interpret the relation $>$ as "has a larger population than." This binary relation is *revealed* by two-alternative-choice tasks of the sort we presented earlier (Hamburg or Cologne?). Now, when we construct a choice function as in one of the definitions above, we go beyond these two-alternative-choice tasks in such a way that one can ask a question such as the following:

Which city has the largest population?

(a) Hamburg
(b) Cologne
(c) Munich

We can thereby represent choices from sets of cardinality greater than two. We believe that this is reasonable extension of Gigerenzer's program. The extension is constructed in terms of an underlying binary preference relation which is determined by a satisficing algorithm.

Of course, in order to construct a choice function for triples and larger sets, we maximize the underlying binary preference relation. One might argue that this does not satisfy the requirement (satisfied by the probabilistic mental models adopted by Gigerenzer) that inductive inference is carried out by a satisficing algorithm (in Simon's sense). Nevertheless, working with maximization seems to be the first obvious step toward extending the algorithm beyond two-alternative-choice tasks.

We will see that this extension requires the introduction of important changes into the traditional theory of choice. It seems that any "fast and frugal" extension of Take The Best (or Original Take The Best) will require (at least) the introduction of identical changes. In this article we therefore limit ourselves to the study of the aforementioned extension by means of maximization. We present, nevertheless, the sketch of a "fast and frugal" extension in the last section of this article (this extension is analyzed in detail in a companion article).

Let's focus now on the structure of the choice functions for take the best we just defined. Obviously, an OTB or TB choice function will not in general be a regular choice function. In fact, even a decisive Original Take The Best model may induce an OTB choice function that is not regular, as we saw in Example 1.1. We must therefore develop a theory of choice functions that does not impose Regularity in order to find a representation of the choice functions induced by Gigerenzer and his colleagues' algorithms.

We can add to the results of the previous section a generalization linking choice functions to Property α and Property γ. Indeed, Sen's result need not require Regularity. To illustrate why this is so, let us first define a somewhat peculiar preference relation.

Definition 1.13 (Preference). Let C be a choice function on a choice space $(X, \mathcal{P}_{fin}(X))$. We define a binary relation $\gtrdot^C$ on X by setting

$$\gtrdot^C := \{(x,y) \in X \times X : y \notin C(\{x,y\})\}$$

We then have the following general result.

Theorem 1.3. *A choice function C on $(X, \mathcal{P}_{fin}(X))$ is M-rational if and only if it satisfies Property α and Property γ.*

Proof.
($\Rightarrow$) This direction proceeds as in the proofs of Theorems 1.1 and 1.2.

($\Leftarrow$) Suppose C satisfies Property α and Property γ. We show that $\gtrdot^C$ M-rationalizes C. Let $S \in \mathcal{P}_{fin}(X)$. We first show that $M(S, \gtrdot^C) \subseteq C(S)$. Suppose $x \in M(S, \gtrdot^C)$. Then $y \gtrdot^C x$ for no $y \in S$ and therefore $x \in \bigcap_{y \in S} C(\{x,y\})$, so by Property γ, $x \in C(S)$. Now to show that $C(S) \subseteq M(S, \gtrdot^C)$, suppose $x \in C(S)$. Then by Property α, for every $y \in S$, $x \in C(\{x,y\})$ and so not $y \gtrdot^C x$, whereby $x \in M(S, \gtrdot^C)$.

One may have observed that this result demands very little of a binary relation that rationalizes a choice function. To see why, consider a choice space $(X, \mathcal{P}_{fin}(X))$, where $X := \{x,y,z\}$, and a choice function C on $(X, \mathcal{P}_{fin}(X))$ defined by setting $C(\{x,y,z\}) := \varnothing$ and

$$\begin{array}{lll} C(\{x\}) := \{x\} & C(\{y\}) := \{y\} & C(\{z\}) := \varnothing \\ C(\{x,z\}) := \{x\} & C(\{y,z\}) := \{y\} & C(\{x,y\}) := \varnothing. \end{array}$$

It is easy check that C thus defined satisfies Property α and Property γ. Observe that among other things, the relation $\gtrdot^C$ is neither asymmetric nor irreflexive.

The lesson to be drawn from this is that although a generalization of the results of the previous section would involve conditions weaker than Regularity, a better generalization would ensure that a binary relation that M-rationalizes a choice function is asymmetric or at least irreflexive. We can guarantee irreflexivity with the following property:

Property ρ. For every $x \in X$ such that $\{x\} \in \mathcal{S}$, $C(\{x\}) = \{x\}$

We can also guarantee asymmetry with the following property:

Property σ. For every $x, y \in X$ such that $\{x,y\} \in \mathcal{S}$, $C(\{x,y\}) \neq \varnothing$.

Observe that as one should expect, Property σ entails Property ρ. We obtain our first better result.

Theorem 1.4. *A choice function C on $(X, \mathcal{P}_{fin}(X))$ is irreflexive M-rational if and only if it satisfies Property α, Property γ, and Property ρ.*

Proof. The proof here proceeds as in the proof of Theorem 1.3. Clearly C is irreflexive M-rational only if it satisfies Property ρ as well as Property α and Property γ. If C satisfies Property ρ, $\gtrdot^C$ is irreflexive, and if C also satisfies Property α and Property γ, $\gtrdot^C$ M-rationalizes C.

We then have a corollary similar to Corollary 1.1.

Corollary 1.2. *A choice function C on $(X, \mathcal{P}_{fin}(X))$ satisfying Property ρ is M-rational if and only if $\gtrdot^C$ uniquely M-rationalizes C.*

But of course, an M-rational choice function satisfying Property ρ is not necessarily rationalized by $>^C$, which guarantees asymmetry. Yet both OTB are TB choice functions are M-rationalized by asymmetric binary relations, so we should seek a representation result that ensures as much. The next result ensures that any binary relation that M-rationalizes C is asymmetric, and of course, for this $>^C$ meets the task.

Theorem 1.5. *A choice function C on $(X, \mathcal{P}_{fin}(X))$ is asymmetric M-rational if and only if it satisfies Property α, Property γ, and Property σ.*

Proof. As before, the proof proceeds as in the proof of Theorem 1.3, and obviously C is asymmetric M-rational only if it satisfies Property σ in addition to Property α and Property γ. For the converse, observe that if C satisfies Property σ, then $x \gtrdot^C y$ if and only if $x \in C(\{x,y\})$ and $y \notin C(\{x,y\})$, and the latter holds just in case $x >^C y$. Of course, if C also satisfies Property α and Property γ, $\gtrdot^C$ and so $>^C$ M-rationalizes C.

We again have a corollary.

Corollary 1.3. *A choice function C on $(X, \mathcal{P}_{fin}(X))$ satisfying Property σ is M-rational if and only if $>^C$ uniquely M-rationalizes C.*

It also follows that a choice function C satisfying Property σ is M-rational just in case $\gtrdot^C$ M-rationalizes C. Moreover, OTB and TB choice functions are at most singleton-valued. That is, if $C_{\mathfrak{H}}$ is an OTB or TB choice function, then for every $S \in \mathcal{P}_{fin}(X)$, if $C_{\mathfrak{H}}(S) \neq \varnothing$, then $|C_{\mathfrak{H}}(S)| = 1$. (For an arbitrary set A, we write $|A|$ to denote the cardinality of A.) Why is this? This is so because every OTB or TB choice function is also connected M-rational, as indicated in Section 5.1 . We can guarantee connectivity with the following condition:

Property π. For every $S \in \mathcal{S}$, if $C(S) \neq \varnothing$, then $|C(S)| = 1$.

As one should expect, Property π is independent of Property σ, and in the presence of Property α, a choice function C on $(X, \mathcal{P}_{fin}(X))$ satisfies Property π if and only if for every $x, y \in X$, $|C(\{x,y\})| = 1$.

Now that we have briefly indicated how to generalize the results of the previous section, we present a result most relevant for our study of OTB and TB choice functions.

Theorem 1.6. *A choice function C is asymmetric, connected M-rational if and only if it satisfies Property α, Property γ, Property σ, and Property π.*

Proof. The direction from left to right is trivial. For the converse, observe if C satisfies Property α, Property γ, and Property σ, $>^C$ M-rationalizes C and is furthermore asymmetric. If C also satisfies Property π, clearly $>^C$ is connected.

Of course, OTB and TB choice functions satisfy Property α, Property γ, Property σ, and Property π. We thereby have the following theorem.

Theorem 1.7. *Every OTB or TB choice function satisfies Property α, Property γ, Property σ, and Property π.*

Indeed, the converse holds for OTB and TB choice functions.

Theorem 1.8. *Every choice function satisfying Property α, Property γ, Property σ, and Property π is both an OTB and TB choice function.*

What about OTB choice functions based on an Original Take The Best model that has no guessing? Somewhat surprisingly, such OTB choice functions can be classified by the same properties of the previous theorem.

Theorem 1.9. *Every choice function C satisfying Property α, Property γ, Property σ, and Property π is an OTB choice function $C_{\mathfrak{H}}$ based on a decisive Original Take The Best model $\mathfrak{H}$.*

Can we produce a similar representation for a Take The Best model? It is easy to see that if $\mathfrak{H}$ is a discriminating Take The Best model, then $\mathfrak{H}$ is asymmetric, connected, and transitive, whereby it follows that $\mathfrak{H}$ is also negatively transitive and acyclic.

Lemma 1.2. *If $\mathfrak{H}$ is a discriminating Take The Best model, then $>$ is transitive.*

Proof. Let $x, y, z \in X$ be such that $x > y$ and $y > z$. Let i be the least index for which $x >_i y$, and let j be the least index for which $y >_j z$. First, assume $i < j$. Then it must be the case that $Cue_i(z) \in \{-, ?\}$, for otherwise, if $Cue_i(z) = +$, then $>_i$ is discriminating between z and y, yielding a contradiction; thus, $Cue_i(z) \in \{-, ?\}$. Furthermore, i is the least index for which Cue_i is discriminating between x and z, for otherwise, either j is not the least index for which Cue_j is discriminating between y and z or i is not the least index for which Cue_i is discriminating between x and y. Thus, $x > z$. A similar argument shows that if $j < i$, then $x > z$.

Choice functions based on a discriminating Take The Best model are "super" rational, and in addition to Property π, satisfy the following conditions:

Property β. For every $S, T \in \mathcal{S}$, if $S \subseteq T$ and $C(S) \cap C(T) \neq \varnothing$, then $C(S) \subseteq C(T)$.
Aizerman's Property. For every $S, T \in \mathcal{S}$, if $S \subseteq T$ and $C(T) \subseteq S$, then $C(T) \subseteq C(S)$.
Property ι. For every $S \in \mathcal{S}$, $|C(S)| = 1$.

We then have the following general result.

Theorem 1.10. *Let C be a choice function. Then the following are equivalent:*

(i) C is asymmetric, transitive, connected M-rational.
(ii) C is asymmetric, negatively transitive, connected M-rational.
(iii) C is acyclic, connected M-rational.
(iv) C satisfies Property α and Property ι.
(v) C is regular and satisfies Property α and Property π.
(vi) C is regular and satisfies Property α, Property γ, and Property π.
(vii) C is regular and satisfies Property α, Property β, and Property π.
(viii) C is regular and satisfies Property α, Aizerman's Property, and Property π.

Proof. Left to the reader.

Such properties characterize a Take The Best model for which there is no guessing.

Theorem 1.11. *Every TB choice function $C_{\mathfrak{H}}$ based on a discriminating Take The Best model $\mathfrak{H}$ satisfies Property α and Property ι.*

Theorem 1.12. *Every choice function C satisfying Property α and Property ι is a TB choice function $C_{\mathfrak{H}}$ based on a discriminating Take The Best model $\mathfrak{H}$.*

Proof. Let C be a choice function satisfying Property α and Property ι. We first define a sequence of sets recursively. In the following, let $n := |X|$. Set $X_0 := C(X)$. Then assuming $X_1, \ldots, X_{m-1}$ are defined, set

$$X_m := \begin{cases} C(\bigcap_{i<m}(X \setminus X_i)) & \text{if } \bigcap_{i<m}(X \setminus X_i) \neq \varnothing \\ \varnothing & \text{otherwise.} \end{cases}$$

Clearly $\{X_i : i < n\}$ is a partition of X. We now define a collection of cues $(Cue_i)_{i<n}$. Define Cue_0 by setting $Cue_0(x) := +$ for every $x \in X$. Then for each i with $0 < i < n$, define Cue_i by setting for every $x \in X$,

$$Cue_i(x) := \begin{cases} + & \text{if } x \in \bigcup_{j<i} X_j \\ ? & \text{otherwise.} \end{cases}$$

Now let $\mathfrak{H} := (X, (Cue_i)_{i<n}, Ran, >)$, where $>$ is defined as in Definition 1.5 and *Ran* is picked arbitrarily. It is easy to verify that $\mathfrak{H}$ is a discriminating Take The Best model and that $> = >^C$, whence $C = C_{\mathfrak{H}}$, as desired.

1.5 Conclusion and Discussion

There is an on-going discussion about the psychological plausibility of probabilistic mental models in general and Take The Best in particular (see [6, 9]). Much of this discussion centers on the methods needed to obtain a cue validity ordering required to implement Take The Best. Some critics have suggested, for example, that cue validities cannot be computed because memory itself does not encode missing information. According to [9], this view is misinformed and the relevant cue orderings can arise from evolutionary, social, or individual learning.

It seem obvious that the plausibility of Take The Best very much depends on finding computationally feasible ways of determining cue validity orderings. In this article we presuppose that Gigerenzer and his colleagues are right and that the relevant cue validity orderings are computable. Then the main remaining problem is how to use cue validity orderings. Of course, it is important to take note that an Original Take The Best model and a Take The Best model are two fundamentally different ways of using the information provided by cues.

We saw that an Original Take The Best model, even without guessing, is compatible with violations of transitivity and the existence of preference cycles. On the other side of the spectrum, we have seen that the discriminating Take The Best model is particularly well behaved. The ordering that this model induces obeys transitivity, and the corresponding choice function obeys not only classical coherence constraints like Property α and Property γ but also other important constraints like Aizerman's Property and Property β.

Gigerenzer compares both heuristics in [7, pp. 194, 195]. He does not recommend one in particular but he mentions that it has been demonstrated that there are systematic intransitivities resulting from incommensurability along one dimension between two biological systems. This is the example he has in mind:

> Imagine three species a, b and c. Species a inhabits both water and land; species b inhabits both water and air. Therefore the two only compete in water, where species a defeats species b. Species c inhabits land and air, so it only competes with b in the air, where it is defeated by b. Finally, when a and c meet, it is only on land, and here, c is in its element and defeats a. A linear model that estimates some value for the combative strength of each species independently of the species with which it is competing would fail to capture this non-transitive cycle.

Gigerenzer seems to suggest that models that do not allow for the possibility of non-transitive cycles would not be rich enough to capture intransitivities exhibited by biological systems. This argument seems to favor the use of the Original Take The Best heuristic. But the argument based on the example is nevertheless notoriously weak. The non-transitive cycle appears only if one adopts a particularly poor representational strategy. Gigerenzer seems to argue that there are intrinsic cycles out there in the world to be captured by our models. Therefore, the imposition of transitivity on a model allegedly limits its representational capacity. But this is not the case. The existence of cycles is relative to a choice of representational language (where the predicate "defeats" is not relativized to air, land or water). The cycle vanishes if the language is rich enough to represent the example more thoroughly.

On the other hand, one might argue that the agent exhibiting non-transitive preferences could be threatened by a smart bookie who can always take advantage of him. These pragmatic arguments, nevertheless, are plagued by all sort of controversial assumptions that limit their use.

Even if one is convinced that there is no reason to use the stricter stopping rule characteristic of the Original Take The Best heuristic,[3] we remind the reader that the use of guessing can reintroduce non-transitive cycles. So it seems that a general theory of choice functions for Take The Best should allow for the possibility of cycles and therefore should abandon Regularity.

Of course, a theory of choice functions of this sort does not have a prescriptive or normative role, but a descriptive role. Its sole function is not to impose rationality constraints on choice but to faithfully register patterns of choice permitted by the algorithm. As we argued above, the theory that thus arises still retains some of the central coherence constraints of normative theories. But the interpretation of these constraints is different in the context of the descriptive theory. The coherence constraints (like Property α) reflect regularities verified by the choice functions that correspond to the algorithm, which, in turn, is taken as an epistemological primitive.

One of the central points that Gigerenzer wants to make is that fast and frugal methods, when adapted, can be as accurate or more accurate than linear models that integrate information. But he seems to suggest as well that speed and frugality cannot be reconciled with the traditional norms of rationality, like transitivity.[4] As we

[3] For example, if one thinks that the stopping rule used in Take The Best is psychologically more adequate.

[4] In fact, Gigerenzer's interest is to question the view – that he sees as common and widespread – that *only* "rational" methods are accurate.

showed in this article, Gigerenzer has failed to make a strong case for questioning this view. Indeed, there are frugal methods that are fairly well behaved from the normative point of view (like a discriminating Take The Best model).

1.5.1 Future Work

In Section 1.3 we noted that it would be nice to have a "fast and frugal" extension of TB. We are working on such an extension in a companion article [3]. Here is the basic idea of the extension: when an agent has to decide what is admissible from a set S of cardinality larger than two the decision involves first an act of *framing* where the set is ordered producing a list. So, if the set has three elements the initial stage consists on producing a list:

$$L(S) = (a_1, a_2, a_3)$$

After the set S has been framed as a list, the list is used to pick an element as follows: one starts with the first pair (a_1, a_2) and compares the two elements (by using the binary relation $<$ that we have used in this article). Say that the dominant element is a_1. Then a_1 is compared with the third element a_3. We take the dominant element from this pair and the process terminates by selecting this dominant element (say a_3).

This type of choice has been studied in a recent article by A. Rubinstein and Y. Salant [16]. The authors call it *successive choice*. As a matter of fact [16] offers an interesting model of choice from lists that is obviously relevant for our extension of TB.

Among other things, Rubinstein and Salant discuss various axioms that apply to choice from lists. A salient one is a version of the independence of irrelevant alternatives (another name of the condition we called α above). They call this axiom "List Independence of Irrelevant Alternatives" (LIIA):

LIIA A choice function from lists C satisfies LIIA for every list $(a_1, \dots a_k)$, if $C(a_1, \dots a_k) = a_i$, then $C(a_1, \dots, a_{j-1}, a_{j+1}, \dots, a_k)$, for all $1 \leq j \leq k$, $j \neq i$.

The axiom says that deleting an element that is not chosen from a list does not alter the choice. They then prove a representation result for choice from lists in terms of this axiom (in fact in terms of a condition that is equivalent to LIIA).

As Rubinstein and Salant remark in [16] not every successive choice function satisfies LIIA. In our case, when the underlying preference relation allows for cycles, it is easy to see that the choice function induced by our extension of the Take The Best algorithm fails to obey LIIA.

When the underlying preference relation does obey transitivity the extended choice function is better behaved and we can use Rubinstein and Salant's main representation result to characterize it. In any case, it is clear that that this extension does meet the requirement of representing inductive inference via a satisficing algorithm (it turns out that satisficing itself is a particular type of choice from lists).

There are many possible areas of application of the theory sketched here. One of these applications is the area of belief change. One can study, for example, the notion of *contraction* of a sentence A from a set of sentences K, which consist on effectively eliminating A from K. Notice that in order to do so it is not enough to just delete A because the sentence might be implied by other sentences in K. So one has to *choose* what elements of K one wants to delete in order to contract A from K. Recently there has been some work studying the general properties that an operator of contraction should have. And there has been some work as well exploring systematic relations between standard axioms in the theory of choice functions (conditions like α, etc) and salient axioms in the theory of contraction. So, it seems that an obvious thing one can do is to consider the constraints on contraction functions imposed by bounded conditions on choice. The corresponding conditions can axiomatize a bounded notion of belief change where one choses among possible contractions in a fast and frugal way.

We consider the application sketched above in a second companion article. We also consider there in more detail some foundational issues related to the model presented in this article (issues related to the reasons for looking at the choice functions induced by fast and frugal methods).

References

1. M. Aizerman. New problems in the general choice theory: Review of a research trend. *Social Choice and Welfare*, 2(4):235–282, December 1985.
2. M. Aizerman and A. Malishevksi. General theory of best varients choice: Some aspects. *IEEE Transactions of Automatic Control*, 26(5):1030–1040, October 1981.
3. H. Arló-Costa and A.P. Pedersen. A fast and frugal extension of Gigerenzer's Take the Best. Manuscript, Carnegie Mellon University, 2008.
4. J.L. Bermúdez. Rationality, logic and fast and frugal heuristics. *Behavioral and Brain Sciences*, 23:744–745, 2000.
5. M.S. Dawkins. *Unravelling Animal Behaviour*. Longman, London Second edition, 1995.
6. A.M. Dougherty, M.R. Franco-Watkins, and R. Thomas. Psychological plausibility of the theory of probabilistic mental models and the fast and frugal heuristics. *Psychological Review*, 115(1):199, 213, 2008.
7. G. Gigerenzer. *Adaptive Thinking: Rationality in the Real World*. Oxford University Press, Oxford, 2000.
8. G. Gigerenzer and D.G. Goldstein. Reasoning the fast and frugal way: Models of bounded rationality. *Psychological Review*, 103:650–669, 1996.
9. U. Gigerenzer, G. Hoffrage, and D.G. Goldstein. Fast and frugal heuristics are plausible models of cognition: Reply to Dougherty, Franco-Watkins, and Thomas (2008). *Psychological Review*, 115(1):230–239, 2008.
10. S.O. Hansson. *A Textbook of Belief Dynamics*. Kluwer Academic Publishers, Dordrecht, 1999.
11. D. Kahneman. A perspective on judgment and choice: Mapping bounded rationality. *American Psychologist*, 58(9):697–720, September 2003.
12. D. Kahneman and A. Tversky. Judgment under uncertainty: Heuristics and biases. *Science*, 185:1124–1131, 1974.
13. H. Moulin. Choice functions over a finite set: A summary. *Social Choice and Welfare*, 2(2):147–160, September 1985.

14. H. Rott. Belief contraction in the context of the general theory of rational choice. *Journal of Symbolic Logic*, 58(4):1426–1450, December 1993.
15. H. Rott. *Change, Choice and Inference: A Study of Belief Revision and Non-monotonic Reasoning*. Oxford University Press, Oxford, 2001.
16. A. Rubinstein and Y. Salant. A model of choice from lists. *Theoretical Economics*, 45:3–17, 2006.
17. A.K. Sen. Choice functions and revealed preference. *The Review of Economic Studies*, 38(3):307–317, July 1971.
18. A.K. Sen. Social choice theory: A re-examination. *Econometrica*, 45(1):53–88, January 1977.
19. H.A. Simon. Invariants of human behavior. *Annual Review of Psychology*, 41:1–20, January 1990.
20. K. Suzumura. *Rational Choice, Collective Decisions, and Social Welfare*. Cambridge University Press, Cambridge, 1983.

Chapter 2
Why Do We Need Justification Logic?

Sergei Artemov*

2.1 Introduction

Since Plato, the notion of justification has been an essential component of epistemic studies (cf. [17, 24, 26, 28, 38, 44, 51], and many others). However, until recently, the notion of justification was conspicuously absent in the mathematical models of knowledge within the epistemic logic framework. Commencing from seminal works [30, 55], the notions of Knowledge and Belief have acquired formalization by means of modal logic with modals *F is known* and *F is believed*. Within this approach, the following analysis was adopted: For a given agent,

$$F \textit{ is known} \quad \sim \quad F \textit{ holds in all epistemically possible situations.}$$

The deficiency of this approach is displayed most prominently, in the *Logical Omniscience* feature of the modal logic of knowledge (cf. [19, 20, 31, 43, 46]). This lack of a justification component has, perhaps, contributed to a certain gap between epistemic logic and mainstream epistemology [28, 29]. We would like to think that Justification Logic is a step towards filling this void.

Justification Logic had been anticipated in [25] (as the logic of explicit mathematical proofs) and in [54] (in epistemology), developed in [2, 3, 36, 42] and other papers (as the Logic of Proofs), and then in [4, 6, 7, 9, 14, 22, 23, 27, 33, 35, 45, 48, 50, 56] and other papers in a broader epistemic context. It introduces a long-anticipated mathematical notion of justification, making epistemic logic more expressive. We now have the capacity to reason about justifications, simple and compound. We can compare different pieces of evidence pertaining to the same fact.

Sergei Artemov
Graduate Center CUNY, New York City, NY 10016, USA,
e-mail: sartemov@gc.cuny.edu

* This work has been partially supported by NSF grant 0830450, CUNY Collaborative Incentive Research Grant CIRG1424, and PSC CUNY Research Grant PSCREG-39-721.

J. van Benthem et al. (eds.), *Games, Norms and Reasons*, Synthese Library 353,
DOI 10.1007/978-94-007-0714-6_2, © Springer Science+Business Media B.V. 2011

We can measure the complexity of justifications, which leads to a coherent theory of logical omniscience [8]. Justification Logic provides a novel mechanism of evidence-tracking which seems to be a key ingredient in the analysis of knowledge. Finally, Justification Logic furnishes a new, evidence-based foundation for the logic of knowledge, according to which

$$F \textit{ is known} \quad \sim \quad F \textit{ has an adequate justification.}$$

Justification assertions have the format $t{:}F$, which is read generically as

t is a justification of F.

There is also a more strict "justificationist" reading in which $t{:}F$ is understood as

t is accepted by the agent as a justification of F.

Justification Logic is general enough to incorporate other semantics. Arithmetical semantics of proofs has been the original motivation for the early Justification Logic systems [2, 3, 7, 13]. Game theoretical semantics for Justification Logic has been developed in [47, 48]. The topological semantics of Justification Logic has been studied in [10, 11]. Furthermore, [5] quotes Tudor Protopopescu, who suggests that $t{:}F$ could also be assigned an externalist, non-justificationist reading, something like

F satisfies conditions t.

In this setting, *t* would be something like a set of causes or counterfactuals.

Justification Logic has been built so far on the simplest base: *classical Boolean logic*, and it is a natural next step to extend these ideas to more elaborate logical models, e.g., intuitionistic and substructural logics, conditionals, relevance logics, and logics of counterfactual reasoning. There are several good reasons for choosing a Boolean logic base for our first meaningful step. At this stage, we are concerned first with *justifications*, which provide a sufficiently serious challenge on even the simplest Boolean base. Once this case is sorted out in a satisfactory way, we can move on to incorporating justifications into other logics. Second, Boolean-based Justification Logic seems to cover known paradigmatic examples, e.g., Russell's and Gettier's examples [6] and Kripke's Red Barn example, which we consider below.

Within the Justification Logic framework, we treat both – **justifications**, which do not necessarily yield the truth of a belief, and **factive justifications**, which yield the truth of the belief. This helps to capture the essence of discussion about these matters in epistemology, where justifications are not generally assumed to be factive.

In this paper, we consider the case of one agent only, although multi-agent Justification Logics have already been studied [4, 9, 56].

Formal logical methods do not directly solve philosophical problems but rather provide a tool for analyzing assumptions and ensuring that we draw correct conclusions. Our hope is that Justification Logic does just that.

2.2 Justifications and Operations

In order to build a formal account of justification, we will make some basic structural assumptions: justifications are abstract objects which have structure, agents do not lose or forget justifications, agents apply the laws of classical logic and accept their conclusions, etc.

We assume two basic operations on justifications, *Application* "·" and *Sum* "+," both having clear epistemic meaning and exact interpretations in relevant mathematical models.

The *Application* operation "·" performs one epistemic action, a one-step deduction according to the *Modus Ponens* rule. Application takes a justification s of an implication $F \rightarrow G$ and a justification t of its antecedent, F, and produces a justification $s \cdot t$ of the succedent, G. Symbolically,

$$s{:}(F \supset G) \supset (t{:}F \supset (s{\cdot}t){:}G). \tag{2.1}$$

This is a basic property of justification-type objects assumed in combinatory logic and λ-calculi (cf. [52]), Brouwer-Heyting-Kolmogorov semantics [53], Kleene realizability [32], the Logic of Proofs LP [3], etc. Application principle (2.1) is related to the epistemic closure principle (cf., for example, [39])

one knows everything that one knows to be implied by what one knows. (2.2)

However, (2.1) does not rely on (10.3), since (2.1) deals with a broader spectrum of justifications not necessarily linked to knowledge. If justifications s and t are formal Hilbert-style proofs, then $s{\cdot}t$ can be understood as a new proof obtained from s and t by a single application of the rule *Modus Ponens* to all possible premises $F \rightarrow G$ from s, and F from t:

$$s{\cdot}t = s * t * \ulcorner G_1 \urcorner * \ldots * \ulcorner G_n \urcorner,$$

where $*$ is concatenation, $\ulcorner X \urcorner$ denotes the Gödel number of X, and G_i's are all formulas from t for which there is a formula $F \rightarrow G_i$ from s.

The second basic operation *Sum* "+" expresses the idea of pooling evidence together without performing any epistemic action. Operation "+" takes justifications s and t and produces $s+t$, which is a justification for everything justified by s or by t.

$$s{:}F \supset (s+t){:}F \quad \text{and} \quad s{:}F \supset (t+s){:}F.$$

In the context of formal proofs, the sum "$s+t$" can be interpreted as a concatenation of proofs s and t

$$s+t = s * t.$$

Such an operation is needed to connect Justification Logic with epistemic modal logic. Justification Logic systems without "+" have been studied in [12, 34, 35].

Justification terms (polynomials) are built from justification variables $x, y, z, \ldots$ and justification constants $a, b, c, \ldots$ by means of the operations "·" and "+." Constants denote atomic justifications which the system no longer analyzes; variables denote unspecified justifications. For the sake of technical convenience, we assume that each constant comes with indices $i = 1, 2, 3\ldots$ which we will omit whenever it is safe.

More elaborate Justification Logic systems use additional operations on justifications, e.g., verifier "!" and negative verifier "?" [3, 6, 45, 49, 50], but we will not need them in this paper.

2.3 Basic Logic of Justifications

Formulas are built from propositional atoms as the usual formulas of Boolean logic, e.g., by means of logical connectives $\wedge, \vee, \rightarrow, \neg$ with the additional formation rule: *whenever t is a justification term and F is a formula, t:F is again a formula.*

The basic Logic of Justifications J_0 contain the following postulates:

A1. *Classical propositional axioms and rule Modus Ponens,*
A2. *Application Axiom* $s{:}(F \supset G) \supset (t{:}F \supset (s{\cdot}t){:}G)$,
A3. *Sum Axiom* $s{:}F \supset (s{+}t){:}F$, $s{:}F \supset (t{+}s){:}F$.

J_0 is the logic of general (not necessarily factive) justifications for an absolutely skeptical agent for whom no formula is provably justified, i.e., J_0 does not derive $t{:}F$ for any t and F. Such an agent is, however, capable of making *relative justification conclusions* of the form

$$\textit{if } x{:}A,\ y{:}B,\ \ldots,\ z{:}C \textit{ hold, then } t{:}F.$$

J_0 is able, with this capacity, to adequately emulate other Justification Logic systems within its language.

Well-known examples of epistemic reasoning reveal that logical axioms are often assumed justified. Justification Logic offers a flexible mechanism of *Constant Specifications* that represents different shades of this kind of logical awareness.

Justification Logic distinguishes between assumptions and justified assumptions. Constants are used to denote justifications of assumptions in situations where we don't analyze these justifications further. Suppose we want to postulate that an axiom A is justified for a given agent. The way to state it in Justification Logic is to postulate

$$e_1{:}A$$

for some justification constant e_1 with index 1. Furthermore, if we want to postulate that this new principle $e_1{:}A$ is also justified, we can postulate

$$e_2{:}(e_1{:}A)$$

for the similar constant e_2 with index 2, then

$$e_3{:}(e_2{:}(e_1{:}A)),$$

etc. Keeping track of indices for "in-depth justifications" is not really necessary, but it is easy and helps in decision procedures (cf. [37]). By $e_n{:}e_{n-1}{:}\ldots{:}e_1{:}A$, we mean $e_n{:}(e_{n-1}{:}\ldots{:}(e_1{:}A)\ldots)$. A set of assumptions of this kind for a given logic is called a *Constant Specification*. Here is a formal definition.

A **Constant Specification** *CS* for a given logic $\mathscr{L}$ is a set of formulas

$$e_n{:}e_{n-1}{:}\ldots{:}e_1{:}A \quad (n \geq 1),$$

in which A is an axiom of $\mathscr{L}$, and $e_1, e_2, \ldots, e_n$ are constants with indices $1, 2, \ldots, n$. We also assume that *CS* contains all intermediate specifications, i.e., whenever $e_n{:}e_{n-1}{:}\ldots{:}e_1{:}A$ is in *CS*, then $e_{n-1}{:}\ldots{:}e_1{:}A$ is in *CS*, too. Here are typical examples of constant specifications:

- *empty*: $CS = \varnothing$. This corresponds to an absolutely skeptical agent (cf. a comment after axioms of J_0).
- *finite*: *CS* is a finite set of formulas. This is a representative case, since any specific derivation in Justification Logic concerns only finite sets of constants and constant specifications.
- *axiomatically appropriate*: For each axiom A, there is a constant e_1 such that $e_1{:}A$ is in *CS*, and if $e_n{:}\ldots{:}e_1{:}A \in CS$, then $e_{n+1}{:}e_n{:}\ldots{:}e_1{:}A \in CS$.
- *total*: For each axiom A and **any** constants $e_1, e_2, \ldots, e_n$,

$$e_n{:}e_{n-1}{:}\ldots{:}e_1{:}A \in CS.$$

Naturally, the total constant specification is axiomatically appropriate.

Logic of Justifications with given Constant Specification

$$\mathsf{J}_{CS} = \mathsf{J}_0 + CS.$$

Logic of Justifications

$$\mathsf{J} = \mathsf{J}_0 + \mathrm{R4},$$

where R4 is the **Axiom Internalization Rule**:

For each axiom A and any constants $e_1, e_2, \ldots, e_n$, *infer* $e_n{:}e_{n-1}{:}\ldots{:}e_1{:}A$.

Note that J_0 is $\mathsf{J}_\varnothing$, and J is J_{CS} with the total Constant Specification *CS*. The latter reflects the idea of the unrestricted logical awareness for J. A similar principle appeared in the Logic of Proofs LP.

For each constant specification *CS*, J_{CS} enjoys the Deduction Theorem because J_0 contains propositional axioms and *Modus Ponens* as the only rule of inference.

Logical awareness expressed by axiomatically appropriate constant specifications ensures an important *Internalization Property* of the system. This property was anticipated by Gödel in [25] for the logic of explicit mathematical proofs, and was first established for the Logic of Proofs LP in [2, 3].

Theorem 2.1. *For each axiomatically appropriate constant specification CS, J_{CS} enjoys the Internalization Property:*

$$\textit{If} \vdash F, \textit{ then} \vdash p{:}F \textit{ for some justification term } p.$$

Proof. Induction on derivation length. If F is an axiom A, then, since CS is axiomatically appropriate, there is a constant e_1 such that $e_1{:}A$ is in CS, hence an axiom of J_{CS}. If F is in CS, then, since CS is axiomatically appropriate, $e_n{:}F$ is in CS for some constant e_n. If F is obtained by *Modus Ponens* from $X \supset F$ and X, then, by the Induction Hypothesis, $\vdash s{:}(X \supset F)$ and $\vdash t{:}X$ for some s, t. By the Application Axiom, $\vdash (s{\cdot}t){:}F$.

Internalization in J is an explicit incarnation of the Necessitation Rule in modal logic K:

$$\vdash F \quad \Rightarrow \vdash \Box F.$$

Let us consider some basic examples of derivations in J. In Examples 2.1 and 2.2, only constants of level 1 have been used; in such situations we skip indices completely.

Example 2.1. This example shows how to build a justification of a conjunction from justifications of the conjuncts. In the traditional modal language, this principle is formalized as

$$\Box A \wedge \Box B \rightarrow \Box(A \wedge B).$$

In J we express this idea in a more precise justification language.

1. $A \supset (B \supset A \wedge B)$, a propositional axiom;
2. $c{:}[A \supset (B \supset A \wedge B)]$, from 1, by R4;
3. $c{:}[A \supset (B \supset A \wedge B)] \supset [x{:}A \supset (c{\cdot}x){:}(B \supset A \wedge B)]$, axiom A2;
4. $x{:}A \supset (c{\cdot}x){:}(B \supset A \wedge B)$, from 2 and 3, by *Modus Ponens*;
5. $(c{\cdot}x){:}(B \supset A \wedge B) \supset (y{:}B \supset ((c{\cdot}x){\cdot}y){:}(A \wedge B))$, axiom A2;
6. $x{:}A \supset (y{:}B \supset ((c{\cdot}x){\cdot}y){:}(A \wedge B))$, from 4, 5, by some propositional reasoning;
7. $x{:}A \wedge y{:}B \supset ((c{\cdot}x){\cdot}y){:}(A \wedge B)$, from 6, by propositional reasoning.

Derived formula 7 contains constant c, which was introduced in line 2, and the complete reading of the result of this derivation is

$$x{:}A \wedge y{:}B \supset ((c{\cdot}x){\cdot}y){:}(A \wedge B), \textit{ given } c{:}[A \supset (B \supset A \wedge B)].$$

Example 2.2. This example shows how to build a justification of a disjunction from justifications of either disjuncts. In the usual modal language, this is represented by

$$\Box A \vee \Box B \rightarrow \Box(A \vee B).$$

Let us see how this would look in J.

1. $A \supset A \vee B$, by A1;
2. $a{:}[A \supset A \vee B]$, from 1, by R4;
3. $x{:}A \supset (a{\cdot}x){:}(A \vee B)$, from 2, by A2 and *Modus Ponens*;
4. $B \supset A \vee B$, by A1;

5. $b{:}[B \supset A \vee B]$, from 4, by R4;
6. $y{:}B \supset (b{\cdot}y){:}(A \vee B)$ from 5, by A2 and *Modus Ponens*;
7. $(a{\cdot}x){:}(A \vee B) \supset (a{\cdot}x+b{\cdot}y){:}(A \vee B)$, by A3;
8. $(b{\cdot}y){:}(A \vee B) \supset (a{\cdot}x+b{\cdot}y){:}(A \vee B)$, by A3;
9. $(x{:}A \vee y{:}B) \supset (a{\cdot}x+b{\cdot}y){:}(A \vee B)$ from 3, 6, 7, 8, by propositional reasoning.

The complete reading of the result of this derivation is

$$(x{:}A \vee y{:}B) \supset (a{\cdot}x+b{\cdot}y){:}(A \vee B),\ \textit{given}\ a{:}[A \supset A \vee B]\ \textit{and}\ b{:}[B \supset A \vee B].$$

These examples, perhaps, leave the (correct) impression that J can emulate derivations in the corresponding modal logic; here it is K, but at the expense of keeping track of specific justifications. A need for such additional bureaucracy requires explanation and illustration, which is the main goal of this paper. Before we proceed to Section 2.4, in which such an example is provided, we briefly list applications of Justification Logic so far:

- intended provability semantics for Gödel's provability logic S4 with the Completeness Theorem [2, 3];
- formalization of Brouwer-Heyting-Kolmogorov semantics for intuitionistic propositional logic with the Completeness Theorem [2, 3];
- a general definition of the Logical Omniscience property and theorems that evidence assertions in Justification Logic are not logically omniscient [8];
- an evidence-based approach to Common Knowledge (so-called Justified Common Knowledge) which provides a rigorous epistemic semantics to McCarthy's "any fool knows" systems [1, 4, 40]. Justified Common Knowledge offers formal systems which are less restrictive than the usual epistemic logics with Common Knowledge [4];
- formalization of Gettier examples in Justification Logic with missing assumptions and redundancy analysis [6], which demonstrates that Justification Logic methods can be applied in formal epistemology;
- analysis of Knower and Knowability paradoxes [15, 16].

The **Correspodence Theorem** [2, 3, 6, 13, 50] is a cumulative result stating that for each of the major epistemic modal logics K, T, K4, S4, K45, KD45, S5, there is a system of justification terms and a corresponding Justification Logic system (called J, JT, J4, LP, J45, JD45, and JT45) capable of recovering explicit justifications for modalities in any theorem of the original modal logic. This theorem is proven by a variety of methods ranging from cut-elimination in modal logics to a semantical proof using Kripke-Fitting models (cf. Section 2.5).

Complexity issues in Justification Logic have been addressed in [14, 33, 35–37, 41].

2.4 Red Barn Example and Tracking Justifications

We illustrate new capabilities of Justification Logic on a paradigmatic Red Barn example which Kripke developed in 1980 (cf. [39], from which we borrow the formulation, with some editing for brevity).

> Suppose I am driving through a neighborhood in which, unbeknownst to me, papier-mâché barns are scattered, and I see that the object in front of me is a barn. Because I have barn-before-me percepts, I believe that the object in front of me is a barn. Our intuitions suggest that I fail to know barn. But now suppose that the neighborhood has no fake red barns, and I also notice that the object in front of me is red, so I know a red barn is there. This juxtaposition, being a red barn, which I know, entails there being a barn, which I do not, "is an embarrassment".[1]

We consider the Red Barn example a test for theories that explain knowledge. From such a theory, we expect a way to represent what is happening here which maintains epistemic closure principle, but also preserves the epistemic structure of the example.

We present formal analysis of the Red Barn example in epistemic modal logic (Sections 2.4.1 and 2.4.2) and in Justification Logic (Sections 2.4.3 and 2.4.4). We will show that epistemic modal logic only indicates that there is a problem, whereas Justification Logic provides resolution.

To make our point, we don't need to formally capture every single detail of the Red Barn story; it suffices to formalize and verify its "entailment" portion. Let

- B be the sentence "the object in front of me is a barn,"
- R be the sentence "the object in front of me is red."

2.4.1 Red Barn in Modal Logic of Belief

In our first formalization, logical derivation will be performed in epistemic modal logic with "my belief" modality $\Box$. We then externally interpret some of the occurrences of $\Box$ as "knowledge" according to the problem's description. In the setting with belief modality $\Box$, episitemic closure principle (10.3) seems to yield

$$\textit{if } \Box F \textit{ and } \Box(F \rightarrow G) \textit{ are both cases of knowledge, then } \Box G \textit{ is also knowledge.} \tag{2.3}$$

The following is a set of natural formal assumptions of the Red Barn example in the language of epistemic modal logic of belief:

1. $\Box B$, "I believe that the object in front of me is a barn";

2. $\Box(B \wedge R)$, "I believe that the object in front of me is a red barn." At the metalevel, we assume that 2 is knowledge, whereas 1 is not knowledge by the problem's description.

[1] Dretske [18].

In the basic modal logic of belief K (hence in other modal logics of belief), the following hold:

3. $B \wedge R \supset B$, as a logical axiom;
4. $\Box(B \wedge R \supset B)$, obtaned from 3 by Necessitation. As a logical truth, this also qualifies as knowledge.

Within this formalization, it appears that the modal epistemic closure principle (2.3) is violated: line 2, $\Box(B \wedge R)$, and line 4, $\Box(B \wedge R \supset B)$ are cases of knowledge whereas $\Box B$ (line 1) is not knowledge. As we see, the modal language here does not help to resolve this issue, but rather obscures its resolution.

2.4.2 Red Barn in Modal Logic of Knowledge

We will now use epistemic modal logic with 'my knowledge' modality **K**. Here is a straightforward formalization of Red Barn example assumptions:

1. $\neg \mathbf{K}B$, "I do not know that the object in front of me is a barn";
2. $\mathbf{K}(B \wedge R)$, "I know that the object in front of me is a red barn."

It is easy to see that these assumptions are inconsistent in the modal logic of knowledge. Indeed,

3. $\mathbf{K}(B \wedge R \rightarrow B)$, by Necessitation of a propositional axiom;
4. $\mathbf{K}(B \wedge R) \rightarrow \mathbf{K}B$, from 3, by modal logic reasoning;
5. $\mathbf{K}B$, from 2 and 4, by *Modus Ponens*.

Lines 1 and 5 formally contradict each other.

Hence, the language of modal logic of knowledge leads to an inconsistent set of formal assumptions and does not reflect the structure of the Red Barn example properly.

2.4.3 Red Barn in Justification Logic of Belief

Justification Logic seems to provide a more fine-grained analysis of the Red Barn example. We formalize the Red Barn example in J where $t{:}F$ is interpreted as

"I **believe** F for reason t."

This interpretation leaves a possibility to conclude, by some meta-level considerations, that certain assertions $t{:}F$ are in fact cases of knowledge. The epistemic closure principle (10.3) can be naturally formulated according to Application principle (2.1) as *if $t{:}F$ and $s{:}(F \to G)$ are both cases of knowledge, then $(s{\cdot}t){:}G$ is also knowledge.* (2.4)

Note that (2.4) is more precise than (2.3). In (2.3) we postulate that whenever $\Box F$ and $\Box(F \to G)$ are knowledge, the conclusion $\Box G$ is also knowledge, regardless of how this conclusion was obtained. In (2.4), we claim that given that both $t{:}F$ and $s{:}(F \to G)$ are knowledge, $(s{\cdot}t){:}G$ is also knowledge **for specific justification** $s{\cdot}t$. This is how the ambiguous modal language fails to represent the epistemic closure principle: one cannot claim (2.3) when justification behind conclusion $\Box G$ is not linked to those behind premises $\Box F$ and $\Box(F \to G)$. This is the essence of the Red Barn example, and a peril which Justification Logic naturally avoids by virtue of its explicit language.

We naturally introduce individual justifications u for belief that B, and v for belief that $B \wedge R$. The list of assumptions is

1. $u{:}B$, "u is the reason to believe that the object in front of me is a barn";
2. $v{:}(B \wedge R)$, "v is the reason to believe that the object in front of me is a red barn." On the metalevel, the description states that 2 is a case of knowledge, and not merely a belief, whereas 1 is belief which is not knowledge.

Let us try to reconstruct the reasoning of the agent in J:

3. $B \wedge R \supset B$, logical axiom;
4. $a{:}[B \wedge R \supset B]$, from 3, by Axiom Internalization. This is also a case of knowledge;
5. $v{:}(B \wedge R) \to (a{\cdot}v){:}B$, from 4, by Application and *Modus Ponens*;
6. $(a{\cdot}v){:}B$, from 2 and 5, by *Modus Ponens*.

Closure holds! By (2.4) principle on the meta-level, we conclude that $(a{\cdot}v){:}B$ is a case of knowledge. The fact that $u{:}B$ is not a case of knowledge does not spoil the closure principle, since the latter claims knowledge specifically for $(a{\cdot}v){:}B$. Hence, after observing a red façade, I indeed know B, but this knowledge has nothing to do with 1, which remains a case of belief rather than of knowledge, and Justification Logic formalization represents this fairly.

2.4.4 Red Barn in Justification Logic of Knowledge

Within this formalization, $t{:}F$ is interpreted as

"I **know** F for reason t."

As in Section 2.4.2, we assume

1. $\neg u{:}B$, "u is not a sufficient reason to know that the object is a barn";
2. $v{:}(B \wedge R)$, "v is a sufficient reason to know that the object is a red barn."

This is a perfectly consistent set of assumptions even in the logic of factive justifications

$$\mathsf{J} + \textit{Factivity Principle}\ (t{:}F \to F).$$

As in Section 2.4.3, we can derive $(a\cdot v){:}B$ where $a{:}[B\wedge R \supset B]$, but this does not lead to a contradiction. Claims $\neg u{:}B$ and $(a\cdot v){:}B$ naturally co-exist. They refer to different justifications u and $a\cdot v$ of the same fact B; one of them insufficient and the other quite sufficient for my knowledge that B.

2.4.5 Red Barn and Formal Epistemic Models

It appears that in Sections 2.4.3 and 2.4.4, Justification Logic represents the structure of the Red Barn example in a reasonable way which was not directly captured by epistemic modal logic.

In all fairness to modal tools, we could imagine a formalization of the Red Barn example in a sort of bi-modal language with distinct modalities for knowledge and belief, where both claims hold: "$\Box B$," by perceptual belief that B, and "$\mathbf{K}B$" for knowledge that B which is logically derived from perceptual knowledge $\mathbf{K}(B\wedge R)$. However, it seems that such a resolution will, intellectually, involve repeating Justification Logic arguments from Sections 2.4.3 to 2.4.4 in a way that obscures, rather than reveals, the truth. Such a bi-modal formalization would distinguish $u{:}B$ from $(a\cdot v){:}B$ not because they have different reasons (which reflects the true epistemic structure of the problem), but rather because the former is labelled "belief" and the latter "knowledge." But what if we need to keep track of a larger number of different unrelated reasons?

By introducing a number of distinct modalities and then imposing various assumptions governing the inter-relationships between these modalities, one would essentially end up with a reformulation of the language of Justification Logic itself (with distinct terms replaced by distinct modalities). This suggests that there may not really be a "halfway point" between the modal language and the language of Justification Logic, at least inasmuch as one tries to capture the essential structure of examples involving the deductive failure of knowledge (e.g., Kripke's Red Barn example). Accordingly, one is either stuck with modal logic and its inferior account of these examples or else moves to Justification Logic and its superior account of these examples. This move can either come about by taking a multi-modal language and imposing inter-dependencies on different modals – ending up with something essentially equivalent to the language of Justification Logic – or else one can use the language of Justification Logic from the start. Either way, all there is to move to is Justification Logic.

2.5 Basic Epistemic Semantics

This section will provide the basics of epistemic semantics for Justification Logic, the main ideas of which have been suggested by Fitting in [22]. The standard epistemic semantics for J (cf. [6]) has been provided by the proper adaptation of Kripke-Fitting models [22] and Mkrtychev models [42].

A Kripke-Fitting J-**model** $\mathcal{M} = (W, R, \mathcal{E}, \Vdash)$ is a Kripke model $(W, R, \Vdash)$ enriched with an **admissible evidence function** $\mathcal{E}$ such that $\mathcal{E}(t, F) \subseteq W$ for any justification t and formula F. Informally, $\mathcal{E}(t, F)$ specifies the set of possible worlds where t is considered admissible evidence for F.[2]

The intended use of $\mathcal{E}$ is in the truth definition for justification assertions:

$u \Vdash t{:}F$ if and only if

(a) F holds for all possible situations, i.e., $v \Vdash F$ for all v such that uRv;
(b) t is an admissible evidence for F at u, i.e., $u \in \mathcal{E}(t, F)$.

An admissible evidence function $\mathcal{E}$ must satisfy the closure conditions with respect to operations "·" and "+":

- *Application*: $\mathcal{E}(s, F \supset G) \cap \mathcal{E}(t, F) \subseteq \mathcal{E}(s{\cdot}t, G)$. This condition states that whenever s is an admissible evidence for $F \supset G$ and t is an admissible evidence for F, their "product," $s{\cdot}t$, is an admissible evidence for G.
- *Sum*: $\mathcal{E}(s, F) \cup \mathcal{E}(t, F) \subseteq \mathcal{E}(s{+}t, F)$. This condition guarantees that $s{+}t$ is an admissible evidence for F whenever either s is an admissible evidence for F or t is an admissible evidence for F.

Given a model $\mathcal{M} = (W, R, \mathcal{E}, \Vdash)$, the forcing relation $\Vdash$ is extended from sentence variables to all formulas as follows: for each $u \in W$,

(a) $\Vdash$ respects Boolean connectives at each world ($u \Vdash F \wedge G$ iff $u \Vdash F$ and $u \Vdash G$, $u \Vdash \neg F$ iff $u \nVdash F$, etc.);
(b) $u \Vdash t{:}F$ iff $u \in \mathcal{E}(t, F)$ and $v \Vdash F$ for every $v \in W$ with uRv.

Note that an admissible evidence function $\mathcal{E}$ may be regarded as a Fagin-Halpern awareness function [10] equipped with the structure of justifications and closure conditions.

A model $\mathcal{M} = (W, R, \mathcal{E}, \Vdash)$ *respects a Constant Specification CS at $u \in W$* if $u \in \mathcal{E}(c, A)$ for all formulas $c{:}A$ from CS. Furthermore, $\mathcal{M} = (W, R, \mathcal{E}, \Vdash)$ *respects a Constant Specification CS* if $\mathcal{M}$ respects CS at each $u \in W$.

Theorem 2.2. *For any Constant Specification CS, J_{CS} is sound and complete for the class of all Kripke-Fitting models respecting CS.*

Mkrtychev semantics is a predecessor of Kripke-Fitting semantics [42]. *Mkrtychev models* are Kripke-Fitting models with a single world.

---

[2] Admissible evidence here is not certain evidence, but rather relevant evidence. Here is an example from [22]. "What might serve as admissible evidence for the statement, 'George Bush is editor of The New York Times'? Clearly the editorial page of any copy of The New York Times would serve, while no page of Mad Magazine would do (although the magazine might very well contain the claim that George Bush does edit the Times). Admissible evidence need not be evidence of a fact, nor need it be decisive – it could happen that The New York Times decides to omit its editor's name, or prints the wrong one by mistake. Nonetheless, what the Times prints would count as evidence, and what Mad prints would not."

Theorem 2.3. *For any Constant Specification CS, J_{CS} is sound and complete for the class of Mkrtychev models respecting CS.*

Theorem 2.3 shows that the information about Kripke structure in Kripke-Fitting models can be completely encoded by the admissible evidence function. Mkrtychev models play an important theoretical role in Justification Logic [14, 33, 36, 41]. On the other hand, Kripke-Fitting models can be useful as counter-models with desirable properties since they take into account both epistemic Kripke structure and evidence structure. Speaking metaphorically, Kripke-Fitting models naturally reflect two reasons why a certain fact F can be unknown to an agent: F fails at some possible world or an agent does not have a sufficient evidence of F.

Another application area of Kripke-Fitting style models is Justification Logic with both epistemic modalities and justification assertions (cf. [4, 9]).

2.6 Adding Factivity

Factivity states that a given justification of F is factive, i.e., sufficient for an agent to conclude that F is true. The corresponding *Factivity Axiom* claims that justifications are factive:

$$t{:}F \supset F,$$

which has a similar motivation to the Truth Axiom in epistemic modal logic

$$\mathbf{K}F \supset F,$$

widely accepted as a basic property of knowledge.

The Factivity Axiom first appeared in the Logic of Proofs LP as a principal feature of mathematical proofs. Indeed, in this setting Factivity is valid: if there is a mathematical proof t of F, then F must be true.

We adopt the Factivity Axiom for justifications that lead to knowledge. However, factivity alone does not warrant knowledge, which has been demonstrated by Gettier examples ([24]).

Logic of Factive Justifications:

$$\mathsf{JT}_0 = \mathsf{J}_0 + \textit{Factivity Axiom},$$

$$\mathsf{JT} = \mathsf{J} + \textit{Factivity Axiom}.$$

Systems JT_{CS} corresponding to Constant Specifications *CS* are defined similarly to J_{CS}.

JT**-models** are J-models with reflexive accessibility relations R. The reflexivity condition makes each possible world accessible from itself, which exactly corresponds to the Factivity Axiom. The direct analogue of Theorem 2.1 hold for JT_{CS} as well.

Theorem 2.4. *For any Constant Specification CS, JT_{CS} is sound and complete with respect to the class of JT-models respecting CS.*

Mkrtychev JT-models are singleton JT-models, i.e., JT-models with singleton W's.

Theorem 2.5. *For any Constant Specification CS, JT_{CS} is sound and complete with respect to the class of Mkrtychev JT-models respecting CS.*

2.7 Conclusions

Modal logic fails to fully represent the epistemic closure principle whereas Justification Logic provides a more adequate formalization.

Justification Logic extends the logic of knowledge by the formal theory of justification. Justification Logic has roots in mainstream epistemology, mathematical logic, computer science, and artificial intelligence. It is capable of formalizing a significant portion of reasoning about justifications.

It remains to be seen to what extent Justification Logic can be useful for analysis of empirical, perceptual, and a priori types of knowledge. From the perspective of Justification Logic, such knowledge may be considered as justified by constants (i.e., atomic justifications). Apparently, further discussion is needed here.

Acknowledgements The author is very grateful to Walter Dean, Mel Fitting, Vladimir Krupski, Roman Kuznets, Elena Nogina, Tudor Protopopescu, and Ruili Ye, whose advice helped with this paper. Many thanks to Karen Kletter for editing this text. The author is also indebted to the anonymous referee whose valuable comments helped to sharpen some of the arguments. In particular, the last paragraph of Section 2.4.5 has been essentially suggested by the referee.

References

1. E. Antonakos. Justified and common knowledge: Limited conservativity. In S. Artemov and A. Nerode, editors, *Logical Foundations of Computer Science. International Symposium, LFCS 2007, New York, NY, USA, June 2007, Proceedings*, volume 4514 of *Lecture Notes in Computer Science*, pages 1–11. Springer, 2007.
2. S. Artemov. Operational modal logic. Technical Report MSI 95-29, Cornell University, 1995.
3. S. Artemov. Explicit provability and constructive semantics. *Bulletin of Symbolic Logic*, 7(1):1–36, 2001.
4. S. Artemov. Justified common knowledge. *Theoretical Computer Science*, 357(1–3):4–22, 2006.
5. S. Artemov. Symmetric logic of proofs. In A. Avron, N. Dershowitz, and A. Rabinovich, editors, *Pillars of Computer Science, Essays Dedicated to Boris (Boaz) Trakhtenbrot on the Occasion of His 85th Birthday*, volume 4800 of *Lecture Notes in Computer Science*, pages 58–71. Springer, Berlin, Germany, February 2008.

6. S. Artemov. The logic of justification. *The Review of Symbolic Logic*, 1(4):477–513, December 2008.
7. S. Artemov, E. Kazakov, and D. Shapiro. Epistemic logic with justifications. Technical Report CFIS 99-12, Cornell University, 1999.
8. S. Artemov and R. Kuznets. Logical omniscience via proof complexity. In *Computer Science Logic 2006*, volume 4207, pages 135–149. Springer Lecture Notes in Computer Science, Berlin, Germany, 2006.
9. S. Artemov and E. Nogina. Introducing justification into epistemic logic. *Journal of Logic and Computation*, 15(6):1059–1073, 2005.
10. S. Artemov and E. Nogina. Topological semantics of justification logic. In E.A. Hirsch, A. Razborov, A. Semenov, and A. Slissenko, editors, *Computer Science – Theory and Applications. Third International Computer Science Symposium in Russia, CSR 2008 Moscow, Russia, June 7–12, 2008 Proceedings*, volume 5010 of *Lecture Notes in Computer Science*, pages 30–39. Springer, Berlin, Germany, 2008.
11. S. Artemov and E. Nogina. The topology of justification. *Journal of Logic and Logical Philosophy*, 17(1–2):58–71, 2008.
12. S. Artemov and T. Strassen. Functionality in the basic logic of proofs. Technical Report IAM 93-004, Department of Computer Science, University of Bern, Switzerland, 1993.
13. V. Brezhnev. On the logic of proofs. In *Proceedings of the Sixth ESSLLI Student Session, Helsinki*, pages 35–46, 2001. `http://www.helsinki.fi/esslli/`
14. V. Brezhnev and R. Kuznets. Making knowledge explicit: How hard it is. *Theoretical Computer Science*, 357(1–3):23–34, 2006.
15. W. Dean and H. Kurokawa. From the knowability paradox to the existence of proofs. *Synthese*, 176(2):177–225, September 2010.
16. W. Dean and H. Kurokawa. The knower paradox and the quantified logic of proofs. In A. Hieke, editor, *Austrian Ludwig Wittgenstein Society*, volume 31, Kirchberg am Wechsel, Austria, August 2008.
17. F. Dretske. Conclusive reasons. *Australasian Journal of Philosophy*, 49:1–22, 1971.
18. F. Dretske. Is knowledge closed under known entailment? The case against closure. In M. Steup, and E. Sosa, editors, *Contemporary Debates in Epistemology*, pages 13–26. Blackwell, Oxford, 2005.
19. R. Fagin and J. Halpern. Belief, awareness, and limited reasoning: Preliminary report. In *Proceedings of the Ninth International Joint Conference on Artificial Intelligence (IJCAI-85)*, pages 491–501. Morgan Kaufmann, Los Angeles, CA, August 1985.
20. R. Fagin and J. Halpern. Belief, awareness, and limited reasoning. *Artificial Intelligence*, 34(1):39–76, 1988.
21. R. Fagin, J. Halpern, Y. Moses, and M. Vardi. *Reasoning About Knowledge*. MIT Press, Cambridge 1995.
22. M. Fitting. The logic of proofs, semantically. *Annals of Pure and Applied Logic*, 132(1):1–25, 2005.
23. M. Fitting. A quantified logic of evidence. *Annals of Pure and Applied Logic*, 152(1–3):67–83, March 2008.
24. E. Gettier. Is justified true belief knowledge? *Analysis*, 23:121–123, 1963.
25. K. Gödel. Vortrag bei Zilsel/Lecture at Zilsel's (1938a). In S. Feferman, J.W. Dawson, Jr., W. Goldfarb, C. Parsons, and R.M. Solovay, editors, *Unpublished Essays and Lectures*, volume III of *Kurt Gödel Collected Works*, pages 86–113. Oxford University Press, Oxford, 1995.
26. A. Goldman. A causal theory of knowing. *The Journal of Philosophy*, 64:335–372, 1967.
27. E. Goris. Feasible operations on proofs: The logic of proofs for bounded arithmetic. *Theory of Computing Systems*, 43(2):185–203, August 2008. Published online in October 2007.
28. V.F. Hendricks. Active agents. *Journal of Logic, Language and Information*, 12(4):469–495, 2003.
29. V.F. Hendricks. *Mainstream and Formal Epistemology*. Cambridge University Press, New York, NY, 2005.
30. J. Hintikka. *Knowledge and Belief*. Cornell University Press, Ithaca, NY, 1962.

31. J. Hintikka. Impossible possible worlds vindicated. *Journal of Philosophical Logic*, 4:475–484, 1975.
32. S. Kleene. On the interpretation of intuitionistic number theory. *The Journal of Symbolic Logic*, 10(4):109–124, 1945.
33. N.V. Krupski. On the complexity of the reflected logic of proofs. *Theoretical Computer Science*, 357(1):136–142, 2006.
34. V.N. Krupski. The single-conclusion proof logic and inference rules specification. *Annals of Pure and Applied Logic*, 113(1–3):181–206, 2001.
35. V.N. Krupski. Referential logic of proofs. *Theoretical Computer Science*, 357(1):143–166, 2006.
36. R. Kuznets. On the complexity of explicit modal logics. In *Computer Science Logic 2000*, volume 1862 of *Lecture Notes in Computer Science*, pages 371–383. Springer, Berlin, Germany, 2000.
37. R. Kuznets. *Complexity Issues in Justification Logic*. PhD thesis, CUNY Graduate Center, 2008. http://kuznets.googlepages.com/PhD.pdf
38. K. Lehrer and T. Paxson. Knowledge: Undefeated justified true belief. *The Journal of Philosophy*, 66:1–22, 1969.
39. S. Luper. The epistemic closure principle. In Edward N. Zalta, editor, *The Stanford Encyclopedia of Philosophy*, Fall 2010 Edition. http://plato.stanford.edu/archives/fall2010/entries/closureepistemic/
40. J. McCarthy, M. Sato, T. Hayashi, and S. Igarishi. On the model theory of knowledge. Technical Report STAN-CS-78-667, Stanford University, 1978.
41. R. Milnikel. Derivability in certain subsystems of the logic of proofs is Π_2^p-complete. *Annals of Pure and Applied Logic*, 145(3):223–239, 2007.
42. A. Mkrtychev. Models for the logic of proofs. In S. Adian and A. Nerode, editors, *Logical Foundations of Computer Science '97, Yaroslavl'*, volume 1234 of *Lecture Notes in Computer Science*, pages 266–275. Springer, Berlin, Germany, 1997.
43. Y. Moses. Resource-bounded knowledge. In M. Vardi, editor, *Proceedings of the Second Conference on Theoretical Aspects of Reasoning about Knowledge, March 7–9, 1988, Pacific Grove, California*, pages 261–276. Morgan Kaufmann Pbl., 1988.
44. R. Nozick. *Philosophical Explanations*. Harvard University Press, Cambridge, 1981.
45. E. Pacuit. A note on some explicit modal logics. Technical Report PP-2006-29, University of Amsterdam. ILLC Publications, 2006.
46. R. Parikh. Knowledge and the problem of logical omniscience. In Z. Ras and M. Zemankova, editors, ISMIS-87 *(International Symposium on Methodology for Intellectual Systems)*, pages 432–439. North-Holland, 1987.
47. B. Renne. Propositional games with explicit strategies. *Electronic Notes on Theoretical Computer Science*, 165:133–144, 1999.
48. B. Renne. *Dynamic Epistemic Logic with Justification*. PhD thesis, CUNY Graduate Center, May 2008.
49. N. Rubtsova. Evidence reconstruction of epistemic modal logic S5. In *Computer Science – Theory and Applications. CSR 2006*, volume 3967 of *Lecture Notes in Computer Science*, pages 313–321. Springer, Berlin 2006.
50. N. Rubtsova. On realization of S5-modality by evidence terms. *Journal of Logic and Computation*, 16:671–684, 2006.
51. R.C. Stalnaker. Knowledge, belief and counterfactual reasoning in games. *Economics and Philosophy*, 12:133–163, 1996.
52. A.S. Troelstra and H. Schwichtenberg. *Basic Proof Theory*. Cambridge University Press, Amsterdam, 1996.
53. A.S. Troelstra and D. van Dalen. *Constructivism in Mathematics*, Volumes 1, 2. North–Holland, Amsterdam, 1988.
54. J. van Benthem. Reflections on epistemic logic. *Logique & Analyse*, 133–134:5–14, 1993.
55. G.H. von Wright. *An Essay in Modal Logic*. North-Holland, Amsterdam, 1951.
56. T. Yavorskaya (Sidon). Multi-agent Explicit knowledge. In D. Grigoriev, J. Harrison, and E.A. Hirsch, editors, *Computer Science – Theory and Applications. CSR 2006*, volume 3967 of *Lecture Notes in Computer Science*, pages 369–380. Springer, Berlin, Germany, 2006.

Chapter 3
Why Meanings Are Not Normative

Akeel Bilgrami

1. This paper is a response prompted by a dissatisfaction with the prevalent discussion of the last many years on the subject of meaning and normativity following the publication of Kripke's book on Wittgenstein's discussion of rule-following. It will present an argument to put into doubt what is a fairly widespread assumption about the normative nature of linguistic meaning by looking at the relation that linguistic meaning bears to an agent's linguistic intentions. The heart of the argument will be constructed using considerations from Frege on the nature of meaning and sense and its relation to questions of self-knowledge. But before presenting the argument, I will need to set the stage with some discussion of Wittgenstein and Grice.

In several passages in his mature work where Wittgenstein discusses the nature of intentional phenomena, focusing most particularly on intentions (as well as expectations), he is keen to distinguish viewing them as mental *processes* and *experiences* from viewing them in terms of the intentions' (or expectations') *fulfillment*. This latter is the idea of elements in the world (including our own actions) that are in *accord* with these intentional states. Thus, just as Rohit Parikh's arriving at an Indian restaurant in midtown Manhattan one evening is a fulfillment of a certain expectation that I have (the expectation that we will have dinner together there that evening) so is my act of taking an umbrella a fulfillment of my intention to do so on a rainy morning. Both are described as being in "accord" with the intentional states in questions.

The terms "fulfillment" and "accord" convey something that is describable as "normative" in a very broad sense of the term. Things are "right" in some sense when there is accord and fulfillment of this kind, wrong if not. Such is the minimal normativity of intentional states. Sticking with "intentions", which will be the particular intentional state that is the focus of my paper, if I were to intend to take an umbrella but took a walking stick instead of an umbrella by mistake, then it would be, well, "a mistake" by these broadly conceived normative lights. So Wittgenstein's view (not explicitly made in these terms, but implicitly very much part of his picture

Akeel Bilgrami
Department of Philosophy, Columbia University, New York, NY 10027, USA,
e-mail: ab41@columbia.edu

J. van Benthem et al. (eds.), *Games, Norms and Reasons*, Synthese Library 353,
DOI 10.1007/978-94-007-0714-6_3,

of intentionality in his mature work) is that the very idea of intention is such that it generates an ideal or norm of correctness, something by the lights of which one can assess one's actions for being correct or wrong, depending on whether they are or fail to be in accord with the intention.

What is the philosophical force behind such talk of the normativity of intentional states? Its force is contrastive: not merely a contrast with the apparently processual and experiential aspects of mentality just mentioned, but also with what Kripke brought to centre stage in his book on Wittgenstein, the dispositional character of mental states. Put most generally, the contrasts are asserted with anti-psychologistic ends in mind: the normative is set against the psychologism of process and of inner experiences as well as of mental tendencies and propensities. Since these contrasts are well known in the discussion of these topics, I will not labour them here beyond saying that normativity, so conceived, is said to be constitutive of intentional states, and if that is so, it puts into doubt that the processual, the inner experiential, and the dispositional, can really be what is primary in our philosophical understanding of intentionality.

There is no gainsaying the centrality of such a normative element in the very idea of intentions, in particular, and intentionality, in general. What I want to question is whether what is true as a general point is true in the case of *linguistic* intentions, in particular the intentions that speakers have regarding the meanings of their words. Might these not be a very special kind of exception to the generality of this truth, providing a sort of limiting or degenerate case of intention and intentionality?

Here is how I have allowed myself to think of it.

2. Let us ask: what are the intentions one has when one says things or means things with one's words (restricting ourselves to assertoric statements for the sake of simplicity and convenience)?

Since Grice's canonical analysis (I should say "analyses" since he fortified his initial analysis in subsequent work) of meaning linked meaning with intention, let us take that as a point of departure.

The first part of his analysis points out that when we say things we have certain nested intentions to have some effect on hearers. In the assertoric case, the intention is to get them to acquire certain beliefs – say, about the world in one's near vicinity. Thus for instance, someone says "That is a snake" with the intention to get someone else to believe that there is a snake in one's path. (In Grice's analysis this intention, as I said, nests with two others – at least – whose point is to ensure that the case is a case of getting someone to believe something by *telling* someone something rather than merely getting it across to them, something that could not be ensured with just that one intention. What prompts these other two nesting intentions that go into the first part of the analysis are not relevant to the concerns of this paper.)

But, in Grice, this analysis invoking these three nested intentions is supposed to be just the beginning of an analysis of meaning. One has to add various things to move from an analysis of *speaker's* meaning, which this analysis provides, to an analysis of *sentence* meaning. The speaker's meaning of the words uttered is analyzed in terms of the specific purpose or intention that the speaker has on that occasion (in the assertoric cases, to get someone to believe something). The sentence meaning is the meaning of the words that the speaker takes his words to have – in Grice's rhetoric – "timelessly". This contrast between what the analysis provides in this first stage with the three nested intentions (i.e., speaker's meaning) and sentence meaning is most explicitly visible when they fail to coincide *even on the surface*, as for instance in metaphors or in indirect speech acts. In a metaphor, one might say some words, such as the familiar, "Man is a wolf" with the intention of getting someone to believe that "Human beings are competitive", in indirect speech acts one might say some words, such as, "The train is about to leave" with the intention to get someone to believe that they ought to walk faster and get on the train. The three intentions of Grice's analysis do not provide the analysis of the sentence meaning, only of what the speaker meant to convey to the hearer on that occasion with the utterance of those words. The speaker does not take the respective *sentences* to mean that human beings are competitive or that someone ought to walk faster. He does intend to get the hearer to believe that human beings are competitive in the one case and that he ought to walk faster in the other, but that is merely speaker's meaning; what he takes the sentences he utters to mean is something quite else.

Grice gave additional analysis of the sentence meaning that the utterance possesses and I will not expound here what that is since it is not Gricean exegesis that I am primarily interested in. In a careful commentary on Grice, Stephen Schiffer was the first to argue that Grice needs to bring in something like a truth-conditional analysis of the sentence meaning – "timeless meaning" – that the speaker takes his words to have, over and above what *he* means on that occasion with the utterance of that sentence. Since truth-conditional analyses of sentence meaning are very familiar by now, let me for the sake of convenience assume that it is they rather than some other analysis which will be the best account of sentence meaning. (If someone were to doubt Schiffer's claim and give some other analysis of sentence-meaning, that should not spoil what I want to say here, since all I want to say, is that even in Grice there is a distinction between speaker's meaning given in his initial analysis with those three nested intentions, and sentence meaning. Which analysis best accounts for the latter makes no difference to my purposes.) In my examples, the truth-conditions of the sentences by no means coincide with the initial Gricean analysis of the speakers' meaning. It would be quite wrong to say that the speaker has in mind that "'Man is a wolf' is true if and only if human beings are competitive" or "'The train is about to leave' is true if and only if the hearer ought to walk faster and get on the train". Rather, he takes it to be the case that "'Man is a wolf' is true if and only if man is a wolf" and "'The train is about to leave' is true if and only if the train is about to leave". These are his sentence-meanings and they depart on the visible surface, in these examples, from the speaker's meaning. And the important point remains that even in cases where there is *no* visible departure of this obvious kind as there is in metaphors or indirect speech acts, one should nevertheless acknowledge the difference between speaker's meaning and sentence meaning. If someone were to say "Human beings are competitive" with the intention to get someone to believe that human beings are competitive that would still leave intact the distinction between speaker's meaning and sentence meaning since the latter would be given by the truth conditions of the sentence, not the intention to get someone to believe something that *happens to coincide* (in *this* but *not* other cases) with the truth-conditions of the sentence.

There is a source of possible confusion here against which we should protect ourselves. I, following Grice and others, have said that when a speaker says something, the sentence meaning is something independent of the intentions he has which are emphasized in Grice's initial analysis, because the initial analysis is only of speaker's meaning, of what he means on that occasion. This may give the quite wrong impression that sentence meaning is not to be something that *he* means *at all*, that it attaches to the words he utters but are not meant by *him*, in the same sense. But it *is* indeed *he* (the *speaker*) who also takes the sentence he utters to have a sentence meaning over and above what he intends someone to believe with the utterance of that sentence.The speaker is not left out of the picture in sentence-meaning. Just because sentence meaning is contrasted with speaker's meaning, it doesn't follow that it is not *speakers* who *take* their utterances to have sentence meaning. It is not as if the sentences uttered by speakers possess meaning in some ulterior way and the speakers who speak them don't take them to have that meaning. (Grice's rhetoric of "timeless" as opposed to "occasional" meaning – clearly echoing the "sentence"/"speaker" meaning distinction – may also mislead in the same way and should be guarded against. Just because so-called "timeless" meaning is contrasted with what a speaker means on an occasion, it doesn't mean that it is not the *speaker* on that *occasion* who takes it to have that timeless meaning.)

Let us now return to the question of the normativity of intentional states as laid out in Wittgenstein's characterization of them, in particular his normative characterization of intentions. Our question, as I said, is the relation between the normative nature of intentions and the normative nature of meaning. In particular if, as Grice shows, intentions are deeply involved in meaning, what I want to explore is the extent to which the normative nature of intentions imparts, or is of a piece with, the alleged normativity of meaning.

What is often said in the literature on this subject is this. Our terms (at any rate many of them) apply to things, and to misapply them is to make a mistake with our terms; and the very possibility of such mistakes amounts to the normativity built into the meanings of our terms. Thus we are right when we apply the term "snake" to snakes but not to any other thing. When related to our intentional utterances of sentences with these terms, such a view of the normativity of meaning amounts, then, to something like this. We intend to say things with the words we utter. Thus – staying, as we have, with assertoric utterances – one might utter, "That is a snake" with the intention of applying those words to a snake in one's path. Now, should it turn out that what is in front of us is, say, a rope and not a snake, we would have gotten things *wrong*; and that possibility of getting things wrong redeems in the *particular* case of meaning things with ones words, Wittgenstein's *general* idea (true of all intentions whether they are in play in meaning or in anything else) that intentions are, in their essence, normative. Such intentions as the one just mentioned with which one utters words such as the ones just cited, are just examples of intentions targeting specifically, not actions such as taking one's umbrella but rather linguistic actions. Just as one might make a mistake and not take one's umbrella (taking a walking stick instead), so also one might make a mistake and say, "That is a snake" in the presence of a rope. In both cases, it is the possibility of such mistakes that reveals the intrinsic normativity of intentions, but in the second case in particular this normativity of intentions captures for meaning (a notion, we have acknowledged to be undeniably tied to intentions) via the intentions with which words are uttered, a special case of this same normativity.

Thus the normativity of the meaning of terms that comes from the idea of the correct or incorrect application of our terms passes over into the normativity of the *intentions with which we use our terms* in utterances. We act *in accord* with these intentions to use words, the intention, say, to use the words "That is a snake" to apply to a snake, only when we do so in the presence of snakes, not in the presence of anything else. That it should pass over in this way might be said to be a very important claim in Wittgenstein because unlike Platonic conceptions of the normativity of meaning, shunned by him, this sort of normativity does not owe to abstractions such as Plato's "forms" or "ideas" but merely to the intentions with which words are used by linguistic agents. Misapplication of a term is not the violation of a norm because it falls afoul of some abstracted ideal (the CONCEPT snake) but because terms are used in utterances with intentions and we can act in ways that fail to be in accord with those intentions.

That is what is often said in the philosophical literature. But there is very good reason to doubt that this picture of the normativity of meaning gets things right. Even a cursory look at what we have been saying about Grice should reveal what the grounds of doubt are, but before I relate it to Grice, let me say what seems obviously wrong with such a view of the normativity of meaning. *What it gets wrong is the intention that is relevant to meaning.* The intention it identifies as being relevant to meaning is in fact relevant to something quite else. The intention relevant to meaning, when one makes assertoric utterances such as the one in the example we are discussing, is not (1) "One utters the words 'That is a snake' with the intention of applying them to a snake in one's path". Rather it is (2) "One utters the words 'That is a snake' with the intention of saying something which is true if and only if that is a snake – or true if and only if there is a snake in one's path." [I mention truth-conditions in (2) for the sake of mere convenience. If someone had another view of meaning than a truth-conditional one – say, one invoking verification or assertibility conditions – one could reformulate (2) accordingly. For the rest of this paper, with that convenience still in mind, I will assume that meaning, i.e., sentence-meaning, is elaborated in terms of truth-conditions.]

Now, let us go back to normativity. We have said, surely uncontroversially, that *the possibility of getting it wrong* is what normativity consists in, in this (or any other) matter. And in linguistic examples (of assertoric utterances in particular) that possibility was supposed to be actualized in cases of the misapplication of terms, cases such as when one says "That is a snake" in the presence of, say, a rope. So let us suppose that one does say those words when there is no snake but a rope in front of one. If one assumes that it is intentions of the form (1) which are relevant to meaning in assertoric utterances, then one is indeed making a mistake. But if one assumes that it is intentions of the form (2) which are relevant to meaning, then no mistake is being made (about meaning) at all when one utters those words in those circumstances. Even if a rope rather than a snake is present, one's intention to say something with certain truth-conditions (something which is true if and only if there is a snake there) is an intention that is impeccably met in these circumstances. The fact that there is a rope and not a snake which is present in the vicinity does not affect in the slightest the aptness of that intention about *meaning*.

Thus it is only by misidentifying the intention relevant to meaning that one is led to think that examples such as these are revealing of the normativity of meaning because they have revealed the built-in possibility of mistakes. These are not mistakes of meaning. They are quite other sorts of mistakes, mistakes due to misperceptions, as in this particular example – in other examples they may be due to quite other causes.

It won't help to say that the idea of mistakes of meaning has to do with the misapplication of terms, so one must find intentions targeting the terms in the sentence uttered and show that *those too* are fulfilled when there are ropes rather than snakes present. It is true that I have only focused on intentions that target the utterances of whole sentences, but the analysis carries over perfectly to intentions that target the terms that compose uttered sentences as well, assuming for the moment that we do have such intentions. Suppose when I utter, "That is a snake", I have a meaning intention that targets, just the word "snake". What shall we identify as the meaning intention for that term? Should it be, "I intend to apply 'snake' to a snake" or should it be "I intend to utter a term, a predicate, that is satisfied if and only if there is a snake there". I claim that it is the latter intention that is properly thought of as a meaning intention. And one has acted in accord with it, even if there is a rope in front of one. It is only the former intention that one has failed to act in accord with, in that circumstance. Misapplication of terms is beside the point (or beside the primary point) as far as meaning intentions are concerned, whether the intentions target the utterance of whole sentences or the terms that compose those sentences.

What I have said about getting the intentions relevant to meaning correctly identified can now be related in obvious ways to the exposition of Grice I presented earlier.

If it is right that speakers who make (assertoric) utterances with specific intentions to get people to believe things, on each such occasion also take their sentence to have a sentence meaning, then it follows strictly that they will also implicitly have *intentions* regarding those sentence meanings. Talk of *taking* the words in one's utterances on occasions to have a sentence meaning makes sense only if one also implicitly has intentions vis a vis what the words are to mean in that sense of (sentence rather than speaker) meaning. And these intentions will be quite different from the three nested intentions in Grice that analyze speaker's meaning since sentence meaning is distinct from speaker's meaning.

Let me sum up and state all of what is involved here very fully. Grice's *initial* analysis captures what it is that the speaker tries to get across to a hearer in a communicative episode. In doing so, it gives the speaker's meaning on that occasion. The analysis captures speaker's meaning by citing three nested intentions of the speaker. These intentions themselves have as their target an effect on the hearer. Essential to the speaker hitting this target is that the words on the speaker's lips have a sentence meaning *as well.* And this kind of meaning needs a *further* analysis. Such analysis is often given by notions such as truth-conditions (or assertibility conditions..). I have said that the speaker must also have *intentions* regarding this sentence meaning since *he takes* the words he utters to have it. It is not as if the sentence meaning is outside the orbit of his mentality. It is right perhaps to describe all this in terms of instrumentality. One tries to get something across to someone, get him or her to have a certain belief, *by* uttering something, some words, with a certain sentence meaning. Both (trying to get someone to believe something and saying something with a certain sentence meaning) are therefore intentional acts. They are done with intentions. No doubt there are other actions involved in such episodes as well, which have further instrumentality. One says something with a certain sentence meaning *by* moving one's throat, etc. This too is an intentional act. It too is done with an intention. Some, perhaps all, of these intentions are highly implicit; they are not in the forefront of the speakers mind. But that does not cast doubt on their being bona fide intentions. It would be quite wrong to think that one's moving one's throat to speak in normal forms of linguistic communication is not an intentional act and not an act done with an intention merely because of its implicitness, in this sense. And the same holds of the intentional act of saying something with a certain sentence meaning. A good schematic description of the instrumentalities of this speech episode, then, is that our protagonist tries to (1) get someone to believe that there is a snake in front of him *by* (2) uttering words which are true if and only if there is a snake in front of him, which he, in turn, does *by* (3) moving his throat and producing the sounds. Each specifies an intentional act, acts done with intentions.

And it has been my point that it is only right to describe the second and intermediate step here as the one that involves the notion of meaning. It is the notion of meaning as it is lodged in (2) that philosophers have long made great and sophisticated efforts to provide some analysis and theory of. Grice was no exception to this, but his is a particularly interesting case because he *began* his analysis by focusing on the sorts of intentions with which (1) is done. He hoped to *build on this basis* and move further to the eventual analysis of meaning as it is found in (2). He called this latter "timeless" meaning (while I am invoking the more standard vocabulary of "sentence meaning") and it was his eventual goal to give an account of it. Whether he was able to do so *on that basis* is not a question I am going to discuss here at all. But it is easy to be misled by something in Grice. In the initial analysis in terms of the intentions that go into the intentional act (1), he used the term "meaning" (sometimes called by him "occasional" meaning to contrast it with "timeless") in describing what he was analyzing in this initial phase. This may give the impression that when we talk of linguistic meaning, (1) is our focus. But it is not. As I said, the subject of meaning is the subject that resides in (2). It is what philosophers (*including* Grice) interested in meaning or semantics have made the object of their theoretical efforts when they study those aspects of language that are not purely syntactic nor purely pragmatic.

That Grice himself was clear about this is obvious from the fact that in subsequent years he provided a detailed supplementary account of certain elements found in (1) by introducing a notion of conversational implicature, and explicitly viewed that as being in the domain of pragmatics, not semantics. The details of that account are not essential here since what I want to really stress is that the supplement is a supplement to Grice's analysis of the phenomenon in (1) and once one notices that something like conversational implicature is relevant to this phenomenon, it becomes explicit that it is to be distinguished from meaning, even for Grice. The idea is that getting something across to someone by telling them things, getting them to form certain beliefs and so on, can be (even if it often is not) achieved by means that may be quite at odds with sentence meanings – as the case of metaphors obviously makes clear. That someone should get someone to believe that human beings are competitive (by saying something manifestly false, i.e., saying words that are true if and only if men are wolves which, they manifestly are not) falls squarely within the domain of (1) above. What Grice does in his work on pragmatics and conversational implicature is to give principles by which one may identify departure of speaker's meaning from meaning (sentence meaning), principles such as "Don't state things that are obvious falsehoods". Violations of such principles are the signs that something like metaphor, for instance, is in play. A speaker intends (in the sense of 1)) to get someone to believe that human beings are competitive by asserting an obvious (zoological) falsehood, thus getting the hearer alerted to the fact that some interesting departure from the sentence meaning must be in play. Thus the theory of conversational implicature or pragmatics is concerned primarily with the initial three fold intentional analysis in (1), making clear that (1) is not the site of semantics or meaning, even though Grice (*in effect* misleadingly, we can say in hindsight) used the term "meaning" to describe what was being analyzed in (1). Meaning is located in (2) and since Grice it has come to be explicitly understood in terms of truth-conditions (or assertibility conditions or). As a result, in the case of metaphors it is very clear that when the world is *manifestly* uncooperative, when the truth conditions have very *obviously* not obtained, then a principle of conversational implicature (Don't speak obvious falsehoods) has been violated. This alerts the hearer to the fact that the speaker (not likely to be making such an obvious mistake or uttering such a blatant and easily discoverable falsehood) intends (in the sense of 1)) something that departs dramatically from what is identified as the meaning or the truth conditions of the words, intends to get across that human beings, far from actually being wolves, are instead merely competitive. And the crucial point remains that the fact that there is often *no* departure (dramatic or otherwise) from (2) in the beliefs that one conveys with one's intentions (in the sense of 1) should not confuse us into thinking that (1) is where semantics or meaning and not pragmatics is in play.

I've spent some time on Grice partly because he was the first philosopher since Wittgenstein to study with some systematic care the relations between intentions and linguistic meaning, which is my particular way of approaching the subject of this paper – the question of the normativity of meaning but partly also to diagnose why philosophers in their claims for the normativity of meaning have misidentified the intentions relevant to linguistic meaning, in the way I mentioned earlier. If my diagnosis is right, one can, if one is not careful enough, be misled by Grice's use of the term 'meaning' to describe the phenomenon in (1), thinking that it is there rather than in (2) where meaning is located. This would account for why one might conclude that someone's use of the words "That is a snake", mistaking a rope for a snake, amounts to a mistake about meaning. How exactly would it account for it?

If one thought that the intention relevant to meaning is that one says something with the intention of conveying something to a hearer or, as Grice puts it in (1), with the intention of getting the hearer to believe something, then perhaps we *would* be able to establish the normativity of meaning along the lines that philosophers have assumed. If someone said "That is a snake" with the intention to get a hearer to believe that there is a snake in front of him, then pointing to a rope rather than a snake might well produce the requisite sort of error in the matter of meaning. One must point to a snake in order to fulfill that particular intention because if the hearer notices that it is a rope, he will not come to believe that that is a snake. Perhaps the idea is better put when it is elaborated more explicitly as follows. If one intends to get someone to believe something which one takes to be true, viz., that there is a snake in front of one, then one way in which one might do it is to *further* intend, in one's literal and sincere assertoric utterance of "That is a snake", to apply those words to a snake in front of one. If there is no snake there, one has *misapplied* the terms, one has *failed* to act in accord with that *further* intention, and so, in turn, quite possibly may fail to fulfill the original intention of producing a certain effect in the hearer.

Notice that I am being careful to say, "quite possibly may" because the hearer, as I said, has to notice that I am perceptually wrong in order to fail to come to have the belief that there is a snake there. Remember Grice's initial analysis is an analysis appealing to the phenomenon of intending to have *effects on hearers*, getting them to have beliefs, and so on. So the hearer would have to fail to acquire the belief that there is a snake there for the intentions, which provide the analysis, to be unfulfilled. It is not therefore *sufficient* for that Gricean intention to be unfulfilled that I say, "That is a snake" when there is a rope present. The hearer, if he is not alert, may not notice that I am wrong and come to believe that there is a snake there, on hearing my words. In that case the Gricean intention would be fulfilled. The only intention that immediately fails to get fulfilled in the presence of a rope is what I referred to as the "further" intention that is tied to what one is applying one's words to. This is the intention that is claimed by many to be directly and immediately tied to the normativity of meaning, when that normativity is viewed in terms of correct and incorrect application of words.

Even so, the important point, if my diagnosis is right, is that this last sort of intention (what I called the "further" intention) is squarely in the region of the intentions to convey things to hearers, the phenomenon that Grice analyzes in his *initial* analysis. It is not at all in the region of the semantic intentions of saying things with certain truth or satisfaction conditions, which analyze the quite different phenomenon of "timeless" or sentence meaning that he wants his analysis to *eventually* work up to. The "further" intention is caught up in the former region of intentions because it is *directly* a part of one's project of conveying things to people; it is directly in the service of getting people to believe things, etc. It is not to be collapsed with or even really deeply associated with sentence meaning and with the latter region of intentions – to say things with certain truth-conditions. And my point is that once we (a) see this and (b) see no *direct* relevance to meaning of any intention in the former region, then whatever else may reveal the possibility of failures to fulfill the intentions relevant to meaning (intentions in the latter region), it won't be such things as mistaking ropes for snakes and, more generally, misapplying one's terms.

So, the issue then becomes: once we *properly* identify the intentions relevant to meaning, what follows about the normativity of meaning? In other words, if such things as mistaking ropes for snakes does not amount to the *failures of accord* with one's meaning intentions that are said to reveal the normativity of meaning in assertoric utterances such as "That is a snake", what sort of thing does reveal it? I pose this question in just this way in order to invite the following suspicion in the reader: can *anything* amount to a failure to act in accord with the intention we have now properly identified as being relevant to the meaning of utterances of that kind? If, as the suspicion is supposed to suggest, the answer to this question *is* "No", then one puts into doubt the idea that meaning is normative, at least to the extent that such normativity is supposed to derive from the (undeniable) normativity of intentionality, in general, and of intentions, in particular. I think it is arguable that the answer to this question is "No".

3. Once properly identified, we have learnt that the intention relevant to meaning targets the truth conditions of one's words. Hence the failure to fulfill that intention would presumably occur only if one failed to get right what their truth conditions are – as opposed to occurring when the truth conditions, which one gets right, fail to *hold* (in our example, when there is no snake but a rope in front of one).

How, then, might one fail to get right what the truth conditions of the sentences one utters, are?

One clear sense in which it might be supposed that one can fail to get them right – or better, one clear *source* for one's failing to get them right – is if one does not *know* what they are. (There is another supposed source, which I will take up a little later.)

The idea here will have to be that I don't know what the truth conditions of my words are, so I intend that they have certain truth conditions, but they are not the correct truth conditions of those words. So a question arises: why should the truth conditions of one's words not always be what one intends them to be? We will return to this question at the end. But first let's ask: how exactly is it that one can intend our words to possess truth conditions they don't in fact possess *as a result of one not knowing* what the truth conditions are? Let's set up an example, a familiar one from the philosophical literature, of such an occurrence. A medical ignoramus intends to say some words that are true if and only if he has a disease either of the joints or ligaments in his thigh. And he says, "I have arthritis in my thigh". He doesn't know that arthritis is a disease of the joints only. So he has said something with truth conditions other than the truth conditions he intended. He has failed to live in accord with his intention. This looks like an example of how, when one does not know what the words one intends to utter mean, one can say something that fails to live up to an intention that (unlike the intention in the example about snakes and ropes) is properly identified as being relevant to meaning.

The crucial task now is to assess this claim that one may not know the meanings of the words one intends to speak Here I do not think the primary question to be asked is: what theoretical account of meaning allows it to be the case that a speaker does not know what he means? It is easy to devise a number of such accounts and they have been devised ever since Plato's highly objectivized notions of meaning understood as given in a heavenly world of "forms" or "ideas", with contemporary versions bringing Plato's heaven down to earth and calling it "society" or "community". Any assessment of the claim needs instead to step back and ask a *prior* question whether we can *tolerate* any theoretical account of meaning in which we breezily allow speakers to fail to know the meanings or truth conditions of their own intended words, and that, in turn, means stepping even further back to ask: by what criteria shall we decide what is and is not tolerable in a theoretical account of meaning thought of in terms of truth conditions?

Responsible stepping back of this sort requires one to at least notice the historical fact that the idea that the meaning of a sentence is given by its truth-conditions was first explicitly formulated in Frege's doctrine of *sense* and so it is perhaps in the notion of sense that we should seek the criteria by which we can assess what seems tolerable or not in an account of meaning. What we will or will not tolerate will depend, therefore, on stating what the notion of sense was introduced to do and see whether it *will* do what it was introduced to do, if we allow that one may not know the senses or meanings of one's words. So let's ask: what is a notion of sense (or meaning) for?

In Frege, as we know, the notion is introduced initially to solve a puzzle about identity. Though that is the occasion in Frege for theoretically motivating the notion of sense, Frege had in mind very large issues in raising the puzzle about identity – the puzzle is a mere surface reflection of one of the most deep and fundamental issues about the relations between language and mind. In fact, Frege's own formulations of the puzzle and his solution to the puzzle don't always make explicit just how deep and fundamental the issue at stake is. One way of putting the point at stake is to describe it as follows: to be misinformed or uninformed is not to be irrational. No account of the mind can confuse these two ways in which a mind can go wrong. Failures of empirical knowledge and logical error are not to be conflated. The puzzle arises precisely because the examples discussed by Frege (and by Kripke, who raises a slightly different version of it) threaten to conflate them. The protagonist in the puzzle who, ex hypothesi, merely lacks worldly knowledge of the identity of a planet (or in Kripke's version, a city) is threatened with a charge of irrationality by precisely such an elementary conflation. And it is Frege's view (though not Kripke's) that introducing the notion of sense will provide the best solution to the puzzle. It is the notion of sense or meaning which makes it clear that no irrationality, no logical contradiction, is entailed by someone saying, for example, that "Hesperus is bright" and "Phosphorus is not bright" or "Londres et jolie" and "London is not pretty". So the puzzle lays down a crucial desideratum: we know the protagonist in the puzzle to be someone who merely lacks knowledge of an a posteriori identity, so we must find a way to characterize his mentality (or this fragment of his mentality) as representing a completely consistent state of affairs. Since it is the positing of senses to his words (over and above their reference) which, according to Frege, helps us achieve such a representation, *nothing* should be tolerated in our understanding of senses that prevents them from decisively carrying out this task. In other words, nothing should be tolerated in the understanding of the notion of sense or meaning, which will prevent senses from doing what they are supposed to do: solve the puzzle and, by doing so, maintain the most fundamental of philosophical distinctions – that between logical error or irrationality and lack of empirical knowledge.

The fact is that senses will *not* decisively solve the Frege style puzzles if it is allowed that we fail to know our own senses or meanings. A failure of transparency in sense will leave it entirely possible that the puzzles about identity can arise one level up and so the puzzles will not be satisfactorily solved; or better, they will not once and for all be *arrested*. Let me explain.

If someone does not know his own senses, he may be in a position to be just as confused as Frege's protagonist in the puzzle, thinking that there are two senses rather than one. Suppose someone wonders, in his ignorance of astronomy: "I wonder if Hesperus is Phosphorus?" To make such a wondering so much as intelligible, a Fregean posits senses. But if the wonderer doesn't know his own senses, he may similarly wonder, one step up, if the sense of "Hesperus" is the same as the sense of "Phosphorus" (or as in Benson Mates pointed out in an ever so slightly different context of discussion, he may wonder whether – or doubt that – the sense of "bachelor" is the sense of "unmarried man".) Thus, there is no genuine arrest of the Frege puzzle (and no eventual solution to it, therefore) if it is not laid down as a basic criterion of senses that they be transparent, i.e., known to their possessors. Without this being laid down, the puzzle can always arise one step up, with protagonists as confused about the identity of their senses as they are about planets and cities.

One implication of this – and a very deep one – is that it amounts to something like a proof that senses are not the sorts of things that we can have multiple perspectives on such that *one* can get their identities confused in the way that we can with planets and cities. Whatever senses are, then, they are not the kind of things that planets and cities are. They are not like any thing which allows multiple perspectives on itself and which therefore allows such confusion to be possible. Things on which we can have more than one perspective are by their nature not transparent, even if they are often in fact known. I suspect that it is, at least partly, because Kripke doesn't quite see this point about the sort of thing senses are that he doesn't follow Frege in invoking senses when he addresses the puzzle.

Those, then, are the considerations that make it intolerable for meanings to not be known by those who speak meaningful words: we will not be guaranteed to solve the Frege puzzles, at least not in a way that arrests them once and for all; and that, in turn, amounts to meanings failing to do the very thing that meanings and senses were introduced to do, viz., allowing one to preserve a fundamental distinction of philosophy between logical error and lack of empirical knowledge. We should therefore regard with suspicion the many accounts of meaning from Plato's down to the secular and mundane versions of Plato in our own time, which allow such an intolerable outcome as a result of prising apart our meanings from our knowledge of them.

The medically ignorant man who says "I have arthritis in my thigh", therefore, though he certainly makes a mistake, makes a mistake about how the term is used in the social linguistic practice, especially among the medically informed experts. His own linguistic practice is not grooving with theirs. *That* is his only mistake, apart from the, ex hypothesi, medical ignorance. But within his own linguistic practice where the words on his lips mean and are intended to mean something that is true if and only if he has a disease of the joints or ligaments in his thigh, he says and thinks something that is both *self-known* to him and something *that* is perfectly *true*. After I set up more conceptual apparatus, I will say a little more on how to represent this idea of his own individual, idiosyncratic practice. But in case, until then, it seems to the reader that this admission of idiosyncrasy fails to keep faith with facts about how individuals show deference to the social, especially the experts in their society, I should point out that there is no need for the view I am presenting to deny that this medically ignorant protagonist will defer to the experts and wishes to defer to them, should he realize *that* mistake. Deference is perfectly compatible with the view. All that deference amounts to on this view is that he will change his linguistic behaviour and adopt theirs. He will start speaking as they do. It will *not* amount to him coming to *know more about what he himself means and thinks*. He always knew what he meant and thought (something he would and *could* only fail to know for *psychological* reasons roughly of the sort Freud studied, not because philosophers have certain secularized Platonist theories about the social aspects of reference and meaning). He has only learnt something about what his fellows, especially the experts, think and how they use words. And because he wishes to defer to the experts he will now start using words as they do and, of course, he will also have become less medically ignorant, more medically knowledgeable.

All this follows as straightforward theoretical implications of meeting the desideratum that we must have a decisive solution to the Frege-style puzzles about identity, where by "decisive" I mean a solution that arrests these puzzles. I have given this argument for the transparency of meaning or sense to block one alleged source for the failure to act in accord with the intention to say something with certain truth conditions. It was claimed that we can fail to act on such an intention if we do not know what the truth conditions of our words are and it is this last claim that the considerations about the Frege puzzle have shown is intolerable for any account of meaning to allow.

But I had also said that that it is not the only supposed source.

If one *has* knowledge of the right truth conditions for one's intended words, can one still get the truth conditions of one's spoken words wrong? How might one intend to say something with certain (correct) truth conditions but say something with some other (incorrect) truth conditions or with no truth conditions? This can happen only if one mis*speaks,* if the sounds one produces do not amount to the words one intends to utter – as for instance in slips of the tongue. So, suppose I say, "I am going towndown" with the intention of saying something that has the truth conditions (something that is true if and only if) I am going downtown. The sounds I make do not, it might be said, amount to words that *in fact* have those truth conditions. (In this particular example, they don't seem to have any truth conditions.) Misspeaking, then, is the second alleged source for failing to live up to our intentions that target truth-conditions.

Is this the best way to analyze such cases of misspeaking – to see them as giving rise to such failures? The issues here are, at bottom, not really different from those already discussed in the cases where the apparent source of the difficulty was an apparent failure to know the meanings of the words one intends to speak. In the present cases, one knows the meanings or truth conditions of the words one intends to speak but not of the words one actually ends up (mis) speaking. But the question is why should the words we actually speak fail to have the truth-conditions we intend them to have, even in these cases of misspeaking?

Once again: is an account of meaning which allows such a failure tolerable? We would only allow it if we were in thrall to accounts of meaning that allow for the possibility of the *meanings* of our words, the words we speak on given occasions with intentions, to be such that we are not aware of what they mean, at the time we utter them. It is only if it is said that I am not aware of the meanings of the words I misspeak that my misspeaking could be seen as a sign that I have uttered something with different truth conditions than the one I intended or, in the particular example I mentioned ("I am going towndown"), something with no truth conditions at all. But why shouldn't misspeakings of this kind get a quite different theoretical treatment, one in which they have just the truth-conditions I intend for them, in which case I am perfectly aware of what my words mean? On this view, the misspeaking is not a case of *meaning* something one doesn't intend, only one of producing sounds one didn't intend to produce. But those sounds mean just what I *intended* to mean by the sounds that I intended to make (but didn't); so I can't fail to know them. One would have thought this theoretical account keeps better faith with the idea of "mis*speaking*" because there is nothing amiss with the meaning at all. The alternative view, which I am opposing and which has it that I ended up meaning something I didn't intend might be capturing something that is better termed mis*meaning*. But it is misspeaking we want to capture, not mismeaning.

There will be a protest: You unfairly foist on the alternative view an attribution of meaning to the misspeaker's utterance that is *his* meaning. But it is not his meaning, it is what the words he utters mean. This move does not help matters at all. The protest, if it is correct, only reveals the true nature of the alternative view. The view amounts to a prising apart of what one's words mean, what truth conditions they have, from the meaning or truth conditions one intends them to have. Such a prising apart has disastrous consequences of a kind that I have already discussed. The reason to see the phenomenon of misspeaking as I am suggesting we should, where the intention to mean something and the meaning of the words one utters are inextricably linked, is quite simply that if they were not so linked, Frege style puzzles would get no solution that arrests them. Fregean puzzles are puzzles that can only be solved if there is no gap between the meanings of the words one utters and the intentions with which we utter them and, therefore, no threat to our self-knowledge of their meaning. By creating a gap between the truth conditions or meaning of the words a speaker utters and what the *speaker means* (by way of the truth conditions he *intends* for his words), the alternative understanding of misspeaking threatens to make it possible for us to be unaware of the meanings of the words we utter (even as it allows us to be aware of the meanings we *intend* to utter). What truth conditions or meanings our words have may now turn on factors that one may be entirely unaware of; and we have already seen that if we allow for speakers to lack self-knowledge of the meaning of what they say, we will have no satisfactory solution to the Fregean puzzles about identity, at any rate no decisive solution which *arrests* those puzzles and prevents them from arising one step up. The important point is that the puzzles arise in a particular conceptual place – at the site of the *meanings* of the words someone utters ("Londres est jolie"/London is not pretty, "Hesperus is bright"/Phosphorus is not bright) and they need a solution *at that site*. Thus it is no good to say that a speaker must know what he intends his words to mean, but he needn't know what his words in fact mean. He needs to know what his words in fact mean, if the puzzle is to get a satisfactory solution because the puzzle is located in the meanings of words. And they won't get this solution if what his words mean are prised apart from what he intends them to mean, because that prising apart is what is responsible for the non-transparency of the meaning or senses of his words that thwarts decisive solutions to the puzzle. This, as they say, is the bottom line. It is, at bottom, the reason to find the second source we are discussing to be dry, and it is the same reason as we had for finding the first source to be fruitless. We need not seek new reasons to deal with claims of the second source once this protest reveals the true nature of the alternative view.

In fact, resisting the view that the protest assumes to be correct is so fundamental not only because such resistance will make possible a satisfying solution to these puzzles via an appeal to senses, as I have been saying, but also because it allows us to understand why the puzzle is the sort of puzzle it is, why it goes so deep. These puzzles about identity, raised by Frege and Kripke, go so deep only because the words we utter *are* inextricably linked to our mentality in a way that the opposing view of misspeaking which prises these two things apart, denies. The puzzles, as I expounded them, are not just raising points about meaning and reference, they are puzzles that reach down to (because they threaten) our most elementary assumptions about the nature of the *rationality* of speakers of a language. And rationality is a property of a speaker's *mind*. So the links between language (or meaning) and mind have to be closer than the opposing view permits, to SO MUCH AS RAISE the puzzles about identity. Kripke himself, despite the fact that he is no friend of Frege's solution (via an appeal to *senses*) to the puzzles about identity, nevertheless sees this importance for the puzzle of the inextricability of the link between meaning and mind. Whatever one may think of his skepticism about "senses" and of his own way of addressing the puzzle, his understanding of how the puzzle arises and what it affects, is deep and instructive. It is precisely because he takes meaning and belief/intention/ (the intentional aspects of the mind, generally) to be inextricably linked, that his puzzle ends up being a "puzzle about *belief*", as his very title makes clear, and therefore about the rationality of his protagonist, Pierre.

Having said that, it also must be said that there are complications in the *specific* way he elaborates the link between meaning and belief that are of special interest when it comes to misspeaking and they ought to be discussed and clarified because they are highly instructive. The link is elaborated by Kripke in what he calls the principle of disquotation. That principle makes the link by saying that whenever a linguistic agent sincerely utters or assents to a sentence (say, "London is not pretty") one may remove the quotation marks and place the sentence in a that-clause specifying the content of the agent's belief (believes that London is not pretty). This principle, along with two other things – something he calls a "principle of translation" (which essentially asserts that translation preserves truth-conditions) and Kripke's own causal, anti-Fregean account of reference – together give rise to his version of the Fregean puzzle. (The principle of translation is only needed for Kripke's own version of the puzzle because it has two languages, French and English.) Suppose, then, that Pierre, brought up in Paris and influenced by his Anglophile nanny, asserts sincerely "Londres est jolie", and then moves to London, without knowing that Londres is London, and lives in some shabby locality and asserts sincerely in his newly acquired language, English, "London is not pretty". With all this supposed, an adherence to the combination of the three things I mentioned (the two principles – of disquotation and of translation – and an anti-Fregean, Kripkean causal account of reference), give rise to the puzzle. We have an agent, who is merely uninformed of the identity of London and Londres, believing two inconsistent things – believing that London is pretty and *believing* that London is not pretty. In that combination of three things, Fregeans would readily adhere to the two principles but will reject the causal doctrine about reference, introducing instead the notion of senses (disallowed by that doctrine) to block this unacceptable implication of Pierre's irrationality. Kripke finds this an unsatisfactory solution to the puzzle, partly because of his prejudices about the notion of sense, some of which quite possibly flow from a failure to understand that senses are in a very deep way not at all like planets and cities – one cannot have multiple perspectives on them, and so they *have* to be the sorts of thing that are self-known to speakers, and therefore they cannot be the subject and occasion of further puzzles one step up. But here I don't want to focus on the dispute between Kripke and the Fregeans. I want to just focus on the principle of disquotation since that is Kripke's particular elaboration of the general idea that meaning and mentality are inextricably linked, the very idea that is essential to my claims about what goes on in misspeaking.

The trouble with this particular elaboration of the general idea is that, at least on the surface, it seems to actually spoil rather than elaborate the inextricability *when it comes to the case of misspeaking*. When someone says, "I am going *town*down", a specific understanding in terms of disquotation of the general idea of the inextricable link would seem to have it that he believes that he is going *town*down. And that is not a very clear and perspicuous attribution of belief. By contrast, in my gloss on misspeaking, I have the agent believing that he is going *down*town (a perfectly clear and perspicuous attribution) because I also have his utterance "I am going towndown" *mean* that he is going downtown. And it is possible to have it mean that because one takes the speaker to intend that his words are true if and only if he is going downtown, and one takes the meanings of his words to be fixed by that semantic intention of his and not by some other factors, often adduced by philosophers, such as social linguistic practice (or the scientific essence of substances, diseases, etc.)

that may in some cases not even by known to the speaker. It is these close links between his meaning and his mentality (i.e., his semantic *intention* and *therefore* also his *belief*) which allow me to make the clear and perspicuous attribution of belief to the speaker. In fact, there is a close circle of links between two mental and one linguistic phenomena: a speaker believes that he is going downtown, he intends the words he utters ("I am going towndown", as it happens) to be true if and only if he is going downtown, and the words he utters are indeed true if and only (they indeed mean that) he is going downtown. So, it looks as if something goes amiss when one elaborates the close links in terms of the principle of disquotation. What the links should attribute in a perfectly clear attribution of a belief, as I have it, gets mangled into a bizarre attribution of the belief that he is going towndown, when the links are seen in strictly disquotational terms.

The fault line here is the *strictly* disquotational elaboration of the close links between meaning and mind

Let me raise the issues first not with the case of misspeaking, which is some distance away from the puzzles about identity, but with a case much closer to the kind of examples discussed in the puzzle by Frege and Kripke. There is a version of the puzzle that Kripke mentions briefly, where the two names uttered in the two seemingly inconsistent sentences generated by the puzzle are not merely in the same language as they are in Frege's original version of the puzzle (unlike in the puzzle about Pierre) but are indeed, on the surface, the *same name*, "Paderewski". There is in fact only one person called "Paderewski", but the protagonist doesn't know that and thinks that there are two different persons, one a statesmen, the other a famous conductor. Thus the apparent inconsistency seems to be starker, with the protagonist saying, not merely, "Hesperus is bright" and "Phosphorus is not bright", but, let's say, "Paderewski is musically accomplished" and "Paderewski is not musically accomplished". Here, elaborating the close link between meaning and mentality in terms of a strict version of the principle of disquotation will land us in a midden. If, as I have been insisting, it is one's semantic intentions that impart meaning's on one's words and one's semantic intentions are to be thought of as having close links with one's beliefs as mentioned in the previous paragraph, and the belief itself is identified via a *disquotational* strategy that links it with the utterance, one will have to say that the speaker semantically intends his first utterance to be true if and only if Paderewski is musically accomplished and the second utterance to be true if and only if Paderewski is not musically accomplished. If that happens, senses or meanings, *even if they are imparted by our semantic intentions*, do not seem to have solved the puzzle at all.

What is obviously needed here to make disquotation line up with the spirit and the underlying point of the close links it elaborates between meaning and mind, is to allow that disquotation comes into play after we make clear what the carriers of the semantics really are. In this example, the term "Paderewski" in each sentence contains an unpronounced (and unwritten) subscript. The sounds and inscriptions as they originally stand are incomplete and they don't properly reveal the *words* that comprise the full and proper semantic items. The semantics (the senses) are carried by neologisms that we will therefore have to introduce, "Paderewski$_1$" and

"Paderewski$_2$". The sounds and inscriptions misleadingly leave the subscripts out. Once we neologize, disquotation can proceed apace. Now, the semantic intentions with which the speaker utters the two sentences can be formulated in a way that allows the meanings or senses to remove the puzzle. Of course, this solution to the puzzle is not Kripke's solution, as I have already said. It is Fregean. It appeals to senses, given in two different semantic intentions, which intend for each of the two (only seemingly inconsistent) sentences, two different truth conditions.

Someone may wish to enter a quarrel here, claiming that this is to abandon the disquotation principle, not rescue it from a strictly rigid reading. But this dispute is paltering and terminological. Whether we say the theoretical move here is "disquotational" or not, hardly matters. What matters is to register that the move fastens on an insight *in Kripke* that the puzzle about identity is a puzzle that reaches all the way down to issues of belief, mind, and rationality, and builds from that insight a notion of sense based on an individual speaker's semantic intentions to characterize the meanings of the words in his speech. Kripke insight that the puzzle goes that deep is just the insight that meaning and mind have inextricably close links, but he then does not see in those close links the possibility of developing this individualist Fregean notion of sense that would decisively resolve the puzzle. And I am claiming that this combination of insight and failure on his part is revealingly present in his own principle of disquotation. The principle itself is an acknowledgement of the close links and is of a piece with (in fact it makes for) the insight that the puzzle reaches all the way down to issues of rationality. The failure to give the principle the less strict and more relaxed reading is of a piece with the failure to build on the insight about those links and provide a plausible version of the Fregean solution. If one sees disquotation in a less strict way than it seems on the surface to be, one has all the ingredients for a convincing Fregean solution to the puzzles, one that would arrest the puzzles once and for all.

And more to the point of our *immediate* interest, we can derive from this more relaxed reading of disquotation an attractive account of what goes in the case of misspeaking. There too, one can say that disquotation should proceed *after* we make clear that "towndown" is really a mispronunciation of "downtown", just as each utterance of "Paderewski" in the two sentences of that version of the puzzle is a mispronunciation of "Paderewski$_1$" and "Paderewski$_2$" respectively. Once this is done, we are landed with no unclear and unintelligible attributions of belief to the person who says "I am going towndown"; and we preserve the close links between meaning and mind which alone allow us to solve the Frege style puzzles decisively, in a way that the alternative view of misspeaking I have been resisting, cannot.

Finally, there is hereabouts another clarification worth making quickly. One should not be under the impression that my way of accounting for misspeaking diminishes the distinction between speakers meaning and sentence meaning. It might seem natural to think that it does. After all, have I not insisted that the meaning of the misspoken *sentence* is exactly what meaning or truth condition the *speaker* intends it to have? And isn't that all it takes to diminish the distinction? It is actually instructive to understand why it does not in any way diminish the general distinction between sentence meaning and speakers meaning. The interesting and important point to emphasize is that misspeaking is a phenomenon quite different from metaphors and indirect speech acts, where that distinction conspicuously presides. In the latter phenomena, speaker's meaning, as I pointed out earlier, comes visibly apart from the sentence meaning. In fact speakers exploit something in the sentence meaning *in order to convey something else* to hearers. They convey that "something else" *by* deploying the sentence meaning of the utterances they make. But in the case of misspeaking, there is only a false impression of speaker's meaning and sentence meaning being visibly apart in the same sense. The idea that the speaker says what she in fact merely happens to say ("I am going towndown") *in order to* deliberately convey something *quite else* – that she is going downtown is completely inappropriate in the analysis of misspeaking. The description "quite else" is entirely out of place as a description of what the speaker is deliberately up to, while it is perfectly correct in describing what the speaker is deliberately up to in metaphors and indirect speech acts. The fact that misspeaking turns on a visible difference between what is uttered and what one is intending to get someone to believe (respectively, "I am going towndown" and "I am going downtown") should not confuse anyone into thinking that the case is similar to the cases of metaphor and indirect speech acts, which also turn on a visible difference between what is uttered and what one is intending to get someone to believe (respectively "Man is a wolf" and "Human beings are competitive" or "The train is about to leave" and "Walk faster to the train".) In the case of misspeaking (the utterance by a speaker of "I am going towndown), the speaker has the following two intentions: to say something that is true if and only if the speaker is going downtown and to get across to the hearer the belief that he is going downtown. There is thus coincidence – rather than departure – of what the speaker intends to get across from what he intends his words to mean whereas in the case of metaphors and indirect speech acts, there is departure rather than coincidence. (I repeat, of course, that to insist on this coincidence is not to say that the distinction between speaker's meaning and sentence meaning is *collapsed* in the case of misspeaking. It is no more collapsed in the case of misspeaking than it is when one says "Human beings are competitive" in order to get someone to believe that human beings are competitive or says "The train is about to leave" in order to get someone to believe that the train is about to leave.)

My claim has been that if we see cases of misspeaking – such as slips of the tongue as having a literal, sentence, meaning that is different from what it *sounds* like – a meaning that is imparted by the semantic intentions of the misspeaker – then we can block misspeaking from becoming a second source for thinking that speakers do not have their own meanings or truth conditions right. And so, just as with the first source discussed earlier, without the possibility of being wrong about our own meanings, we lose our grip on the very idea of a notion of norm that holds of meaning since no one would want to say that talk of normativity is apt when there is no possibility of being wrong or mistaken.

Let's return to the Wittgensteinian notion of the normativity that resides in acting in accord or out of accord with our intentions, the subject that we began with. This paper's brief has been that we cannot fail to act in accord with the intention relevant to meaning. I argued (in the previous section) that that intention is not the intention to apply particular words to particular things rather than others, but the intention to say particular words with particular truth conditions and satisfaction conditions. The subsequent discussion (in the present section) of the phenomenon of misspeaking helps me to stress a point that I have tried to be careful about in my various formulations of this intention relevant to meaning. The intention relevant to meaning is best formulated by saying that a speaker intends with an (assertoric) utterance to say *something* which has particular truth conditions. The word "something" in this formulation has the right generality. Sometimes speakers do not produce the exact sounds they intend to produce, as when they misspeak. Thus when the intention is described with the right generality, such cases will not spoil the efficacy of that intention. Our protagonist who utters, "I am going towndown" does indeed intend to say *something* that is true if and only if he is going downtown. That intention, formulated with that *generality*, is perfectly well fulfilled when he misspeaks, even if another intention formulated without that generality (to utter "I am going downtown" *in particular*) is not. And it is the former intention that is his semantic intention which targets the *sentence* meaning of his utterance.

What I will concede is that when the intention relevant to meaning gets such a general description as I am proposing ("I intend to say *something* which is true if and only if"), we may sometimes find that what a linguistic agent intends as the truth conditions for his utterance may be rather idiosyncratic. A slip of the tongue is proof of that as is the medically ignorant utterance "I have arthritis in my thigh". If the formulation of the intention relevant to meaning is that I intend to say something which is true if and only if then, I can say "I have arthritis in my thigh" with the intention of saying *something* that is true if and only if I have a disease either of the joints or the ligaments in my thigh; or I can say "I am going towndown" with the intention of saying something that is true if and only I am going downtown. The formulation leaves things general enough to allow this sort of leeway for idiosyncratic (individualistic) semantic intentions.

Is there any shortcoming to allowing this sort of idiosyncrasy into literal, sentence, meaning, and not restricting such idiosyncrasy to non-lietral phenomena like metaphors and other figures of speech? The answer to this is, "yes". Idiosyncratic *semantic* intentions for one's words put our hearers to some strain in interpreting the words correctly. Unlike metaphors which, at least in poetry and other creative writing, are *intended* to strain and surprise the reader (in a pleasurable way), the routine utterances that may be the product of misspeaking, presumably are not so intended. What this means is that even if a speaker cannot fail to act in accord with his semantic intentions, as I am insisting, there may be another intention that a speaker has that he does fail to act in accord with, which is the intention to say something that will be *easily* understood by others, understood without strain or without surprise. One assumes that speakers have such an intention in their ordinary speech most of the time, and misspoken utterances or utterances made in medical (or other forms of) ignorance would be actions that fail to be in accord with such an intention. So I am not denying that various intentions, such as the one to be easily understood, are not always fulfilled in the theoretical treatment of meaning I am proposing. But these are not the failures of fulfillment that reflect any *intrinsic* normativity of meaning.

"Speak so as to avoid hearers strain in understanding what you have to say" is not an intention one may have towards one's speech that reflects a norm that is intrinsic to language. It is a purely utilitarian norm. It is not the assumption of normativity that philosophers have made so central to meaning. That assumption of normativity is said by those philosophers to be intrinsic to meaning rather than merely utilitarian.

If the assumption were true, such normativity would reside in intentions which, when properly identified, are intentions that target the sentence meaning or truth conditions of one's words (as I argued in the last section) and which when formulated correctly have an appropriate generality (as I have just argued in this section). If this paper's argument is convincing, intentions, so identified and so formulated, cannot fail to get fulfilled in the speech of those who possess them. And if they cannot fail to get fulfilled, they cannot reflect any genuine normativity.

Meaning intentions, then, are exceptions to Wittgenstein's insight about the nature of intention. They are *not refutations* of his insight and it would be a misunderstanding of the argument of this paper to think it was presented with a view to providing a *counter* example to his claim about the normativity built into the very idea of intention. As a generality, it is indeed true that intentions do have the normativity built into them that Wittgenstein first profoundly brought to our notice. The point rather is that intentions regarding meanings are a degenerate species of intentions and the deepest reasons for this, which I cannot explore here in what is already a very long paper, have to do with the fact that meaning something is a rather unique kind of thing in that *intending* a meaning and *living up* to that intention are – to put it flamboyantly and perhaps a little perversely – more like one thing rather than two, and so failures are not really possible. This remarkable fact is one that I have pursued only briefly elsewhere and intend to explore at much greater length in the future. The present paper's conclusion is, accordingly, relatively modest. Without fully spelling out this unique nature of meaning intentions, it has given an argument to show why they must be viewed, in a very specific sense, as degenerate. It is also modest in another sense. It has not argued that meaning is not normative in any interesting sense, though I believe that to be true and have argued it elsewhere. The conclusion is merely skeptical about the normativity of meaning owing to the normativity of intentions. It argues only that meaning is not normative because, despite its intimate link with intention, it does not inherit the normativity that intentions possess; and the argument is that the normativity that intentions possess *lapse* when intentions target meanings.

Should it be a cause for concern that normativity of this kind goes missing when it comes to meaning? In the passage of this essay's argument, we have seen the extent to which there would have to be a loss of self-knowledge of meaning in order for meaning to be normative and I have hinted at the extent to which that loss would itself owe to a location of meaning in the social realm or in the objective realm of scientific essences. I described these as the mundane versions of Plato's more metaphysically abstract locations. I reckon, then, that any concern we feel at such an absence of norms reflects a residual, secularized yearning for Platonist forms of objectivity, something that Wittgenstein would have seen as calling for therapy, not philosophical satisfaction.

Chapter 4
The Realization Theorem for S5
A Simple, Constructive Proof

Melvin Fitting

4.1 Introduction

Ten years ago I wrote a paper in honor of Rohit Parikh's 60th birthday, [4]. Now I am honored to write another for his 70th. In this paper I will make use of my paper from 10 years ago, to help provide a simple, constructive proof of the Realization Theorem for the modal logic S5. The Realization Theorem is a fundamental result in a developing area known as *justification logics*. Since these are not (yet) standard in the way that modal logics are, I will begin by sketching what justification logics are, saying why they are significant, and saying what the Realization Theorem is about. Then we can get to the more technical material, which essentially amounts to combining work from two of my earlier papers, [4] and [9].

4.2 Justification Logics

Modal logics are familiar things and, just as Hintikka told us, many of them can be interpreted naturally as logics of knowledge – one reads $\Box X$ as "X is known." Thus we want $\Box(X \supset Y) \supset (\Box X \supset \Box Y)$ so that we can draw conclusions from our knowledge, and we want $\Box X \supset X$ to guarantee that what we know is so. This much gives us the logic T. Or we might want to drop the latter condition and study belief instead of knowledge – the modal logic K. Then again, we might add *positive introspection*, $\Box X \supset \Box\Box X$, getting S4, or also *negative introspection*, $\neg\Box X \supset \Box\neg\Box X$, getting S5. But while this approach to (monomodal) logics of knowledge has served well these many years, logics of knowledge are somewhat blunt instruments. Typically we don't just know something, but we know it for a reason – *justifications* are

Melvin Fitting
Department of Mathematics and Computer Science, Lehman College (CUNY), Bronx, NY 10468-1589, USA, e-mail: `melvin.fitting@lehman.cuny.edu`

J. van Benthem et al. (eds.), *Games, Norms and Reasons*, Synthese Library 353,
DOI 10.1007/978-94-007-0714-6_4,

involved. We can think of $\Box$ as a kind of existential quantifier, $\Box X$ says that there exists a justification for X, but the justification itself has been abstracted away.

Justification logics are modal-like logics, having a small calculus of explicit justifications as a formal part of the machinery. *Justification terms* are built up from constants and variables using certain operations. For variables we'll use x, y, x_1, x_2, ... Think of these informally as standing for information from the 'real' world, justifying knowledge we have acquired about things and events, which the logic does not further analyze. There are also basic logical truths – classical tautologies for instance – which we simply accept. We use constants to represent justifications for them, c, d, c_1, c_2, ... Here considerable flexibility is possible. We might want a single justification constant for all tautologies, or for tautology schemes, or for individual tautologies. The setup allows for all these possibilities and more.

Then we have the operation symbols. Two are standard in this area. First, we have the binary symbol $\cdot$ of *application*. The idea is that if term t justifies $X \supset Y$ and term u justifies X then $t \cdot u$ is a justification of Y. Second, we have the binary symbol $+$, representing a kind of weakening or monotonicity. The idea is that $t+u$ justifies anything that t justifies, and also anything that u justifies. We could assume $+$ is commutative, or associative, these would be reasonable assumptions, but we will not make them. We keep things as general as possible.

Finally we have operations corresponding to positive and to negative introspection, ! and ?. If t justifies X, $!t$ should justify the fact that t justifies X. And if t does not justify X, $?t$ should justify the fact that t does not justify X.

The notation that is standard here is, $t{:}X$ is a formula if t is a justification term and X is a formula, and is intended to be read "t justifies X." Here are axiomatic formulations of basic justification logics, corresponding to the modal logics mentioned above.

Axiom Schemes

(a) tautologies (or enough of them)
(b) $t{:}(X \supset Y) \supset (u{:}X \supset (t \cdot u){:}Y)$
(c) $t{:}X \supset (t+u){:}X$ and $u{:}X \supset (t+u){:}X$
(d) $t{:}X \supset X$
(e) $t{:}X \supset !t{:}t{:}X$
(f) $\neg t{:}X \supset ?t{:}\neg t{:}X$

Rules of Inference

(a) Modus ponens, $X, X \supset Y \vdash Y$
(b) Axiom necessitation, if X is an axiom and c is a constant, $\vdash c{:}X$

A *constant specification* is an assignment of axioms to constants. Each proof using the axiom system above, or a subset of it, generates a constant specification – just see what use we made of Axiom necessitation in the course of the proof. Alternately we could start with a constant specification and require that all applications of the Axiom necessitation rule be in accordance with it. Various special conditions can be put on constant specifications, but the only one we are interested in here is *injectivity*. A constant specification is *injective* if at most one axiom is associated with each constant.

The idea of justification logics, and the oldest of them, are due to Artemov [1]. The first of them was called LP, standing for *logic of proofs*. His axiomatization uses the system above without axiom 6. It was referred to as a logic of proofs because justification terms in it could be understood as representing explicit proofs in a logical system, and interpreted arithmetically as such in formal arithmetic. This made it possible to provide a constructive, arithmetic semantics for intuitionistic logic, completing a program begun by Gödel, [10]. That LP can be embedded into formal arithmetic is Artemov's *Arithmetic Completeness Theorem*, which will not concern us here.

If we take any formula of a justification logic and replace every justification term with $\Box$, we get a standard modal formula. This is called the *forgetful functor*. It is a simple matter to check that the forgetful functor turns each axiom of LP into an axiom of S4, and turns every rule application of LP into a rule application of S4. It follows that the forgetful functor turns every theorem of LP into a theorem of S4. Much more difficult to prove is that the converse also holds. Every theorem of S4 is the result of applying the forgetful functor to some theorem of LP. Actually, this holds in a stronger form, which we now state formally. Like the Arithmetic Completeness Theorem, it too is due to Artemov, [1].

Theorem 4.1 (Realization for S4). *Let X be a theorem of S4. Then there is some way of replacing $\Box$ operators of X with justification terms, with negative occurrences being replaced with distinct variables, and positive occurrences by terms that may involve those variables, so that the result is a theorem of LP, provable using an injective constant specification.*

The Realization Theorem plays a key role in Artemov's fulfillment of Gödel's program. Intuitionistic logic embeds in S4 via the well-known mapping that inserts $\Box$ before every subformula. Then S4 embeds in LP; the Realization Theorem is needed here. Finally, LP embeds in formal arithmetic, using the Artemov Arithmetic Completeness Theorem. But of all this, here we are only concerned with the forgetful functor and the Realization Theorem, telling us that LP serves as an explicit version of S4. The Realization Theorem, in effect, Skolemizes the existential quantifiers that are tacit in $\Box$, providing us with an analysis of the reasoning behind the validities of the logic of knowledge S4.

The logic LP can be weakened by omitting positive introspection, axiom 5 (and also modifying the Axiom Necessitation Rule, but we omit details here). Further we can create an analog of a logic of belief, by omitting axiom 4. Today these logics are known as JT and JK. Presumably LP could also be known as JS4, but the original name has been retained for historical reasons. JT and JK also have their Realization Theorems connecting them with T and K, as was also established by Artemov. The proof basically amounts to omitting parts of the full S4/LP proof. By now there are several algorithmic proofs of Realization, [1, 3, 7], and a semantic proof as well, [5].

If we have the full axiom set above, including negative introspection, axiom 6, we have a justification logic known as JS5. There is a Realization Theorem connecting it with the modal logic S5, but now the proof is more than a straightforward modification of Artemov's version for S4. The result has been established using a semantic argument, in [12], with a constructive proof in [2]. It is our intention here to give a simpler constructive argument.

To conclude this section we give a few basic results concerning justification logics. They will play a significant role later on. The first concerns substitution.

Definition 4.1. A *substitution* is a map from justification variables to justification terms. If σ is a substitution and X is a formula, we write $X\sigma$ for the result of replacing each variable x in X with the term $x\sigma$. Similarly for substitution in justification terms themselves.

The following is shown in [1] for LP, and applies, with exactly the same argument, to JS5.

Theorem 4.2 (Substitution Lemma). *If X is a theorem of LP, so is $X\sigma$. Further, if X has an injective proof, so does $X\sigma$.*

The constant specification used for proving X and that used for proving $X\sigma$ will, in general, be different, but that does not matter for our purposes.

The second fundamental result is the Lifting Lemma, also from [1], saying justification logics internalize their own proofs. Let us take $X_1,\ldots,X_n \vdash_{\mathsf{J}} Y_1,\ldots,Y_m$ to mean that $(X_1 \wedge \ldots \wedge X_n) \supset (Y_1 \vee \ldots \vee Y_m)$ is a theorem of the justification logic J. The Lifting Lemma for LP, due to Artemov [1], says that if $s_1{:}X_1,\ldots,s_n{:}X_n,Y_1,\ldots,Y_k \vdash_{\mathsf{LP}} W$, then there is a justification term $u(s_1,\ldots,s_n,y_1,\ldots,y_k)$ (where the y_i are variables) such that $s_1{:}X_1,\ldots,s_n{:}X_n,y_1{:}Y_1,\ldots,y_k{:}Y_k \vdash_{\mathsf{LP}} u(s_1,\ldots,s_n,y_1,\ldots,y_k){:}W$. The version for JS5 is broader in that the right side of the turnstyle can have multiple formulas.

Theorem 4.3 (JS5 Lifting Lemma). *Suppose*

$$s_1{:}X_1,\ldots,s_n{:}X_n,Y_1,\ldots,Y_k \vdash_{JS5} t_1{:}Z_1,\ldots,t_m{:}Z_m,W$$

then there is a proof polynomial $u(s_1,\ldots,s_n,t_1,\ldots,t_m,y_1,\ldots,y_k)$ such that

$$s_1{:}X_1,\ldots,s_n{:}X_n,y_1{:}Y_1,\ldots,y_k{:}Y_k \vdash_{JS5}$$
$$t_1{:}Z_1,\ldots,t_m{:}Z_m,u(s_1,\ldots,s_n,t_1,\ldots,t_m,y_1,\ldots,y_k){:}W.$$

Moreover, if the original derivation was injective, the same is the case for the later derivation.

We omit the (constructive) proof of this, which is similar to that for LP except that axiom 6 plays a role. If one moves formulas $t_i{:}Z_i$ from the right of the turnstyle to the left, as $\neg t_i{:}Z_i$, things are straightforward.

4.3 An S5 Gentzen System

To date, all constructive proofs of Realization Theorems make use of cut-free Gentzen system (or tableau system) proofs. The logic S5 is an anomaly among the most common modal logics, in that it does not seem to have a simple cut-free Gentzen system. The constructive S5 Realization Theorem proof in [2] was based on a hypersequent calculus from [11]. But in [4] we gave a cut-free tableau system for S5 that seems to be as simple as anything in the literature. We will make use of the corresponding Gentzen system here, so in this section we present it. We begin with a standard Gentzen system for S4. Then we explain how to modify it for S5. We assume formulas are built up from propositional letters and $\bot$ using $\supset$ and $\Box$.

A *sequent* for S4 is a pair of finite multisets of modal formulas, where the pair is written $\Gamma \longrightarrow \Delta$, with Γ and Δ being multisets. Using multisets avoids the need for explicit permutation rules. Axioms are the following sequents, where P is any propositional letter.

$$P \longrightarrow P \qquad\qquad \bot \longrightarrow$$

Then the rules of derivation are as follows. In stating them, Γ and Δ are multisets, X and Y are formulas, and if $\Gamma = \{Y_1, \ldots, Y_k\}$ then $\Box\Gamma = \{\Box Y_1, \ldots, \Box Y_k\}$.

$$LW \ \frac{\Gamma \longrightarrow \Delta}{\Gamma, X \longrightarrow \Delta} \qquad\qquad RW \ \frac{\Gamma \longrightarrow \Delta}{\Gamma \longrightarrow \Delta, X}$$

$$LC \ \frac{\Gamma, X, X \longrightarrow \Delta}{\Gamma, X \longrightarrow \Delta} \qquad\qquad RC \ \frac{\Gamma \longrightarrow \Delta, X, X}{\Gamma \longrightarrow \Delta, X}$$

$$L\supset \ \frac{\Gamma, Y \longrightarrow \Delta \quad \Gamma \longrightarrow \Delta, X}{\Gamma, X \supset Y \longrightarrow \Delta} \qquad\qquad R\supset \ \frac{\Gamma, X \longrightarrow \Delta, Y}{\Gamma \longrightarrow \Delta, X \supset Y}$$

$$L\Box \ \frac{\Gamma, X \longrightarrow \Delta}{\Gamma, \Box X \longrightarrow \Delta} \qquad\qquad R\Box \ \frac{\Box\Gamma \longrightarrow X}{\Box\Gamma \longrightarrow \Box X}$$

As usual, a proof of a formula X in this calculus is a proof of the sequent $\longrightarrow X$. This is a standard sequent calculus for S4, and soundness and completeness arguments are well-known in the literature.

To turn this into a proof system for propositional S5, two simple changes are needed. First, the rule $R\Box$ is replaced by a stronger version allowing multiple formulas on the right. Here is the rule we will be using for S5.

$$R\Box \frac{\Box\Gamma \longrightarrow \Box\Delta, X}{\Box\Gamma \longrightarrow \Box\Delta, \Box X}$$

The system, with this new $R\Box$ rule, is sound for S5, but it is not complete when used directly. However we do have completeness in the following odd sense. *If X is a valid formula of S5, there will be a sequent proof of $\longrightarrow \Box X$.*

In order to give an example of a proof, it will be convenient to introduce negation in the usual way, $\neg X$ stands for $X \supset \bot$. It is easy to show the following are derived rules, and we will make use of them in the example. This example will be continued in Section 4.6.

$$L\neg \frac{\Gamma \longrightarrow \Delta, X}{\Gamma, \neg X \longrightarrow \Delta} \qquad\qquad R\neg \frac{\Gamma, X \longrightarrow \Delta}{\Gamma \longrightarrow \Delta, \neg X}$$

Example 4.1. We prove $P \supset \Box\neg\Box\neg P$ in this system by giving a Gentzen derivation of the sequent $\longrightarrow \Box(P \supset \Box\neg\Box\neg P)$.

$$
\begin{array}{ll}
1 & P \longrightarrow P \\
2 & P, \neg P \longrightarrow \\
3 & P, \neg P \longrightarrow \Box\neg\Box\neg P \\
4 & \neg P \longrightarrow P \supset \Box\neg\Box\neg P \\
5 & \Box\neg P \longrightarrow P \supset \Box\neg\Box\neg P \\
6 & \Box\neg P \longrightarrow \Box(P \supset \Box\neg\Box\neg P) \\
7 & \longrightarrow \neg\Box\neg P, \Box(P \supset \Box\neg\Box\neg P) \\
8 & \longrightarrow \Box\neg\Box\neg P, \Box(P \supset \Box\neg\Box\neg P) \\
9 & P \longrightarrow \Box\neg\Box\neg P, \Box(P \supset \Box\neg\Box\neg P) \\
10 & \longrightarrow (P \supset \Box\neg\Box\neg P), \Box(P \supset \Box\neg\Box\neg P) \\
11 & \longrightarrow \Box(P \supset \Box\neg\Box\neg P), \Box(P \supset \Box\neg\Box\neg P) \\
12 & \longrightarrow \Box(P \supset \Box\neg\Box\neg P)
\end{array}
$$

In this: 1 is an axiom, 2 is from 1 by $L\neg$, 3 is from 2 by RW, 4 is from 3 by $R\supset$, 5 is from 4 by $L\Box$, 6 is from 5 by $R\Box$, 7 is from 6 by $R\neg$, 8 is from 7 by $R\Box$, 9 is from 8 by LW, 10 is from 9 by $R\supset$, 11 is from 10 by $R\Box$, and 12 is from 11 by RC.

Soundness and completeness are shown in [4] for a tableau version of this system. That transfers to the present Gentzen system either by adapting the proof, or by showing that tableau proofs translate into sequent calculus proofs. Details are omitted here.

4.4 Annotations and Realizations

The Realization Theorem deals with *occurrences* of modal operators, and treats positive and negative occurrences differently; negatives become proof variables while positives need not. To make this formal, in [6, 7, 9] I introduced *annotated* formulas – providing syntactic machinery to keep track of $\Box$ occurrences.

For annotated formulas, instead of a single operator $\Box$ there is an infinite family, $\Box_1, \Box_2, \ldots$. These are called *indexed* modal operators. Annotated formulas are built up as usual, but using indexed modal operators instead of $\Box$. If X is an annotated formula, and X' is the result of replacing all indexed modal operators, $\Box_n$, with $\Box$, we say X is an *annotated version* of X', and X' is an *unannotated version* of X.

A *properly* annotated formula is an annotated formula meeting the conditions that: no indexed modal operator occurs twice, and if $\Box_n$ occurs in a negative position n is even, and if it occurs in a positive position n is odd.

Annotations are simply a bookkeeping device to keep track of occurrences of modal operators and their polarities – negative occurrences are even, positive occurrences are odd. Properly annotated formulas are fundamental, but formulas that are annotated but not properly so also arise. For instance, if $X \supset Y$ is properly annotated, the subformula Y is also, but X is not (though $\neg X$ is). Generally we will fix a properly annotated formula X and work with subformulas of it, where these may not be properly annotated.

It is easy to see that every modal formula has many properly annotated versions. But further, it is also easy to give an annotated version of the S5 proof system in Section 4.3. Except for the two modal rules, all the axioms and rules have exactly the same form *but formulas are now annotated.* Annotations must be preserved in moving from sequents above the line to sequents below the line. The two modal rules become the following.

$$L\Box \ \frac{\Gamma, X \longrightarrow \Delta}{\Gamma, \Box_{2n} X \longrightarrow \Delta}$$

$$R\Box \ \frac{\Box_{2n_1} Y_1, \ldots, \Box_{2n_j} Y_j \longrightarrow \Box_{2m_1+1} Z_1, \ldots, \Box_{2m_k+1} Z_k, X}{\Box_{2n_1} Y_1, \ldots, \Box_{2n_j} Y_j \longrightarrow \Box_{2m_1+1} Z_1, \ldots, \Box_{2m_k+1} Z_k, \Box_{2p+1} X}$$

If there is a sequent proof ending with $\longrightarrow X$ (annotated or unannotated), it has one in which every formula that appears is a subformula of X; more strongly, it has one in which every formula on the left of an arrow is a negative subformula of X and every formula on the right of an arrow is a positive subformula of X.

Proposition 4.1. *Let Z be an unannotated modal formula and X be any properly annotated version of Z. If there is an unannotated sequent proof of $\longrightarrow Z$ then there is an annotated sequent proof of $\longrightarrow X$.*

Briefly, take a proof of $\longrightarrow Z$ in the unannotated sequent calculus, and use this to construct an annotated proof of $\longrightarrow X$. Replace the final $\longrightarrow Z$ with $\longrightarrow X$, then propagate the annotations upward, from conclusions of rules to premises, until the entire sequent construction has been annotated. A formal version of this verification amounts to an induction on the number of sequents in the unannotated proof, and is omitted. An example can be found in Section 4.6.

Now that we have annotated formulas, realizations can be defined functionally in a natural way. A *realization function* is a mapping from positive integers to proof polynomials that maps even integers to justification variables. Moreover it is assumed that all realization functions behave the same on the even integers, specifically, if r is any realization function, $r(2n) = x_n$, where $x_1, x_2, \ldots$ is the list of justification variables arranged in a standardized order. If X is an annotated formulala, and r is a realization function, by $r(X)$ is meant the result of replacing each modal operator $\Box_i$ in X with the proof polynomial $r(i)$. The result, $r(X)$ is formula of justification logic. Now, here is a statement of the theorem we are after.

Theorem 4.4. *If Z is a theorem of S5, then for any properly annotated version X of Z there is a realization function r such that $r(X)$ is injectively provable in JS5.*

4.5 Modifying Realizations

Realizations, treated as functions, can be combined and modified in various ways, though this is not a simple process. In [6, 7, 9] several ways of doing so were presented, with the work continued in [8]. We need two of these.

The first of our results from [7, 9] has to do with the merging of different realizations for the same formula. As originally stated the theorem was for LP, but the proof only makes use of the fact that $+$ and $\cdot$ are among the available operations, and an Internalization Lemma is provable, and hence it holds for JS5 as well. Also, it applies to the merging of many realization functions – we will only need it for two of them.

Definition 4.2. Let X be an annotated formula and r_1 and r_2 be realization functions. We say a pair $\langle r, \sigma \rangle$ consisting of a realization function and a substitution *hereditarily merges* r_1 and r_2 on on X provided, for each subformula φ of X:

(a) if φ is a positive subformula of X then both $r_1(\varphi)\sigma \supset r(\varphi)$ and $r_2(\varphi)\sigma \supset r(\varphi)$ are theorems, provable with an injective constant specification;
(b) if φ is a negative subformula of X then both $r(\varphi) \supset r_1(\varphi)\sigma$ and $r(\varphi) \supset r_2(\varphi)\sigma$ are theorems, provable with an injective constant specification.

Theorem 4.5 (Realization Merging). *Let X be a properly annotated formula, and r_1 and r_2 be realization functions. Then there is a realization/substitution pair $\langle r, \sigma \rangle$ that hereditarily merges r_1 and r_2 on X.*

The proof of this Theorem is entirely algorithmic, but it is complex and is omitted here, as is the algorithm and verification for the next result – they can be found in detail in [9]. The second item from that paper is an analog of the replacement property of classical logic. Suppose $\psi(P)$ is a classical formula and P is a propositional letter, we write $\psi(A)$ for the result of replacing all occurrences of P in $\psi(P)$ with occurrences of the formula A. Suppose P has only positive occurrences in $\psi(P)$. Then, if $A \supset B$ is provable so is $\psi(A) \supset \psi(B)$. If P has only negative occurrences then $A \supset B$ provable yields that $\psi(B) \supset \psi(A)$ is provable. We gave a corresponding result for the justification logic LP, but it too applies to JS5. Further, in [9] the replacement result was narrowed to be more directly applicable to the construction of realizations. In fact, we did not narrow it enough, and the realization algorithm given there appeared to be more complicated than was needed. Here we first state the appropriate item from [9], then use it to derive the narrower thing we actually need.

Definition 4.3. Let $X(P)$ be an annotated formula in which the propositional letter P has at most one positive occurrence, let A and B be annotated formulas, and let r_1 be a realization function. We say the realization/substitution pair $\langle r, \sigma \rangle$ *hereditarily replaces* $r_1(A)$ with $r_1(B)$ at P in $X(P)$ provided, for each subformula $\varphi(P)$ of $X(P)$:

(a) if $\varphi(P)$ is a positive subformula of $X(P)$ then $r_1(\varphi(A))\sigma \supset r(\varphi(B))$ has a proof with an injective constant specification;
(b) if $\varphi(P)$ is a negative subformula of $X(P)$ then $r(\varphi(B)) \supset r_1(\varphi(A))\sigma$ has a proof with an injective constant specification.

We actually need a very simple version of replacement, in which we jointly replace $t{:}F$ and $u{:}F$ with the weaker $(t+u){:}F$. Here is a formulation using the machinery of realization functions.

Theorem 4.6 (Realization Weakening). *Assume the following.*

S–1. $X(P)$ *is a properly annotated formula in which the propositional letter* P *has at most one positive occurrence;*
S–2. $\Box_p K$ *and* $\Box_q K$ *are both are properly annotated, there is no annotation overlap between* $X(P)$ *and* $\Box_p K$*, and* $X(P)$ *and* $\Box_q K$*, and* p *and* q *are different;*
S–3. r_1 *and* r_2 *are realization functions with* $r_1(K) = r_2(K)$*;*
S–4. $r_1(q) = r_2(q) = r_1(p) + r_2(p)$.

Then there is a realization/substitution pair $\langle r, \sigma \rangle$ *that hereditarily replaces* $r_1(\Box_p K)$ *with* $r_1(\Box_q K)$ *at* P *in* $X(P)$*, and hereditarily replaces* $r_2(\Box_p K)$ *with* $r_2(\Box_q K)$ *at* P *in* $X(P)$.

If we set $r_1(K) = r_2(K) = F$, $r_1(p) = t$, $r_2(p) = u$, and $r_1(q) = r_2(q) = t+u$, the theorem above provides a replacement of $t{:}F$ and $u{:}F$ with $(t+u){:}F$, as promised.

In [9] as it has been available in pre-publication form, the proof of the Realization Theorem made use of the Realization Weakening Theorem directly. In fact, only a very narrow special case of it is needed, and this is embodied in the following Corollary. The proof of this Corollary was implicit in [9], and is made explicit here. Hopefully, the paper [9] itself can be revised before publication by incorporating the present approach into it.

Corollary 4.1. *Suppose X is a properly annotated formula with $\Box_p K$ as a positive subformula. Let r_1 be a realization function, and u be a justification term. There is a realization/substitution pair $\langle r, \sigma \rangle$ such that:*

(a) if φ is a positive subformula of X then $r_1(\varphi)\sigma \supset r(\varphi)$ is provable using an injective constant specification;
(b) if φ is a negative subformula of X then $r(\varphi) \supset r_1(\varphi)\sigma$ is provable using an injective constant specification;
(c) $u{:}r_1(K)\sigma \supset r(\Box_p K)$ is an injective theorem.

The corollary above can be given an intuitive meaning consistent with what we have been saying. Suppose we write t for $r_1(p)$. Since $\Box_p K$ is a positive subformula of X conclusion 1 has, as a special case, the injective provability of

$$t{:}r_1(K)\sigma \supset r(\Box_p K)$$

while conclusion 3 asserts the injective provability of

$$u{:}r_1(K)\sigma \supset r(\Box_p K)$$

and thus we might loosely describe what is happening as: $\langle r, \sigma \rangle$ *weakens t to $t + u$ at p*.

Proof. We will derive this from Theorem 4.6. We first introduce a new index q and a second realization function, r_2. Then we apply Theorem 4.6, and eliminate the index q at the end.

The formula $\Box_p K$ occurs as a positive subformula of X so it must occur exactly once, since X is properly annotated and so the index p can occur only once. Let P be a propositional letter that does not occur in X, and let $X(P)$ be like X except that the subformula $\Box_p K$ has been replaced with P. Then P must have a single positive occurrence in $X(P)$, and X is the same as $X(\Box_p K)$. Also, no index in $\Box_p K$ can occur in $X(P)$, again since $X = X(\Box_p K)$ is properly annotated.

Let q be an odd index that does not occur in X (and hence it is different than p). Then $X(P)$, $\Box_p K$, and $\Box_q K$ are properly annotated, $X(P)$ and $\Box_p K$ have no annotation overlap, and $X(P)$ and $\Box_q K$ have no annotation overlap.

Modify the definition of r_1 so that $r_1(q) = r_1(p) + u$. Since q does not occur in X this does not change the behavior of r_1 on X or its subformulas.

Define a second realization function r_2 to be the same as r_1, except that $r_2(p) = u$. Since $\Box_p K$ is a subformula of X, and X is properly annotated, p does not occur in K and hence $r_1(K) = r_2(K)$. Also by definition, $r_2(q) = r_1(q) = r_1(p) + u = r_1(p) + r_2(p)$.

Now we can apply Theorem 4.6, Realization Weakening. There is a realization/substitution pair $\langle r^*, \sigma^* \rangle$ that hereditarily replaces $r_1(\Box_p K)$ with $r_1(\Box_q K)$ at P in $X(P)$ and hereditarily replaces $r_2(\Box_p K)$ with $r_2(\Box_q K)$ at P in $X(P)$. The realization function r^* is almost the one we want – it needs one modification. Let r be like r^* except that $r(p) = r^*(q)$. We will show the realization/substitution pair $\langle r, \sigma^* \rangle$ does the job.

Let φ be an arbitrary subformula of X. Recall that $X(P)$ was like X but with $\Box_p K$ replaced with P, and so $X = X(\Box_p K)$. Since φ is a subformula of $X(\Box_p K)$ there are three possibilities. It could be that φ is a subformula of K. Otherwise φ is not a subformula of K, in which case either $\Box_p K$ is a subformula of φ (possibly not proper), or $\Box_p K$ and φ are disjoint. If $\Box_p K$ is a subformula of φ then there is a subformula $\varphi(P)$ of $X(P)$ with $\varphi = \varphi(\Box_p K)$. We can treat the case where φ and $\Box_p K$ are disjoint the same way by allowing P to occur vacuously in $\varphi(P)$. Thus we really have two cases to consider. We handle each case separately to establish conclusions 1 and 2 of the Corollary. Conclusion 3 is shown the same way in either case.

Suppose first that φ is a subformula of K. Since $r^*(K) = r_1(K)\sigma^*$ we must have $r^*(\varphi) = r_1(\varphi)\sigma^*$ Since r and r^* only differ on q, and q cannot occur in K, we have $r(\varphi) = r_1(\varphi)\sigma^*$. This gives us conclusions 1 and 2 for this case.

For the second case, let $\varphi(P)$ be a subformula of $X(P)$. Let us say it is a positive subformula – the negative subformula case is handled similarly. Since $\langle r^*, \sigma^* \rangle$ hereditarily replaces $r_1(\Box_p K)$ with $r_1(\Box_q K)$ at P in $X(P)$, we have the provability of $r_1(\varphi(\Box_p K))\sigma^* \supset r^*(\varphi(\Box_q K))$. Since $r(p) = r^*(q)$, this says we have the provability of $r_1(\varphi(\Box_p K))\sigma^* \supset r(\varphi(\Box_p K))$, and this is conclusion 1 of the Corollary.

Finally, P is a positive subformula of $X(P)$, so since $\langle r^*, \sigma^* \rangle$ also hereditarily replaces $r_2(\Box_p K)$ with $r_2(\Box_q K)$ at P in $X(P)$, we have provability of $r_2(\Box_p K)\sigma^* \supset r^*(\Box_q K)$. Again since $r^*(q) = r(p)$, this give provability of $r_2(\Box_p K)\sigma^* \supset r(\Box_p K)$. And since $r_2(p) = u$ and $r_2(K) = r_1(K)$ we have provability of $u{:}r_1(K)\sigma^* \supset r(\Box_p K)$, and this is conclusion 3 of the Corollary.

Now we combine the machinery presented so far, and give a proof of the Realization Theorem for S5, along the same lines as the one for S4 in [9].

Theorem 4.7. *If Z_0 is a theorem of S5, there is a realization of Z_0 that is an injectively provable theorem of JS5. More precisely, if Z_0 is a theorem of S5, then for any properly annotated version Z of Z_0 there is a realization function r such that $r(Z)$ is injectively provable in JS5.*

Proof. Assume Z_0 is a theorem of S5, and Z is a properly annotated version of Z_0. There is a sequent proof $\mathscr{P}$ of $\rightarrow \Box Z$ in the annotated Gentzen calculus of Section 4.3. We will show that, in a suitable sense, every sequent in $\mathscr{P}$ is realizable. It will follow that $\Box Z$ is realizable, and then so is Z, using Axiom Scheme 4 from Section 4.2.

There is a standard connection between sequents and formulas. Here it is, for the record. For each sequent S of annotated formulas, an annotated formula $\|S\|$ is defined.

(a) $\|X_1, \ldots, X_n \longrightarrow Y_1, \ldots, Y_m\| = [(X_1 \wedge \ldots \wedge X_n) \supset (Y_1 \vee \ldots \vee Y_m)]$.
(b) $\|X_1, \ldots, X_n \rightarrow \| = [(X_1 \wedge \ldots \wedge X_n) \supset \bot]$.
(c) $\| \longrightarrow Y_1, \ldots, Y_k\| = [(Y_1 \vee \ldots \vee Y_k)]$.

We now show that for every sequent S in proof $\mathscr{P}$, there is a realization function r such that $r(\|S\|)$ is injectively provable in JS5. The final sequent is $\longrightarrow \Box Z$, and

$\| \longrightarrow \Box Z\|$ is simply $\Box Z$, and the existence of a realization function for Z follows. We show axioms in $\mathscr{P}$ are realized, and that realization is preserved by the rules of derivation.

Sequents that are axioms have no modal operators, so for these we can take any realization function.

With two exceptions, if r realizes the premise of a rule, it also realizes the conclusion. The two exceptions are $L\supset$ (which has two premises), and $R\Box$. We concentrate on these two cases, and leave the others to you – the arguments are straightforward.

The two hard cases are handled more-or-less the same way that worked for the S4 Realization Theorem in [9], so here we only sketch the ideas.

We begin with $L\supset$. Suppose both $\Gamma, Y \longrightarrow \Delta$ and $\Gamma \longrightarrow \Delta, X$ are realized, though different realization functions may be involved. Say $r_1(\|\Gamma, Y \longrightarrow \Delta\|)$ and $r_2(\|\Gamma \longrightarrow \Delta, X\|)$ both have injective S5 proofs. By the Realization Merging Theorem 4.5, there is a realization/substitution pair $\langle r, \sigma\rangle$ that hereditarily merges r_1 and r_2 on Z. Then $r(\|\Gamma, X \supset Y \longrightarrow \Delta\|)$ is injectively provable, as we now verify.

Suppose φ is an annotated formula on the left in one of the premise sequents, so φ is in Γ or is Y itself. Then φ is a negative subformula of Z, so both $r(\varphi) \supset r_1(\varphi)\sigma$ and $r(\varphi) \supset r_2(\varphi)\sigma$ are injectively provable. Similarly if φ on the right, a member of Δ or X itself, it is a positive subformula of Z and so both $r_1(\varphi)\sigma \supset r(\varphi)$ and $r_2(\varphi)\sigma \supset r(\varphi)$ are injectively provable. Either way, it is easy to see that $r_1(\|\Gamma, Y \to \Delta\|)\sigma \supset r(\|\Gamma, Y \to \Delta\|)$ and $r_2(\|\Gamma \to \Delta, X\|)\sigma \supset r(\|\Gamma \to \Delta, X\|)$ are injectively provable. Since $r_1(\|\Gamma, Y \longrightarrow \Delta\|)$ is injectively provable, by the Substitution Lemma 4.2 so is $r_1(\|\Gamma, Y \longrightarrow \Delta\|)\sigma$. Similarly $r_2(\|\Gamma \longrightarrow \Delta, X\|)\sigma$ is injectively provable. It follows that both $r(\|\Gamma, Y \to \Delta\|)$ and $r(\|\Gamma \to \Delta, X\|)$ are injectively provable. Now we have a single realization function, r, for both sequents and it is easy to show that $r(\|\Gamma, X \supset Y \longrightarrow \Delta\|)$ is injectively provable.

Finally we consider the case $R\Box$. Suppose

$$\Box_{n_1} Y_1, \ldots, \Box_{n_j} Y_j \longrightarrow \Box_{p_1} Z_1, \ldots, \Box_{p_k} Z_k, X \tag{4.1}$$

is realized, say using the realization function r_1. In this each n_1 is even, being in a negative position, and each p_i is odd, being in a positive position. We show

$$\Box_{n_1} Y_1, \ldots, \Box_{n_j} Y_j \longrightarrow \Box_{p_1} Z_1, \ldots, \Box_{p_k} Z_k, \Box_p X \tag{4.2}$$

is realized, where p is odd. It is assumed that (4.1) and (4.2) occur in proof $\mathscr{P}$.

Since r_1 realizes (4.1), there is a JS5 proof of the following (where $r_1(n_i)$ is a variable for each n_i).

$$[r_1(\Box_{n_1} Y_1) \wedge \ldots \wedge r_1(\Box_{n_j} Y_j)] \supset [r_1(\Box_{p_1} Z_1) \wedge \ldots \wedge r_1(\Box_{p_k} Z_k) \wedge r_1(X)] \tag{4.3}$$

Using (4.3) and the Lifting Lemma, Theorem 4.3, there is a justification term, call it u, such that

$$[r_1(\Box_{n_1}Y_1)\wedge\ldots\wedge r_1(\Box_{n_j}Y_j)]\supset[r_1(\Box_{p_1}Z_1)\wedge\ldots\wedge r_1(\Box_{p_k}Z_k)\wedge u{:}r_1(X)] \tag{4.4}$$

is injectively provable.

If the index p has no occurrences in the sequent (4.1) then things are simple since what we do with p has no effect on what we have already done. Define a realization function r to be like r_1 except that $r(p)=u$. Then it follows immediately from (4.4) that r realizes (4.2).

The simple case just discussed may not always happen – the index p may already occur in (4.1) and so $r_1(p)$ may already be playing an important role in realizing (4.1). In such event, in realizing (4.2) we realize $\Box_p$ not with u itself, but with $r_1(p)+u$. We now establish that this works.

Apply Corollary 4.1 of the Realization Weakening Theorem. (Recall, $\Box_pX$ must be a positive subformula of Z since it occurs on the right of the arrow in the Gentzen proof $\mathscr{P}$ of Z.) There is a realization/substitution pair $\langle r,\sigma\rangle$ that meets the following conditions

(a) if φ is a positive subformula of Z then $r_1(\varphi)\sigma\supset r(\varphi)$ is an injective theorem;
(b) if φ is a negative subformula of Z then $r(\varphi)\supset r_1(\varphi)\sigma$ is an injective theorem;
(c) $u{:}r_1(X)\sigma\supset r(\Box_pX)$ is an injective theorem.

From (4.4) and the Substitution Lemma, Theorem 4.2, we have provability of the following.

$$[r_1(\Box_{n_1}Y_1)\sigma\wedge\ldots\wedge r_1(\Box_{n_j}Y_j)\sigma]\supset[r_1(\Box_{p_1}Z_1)\sigma\wedge\ldots\wedge r_1(\Box_{p_k}Z_k)\sigma\wedge u{:}r_1(X)\sigma] \tag{4.5}$$

Using this, we will now show that

$$[r(\Box_{n_1}Y_1)\wedge\ldots\wedge r(\Box_{n_j}Y_j)]\supset[r(\Box_{p_1}Z_1)\wedge\ldots\wedge r(\Box_{p_k}Z_k)\wedge r(\Box_pX)] \tag{4.6}$$

has a proof, and thus (4.2) is realized.

Consider one of the formulas on the left of sequent (4.2), say $\Box_{n_i}Y_i$. This is a negative subformula of Z and hence $r(\Box_{n_i}Y_i)\supset r_1(\Box_{n_i}Y_i)\sigma$ is provable. We thus have provability of the following.

$$[r(\Box_{n_1}Y_1)\wedge\ldots\wedge r(\Box_{n_j}Y_j)]\supset[r_1(\Box_{n_1}Y_1)\sigma\wedge\ldots\wedge r_1(\Box_{n_j}Y_j)\sigma] \tag{4.7}$$

Next consider a formula on the right of sequent (4.2) of the form $\Box_{p_i}Z_i$. This is a positive subformula of Z and so $r_1(\Box_{p_i}Z_i)\sigma\supset r(\Box_{p_i}Z_i)$ is provable. So we have provability of the following.

$$[r_1(\Box_{p_1}Z_1)\sigma\wedge\ldots\wedge r_1(\Box_{p_k}Z_k)\sigma]\supset[r(\Box_{p_1}Z_1)\wedge\ldots\wedge r(\Box_{p_k}Z_k)] \tag{4.8}$$

Finally, by conclusion 3 of the Corollary,

$$u{:}r_1(X)\sigma\supset r(\Box_pX) \tag{4.9}$$

is provable.

Now, simply combining (4.7), (4.5), (4.8), and (4.9), we immediately have (4.6).

4.6 A Realization Example

We produce a realization for the S5 theorem $P \supset \Box\neg\Box P$, where P is a propositional letter. In Section 4.3 we gave a Gentzen system proof for this, which amounted to constructing a sequent proof for $\Box(P \supset \Box\neg\Box P)$. We build on that.

To begin, we need a properly annotated version of $\Box(P \supset \Box\neg\Box P)$. We use $\Box_1(P \supset \Box_3\neg\Box_2\neg P)$. And next we need an annnotated sequent proof of this. One can be created by starting with the unannotated proof, annotating the last sequent, and propagating the annotations upward. The result is the following.

1	$P \longrightarrow P$
2	$P, \neg P \longrightarrow$
3	$P, \neg P \longrightarrow \Box_3\neg\Box_2\neg P$
4	$\neg P \longrightarrow P \supset \Box_3\neg\Box_2\neg P$
5	$\Box_2\neg P \longrightarrow P \supset \Box_3\neg\Box_2\neg P$
6	$\Box_2\neg P \longrightarrow \Box_1(P \supset \Box_3\neg\Box_2\neg P)$
7	$\longrightarrow \neg\Box_2\neg P, \Box_1(P \supset \Box_3\neg\Box_2\neg P)$
8	$\longrightarrow \Box_3\neg\Box_2\neg P, \Box_1(P \supset \Box_3\neg\Box_2\neg P)$
9	$P \longrightarrow \Box_3\neg\Box_2\neg P, \Box_1(P \supset \Box_3\neg\Box_2\neg P)$
10	$\longrightarrow (P \supset \Box_3\neg\Box_2\neg P), \Box_1(P \supset \Box_3\neg\Box_2\neg P)$
11	$\longrightarrow \Box_1(P \supset \Box_3\neg\Box_2\neg P), \Box_1(P \supset \Box_3\neg\Box_2\neg P)$
12	$\longrightarrow \Box_1(P \supset \Box_3\neg\Box_2\neg P)$

Next, we follow the algorithm embodied in the proof of Theorem 4.7, and show each sequent above is realizable in JS5. In doing this we make use of Corollary 4.1 which, in turn, uses the Realization Weakening Theorem 4.6. The proof of this theorem is algorithmic, and the algorithm can be found in [9] – here we simply give the results of the application of the algorithm.

Sequents 1, 2, 3, 4, 5 To realize 1, any realization function will do. Then this is also the case for 2, 3, 4, and 5, which follow using rules $L\neg$, RW, $R\supset$, and $L\Box$ respectively. We must have $r_1(2) = x_1$ and, to be specific, let us say $r_1(3) = a$. Then 5 is realized by the following justification formula, provable in JS5.

$$x_1{:}\neg P \supset [P \supset a{:}\neg x_1{:}\neg P]$$

Sequents 6, 7 Sequent 6 follows from 5 by $R\Box$, and since the index 1 does not occur in sequent 5 things are simple. By the Lifting Lemma there is a justification term q such that $x_1{:}\neg P \supset q{:}[P \supset a{:}\neg x_1{:}\neg P]$ is a theorem of JS5. The proof of the Lifting Lemma is constructive, but here we will not explicitly produce q. We simply note that such a term can be produced. Now, a realization function for 6 is r_2 where $r_2(1) = q$, $r_2(2) = x_1$, and $r_2(3) = a$. The same realization function works for sequent 7 since the rule $R\neg$ is involved. Thus we have JS5 provability of the following.

$$\neg x_1{:}\neg P \vee q{:}[P \supset a{:}\neg x_1{:}\neg P]$$

Sequents 8, 9, 10 Sequent 8 follows from 7 by $R\Box$, and index 3 already occurs in 7 so we are in a more complex situation. Since we have JS5 provability of the formula given above, realizing sequent 7, by the Lifting Lemma there is a justification term r such that $r{:}\neg x_1{:}\neg P \wedge q{:}(P \supset a{:}\neg x_1{:}\neg P)$ is provable. The idea is to have realization function r_3 come from r_2 using Corollary 4.1, weakening $a = r_2(3)$ to $a + r$. Working this out produces a realization function such that $r_3(1) = s \cdot q$, $r_3(2) = x_1$, and $r_3(3) = a + r$, where s internalizes a proof of the JS5 theorem $(P \supset a{:}\neg x_1{:}\neg P) \supset (P \supset (a+r){:}\neg x_1{:}\neg P)$. That is, we have JS5 provability of $s{:}[(P \supset a{:}\neg x_1{:}\neg P) \supset (P \supset (a+r){:}\neg x_1{:}\neg P)]$. The same realization function works for sequents 9 and 10. Then, sequent 10 has the following realization.

$$(P \supset (a+r){:}\neg x_1{:}\neg P) \vee (s \cdot q){:}(P \supset (a+r){:}\neg x_1{:}\neg P)$$

Sequents 11, 12 Sequent 11 is from 10 by $R\Box$, and the index 1 occurs in 10, so we again have a complex situation. The formula above, realizing sequent 10, is provable so by the Lifting Lemma, again, there is a justification term t such that $t{:}(P \supset (a+r){:}\neg x_1{:}\neg P) \vee (s \cdot q){:}(P \supset (a+r){:}\neg x_1{:}\neg P)$ is provable. Now Corollary 4.1 comes into play again, and we arrive at the following. Let r_4 be like r_3 except that $r_4(1) = r_3(1) + t$. That is, $r_4(1) = (s \cdot q) + t$, $r_4(2) = x_1$, and $r_4(3) = a + r$. This realizes sequent 11, and also sequent 12. We thus have the JS5 provability of the following.

$$(s \cdot q + t){:}(P \supset (a+r){:}\neg x_1{:}\neg P)$$

Conclusion Now we use axiom 4 and modus ponens with the formula above. The following is provable in JS5:

$$P \supset (a+r){:}\neg x_1{:}\neg P \tag{4.10}$$

where a is arbitrary, and r is a justification term such that $r{:}\neg x_1{:}\neg P \wedge q{:}(P \supset a{:}\neg x_1{:}\neg P)$ is provable, where q in turn is a justification term such that $x_1{:}\neg P \supset q{:}[P \supset a{:}\neg x_1{:}\neg P]$ is provable. Notice that q itself has disappeared from (4.10) except insofar as it has been incorporated into r. It is (4.10) that realizes the S5 theorem $P \supset \Box\neg\Box\neg P$.

References

1. S. Artemov. Explicit provability and constructive semantics. *The Bulletin for Symbolic Logic*, 7(1):1–36, 2001.
2. S. Artemov, E. Kazakov, and D. Shapiro. On logic of knowledge with justifications. Technical Report CFIS 99–12, Cornell University, Cornell, New York, 1999.
3. V. Brezhnev and R. Kuznets. Making knowledge explicit: How hard it is. *Theoretical Computer Science*, 357(1–3):23–34, 2006.
4. M.C. Fitting. A simple propositional S5 tableau system. *Annals of Pure and Applied Logic*, 96:107–115, 1999. Originally in *The Parikh Project, Seven papers in honour of Rohit*, Uppsala Prints and Reprints in Philosophy, Uppsala, Sweden, 1996 Number 18.
5. M.C. Fitting. The logic of proofs, semantically. *Annals of Pure and Applied Logic*, 132:1–25, 2005.
6. M.C. Fitting. A replacement theorem for LP. Technical Report TR-2006002, CUNY Ph.D. Program in Computer Science, 2006. `http://www.cs.gc.cuny.edu/tr/`
7. M.C. Fitting. Realizations and LP. In S. Artemov and A. Nerode, editors, *Logical Foundations of Computer Science – New York '07*, pages 212–223. Springer, 2007. Lecture Notes in Computer Science, 4514.
8. M.C. Fitting. Realizing substitution instances of modal theorems. Technical Report TR-2007006, CUNY Ph.D. Program in Computer Science, 2007. `http://www.cs.gc.cuny.edu/tr/`
9. M.C. Fitting. Realizations and LP. *Annals of Pure and Applied Logic*, 161(3):368–387, December 2009.
10. K. Gödel. Eine Interpretation des Intuitionistischen Aussagenkalküls. *Ergebnisse eines Mathematischen Kolloquiums*, 4:39–40, 1933; Translation "An interpretation of the intuitionistic propositional calculus," In K. Gödel, Collected Works, S. Feferman et al., editors, volume 3, pages 296–302, Oxford University Press, Oxford and New York, 1995.
11. G. Mints. Lewis' systems and system T. In *Selected papers in proof theory (1965–1973)*. Bibliopolis, Pittisburg, P.A, 1992.
12. N. Rubtsova. Evidence reconstruction of epistemic modal logic S5. In D. Grigoriev, J. Harrison, and E.A. Hirsch, editors, *Computer Science – Theory and Applications*, Lecture Notes in Computer Science, volume 3967, pages 313–321. Springer, Berlin, Germany, 2006.

Chapter 5
Merging Information

Sujata Ghosh* and Fernando R. Velázquez-Quesada**

5.1 Introduction: The Milieu

On innumerous occasions in our everyday life we are forced to make decisions on the basis of incomplete information that we acquire regarding the current state of affairs. While playing poker, we are forced to decide whether to bet without being sure about the opponents' hands; we have *imperfect information* about the situation of the game. Scheduling cricket matches in the nor'western season is just like a game of chance; there is no guarantee that a match would be played on the scheduled day because of the possibility of the sudden storms, and one has to depend on weather forecasts which are invariably incomplete in terms of their information content.

Thus, more often than not we are in imperfect information situations, where we do not know whether a relevant fact is the case or not. But even though we do not have the precise information, decisions have to be taken. Then, we rely not only in what we know, but also in what we *believe*. Though we do not know if it will rain this afternoon or not, we usually find one case more plausible than the other, and we act based on this assumption.

Evidently, "belief" is rather a dynamic notion: we may believe now that it will not rain this afternoon, but this belief will change if we see the sky getting darker. Studying the dynamic nature of beliefs is the main motivation behind *Belief Revision*, the field focussed on the process of updating beliefs in order to accept con-

Sujata Ghosh
Institute of Artificial Intelligence, University of Groningen, Groningen, The Netherlands, e-mail: `sujata@ai.rug.nl`
Fernando R. Velázquez-Quesada
Institute for Logic, Language and Computation, University of Amsterdam, Amsterdam, The Netherlands, e-mail: `F.R.VelazquezQuesada@uva.nl`

* Acknowledges NWO grant # 600.065.120.08N201

** Acknowledges a scholarship by **Consejo Nacional de Ciencia y Tecnología (CONACyT)** from México. Scholarship holder # 167693

J. van Benthem et al. (eds.), *Games, Norms and Reasons*, Synthese Library 353,
DOI 10.1007/978-94-007-0714-6_5, © Springer Science+Business Media B.V. 2011

sistently a new piece of information. It is basically the product of two converging research traditions. The first one has a computer science flavor, with its origins in the development of procedures to update databases (for example, the truth maintenance systems of [18] and, more recently, [13] and [14]); the second one is more philosophical, and has its origins in the discussion of the mechanisms by which scientific theories develop, proposing requirements for *rational* belief change (for example, the studies of [40, 41] and [32]).

But in general, when new information arrives, it does not show up just from one source: we can get information about the weather by listening to the radio, looking at some webpage, asking a friend and even by looking at how cloudy the sky is. Of course, not all the sources are equally trustworthy: we find a rain prediction of the forecast on internet more reliable than the sunny day prediction of an enthusiastic friend. Rather than revising our beliefs to accept a single incoming piece of information, we *merge* the information coming from the different sources (including our current beliefs), taking into account the different reliability of each one. This is the main motivation behind the field appropriately called *Belief Merging*.

And we can imagine an even more realistic case, where the different sources of information indicate not only opinions about the facts in discussion but also opinions about themselves. It is not strange to find a radio broadcasting stating not only that it will rain this afternoon but also that we should not trust in anyone who says the opposite.

In the present work we propose a further extension of the formal studies of belief merging, with an aim of making it closer to the reality. We consider situations where the different sources of information (considered as agents) also have opinions about each other. In Section 5.2 we present a brief survey of the main traditions in the formal studies on merging of information. Section 5.3 comprises the main contribution of this work, presenting a logic in which we can express agent's opinions about facts and other agents (LO) and then extending it to express beliefs and also preferences of agents over other agents (LOB^-). Some discussions are provided, pointing towards the interactive nature of these epistemic attitudes. Finally, in Section 5.4 we discuss our general conclusions and and give pointers for further work.

5.2 The Vast Realm of Approaches

In the literature one can find many proposals for postulates and procedures to revise and merge beliefs. Here we provide a brief description of the main ideas that are being nurtured in these fields. Over the years, different approaches to the problem have been proposed, some from a logical point of view but some others from a connectionist perspective. We give a small survey of the most relevant works in both the areas.

5.2.1 Revising vs Merging

Belief Revision focusses on the process of modifying beliefs so as to accept a new piece information in a consistent manner. One of the most important traditions in the field is the AGM model introduced in [1]. Following the philosophical origins of the field, the authors discussed a set of postulates for rational revision – properties that an operator that performs revision should satisfy in order to being considered rational. The AGM approach *"analyzes belief change without committing to any particular mechanism, providing just abstract postulates on the process"* ([51]).

This approach contains two asymmetries in its formulation. The first one is the precedence of incoming information over current beliefs (current beliefs should change to *accept* the new information consistently); the second one is the richer structure assigned to the belief set (a collection of formulas, sometimes with an underlying structure) compared with that of the incoming information (just a formula). While theories of non-prioritized belief change have emerged to tackle the first issue [31, 44], other kinds of generalizations have surfaced with works considering the aggregation of finite sets of information, all of them with similar structure and arbitrary precedence, into a collective one. Instead of *revising* beliefs, we *merge* all available information: this is the main idea behind *Belief Merging* [34–36].

It is also interesting to notice that the aggregation procedure in *Belief Merging* faces problems similar to those addressed in *Social Choice Theory*. Both fields consider several sources of information with different precedence (based on reliability in *Belief Merging* and priority in *Social Choice*), which provide an order over the relevant entities (beliefs about the current state of affairs in *Belief Merging*, preferences over a set of options in *Social Choice*). Links between these two disciplines have been investigated in [19, 21, 34, 37], among others.

5.2.2 Different Approaches

As various other analogous research areas, revising and merging beliefs can be studied from two perspectives: an abstract one providing and discussing properties of ideal solutions, or a more practical one providing solutions and verifying which properties they satisfy. When providing a specific procedure to get solutions, several approaches may be considered. In the area of revising and merging of beliefs, two have been used most extensively: logical approaches, providing models and formal languages to describe the relation between the input and output of the revising and merging mechanisms, and connectionist approaches, considering the phenomena as the emergent processes of interconnected networks of simple units (usually, neural network models).

5.2.2.1 Logical Approaches

There are several approaches that use modal logic tools. Authors like André Fuhrmann [20], Johan van Benthem [50] and Maarten de Rijke [15, 16] showed how theories of change can be analyzed with the help of *"dynamic modal logic"*. By using a multi-modal language, dynamic logic allows to express the effect of actions: formulas of the form $[a]\varphi$ indicate that φ is the case after every execution of the action a. Moreover, it allows us to build more complex actions from the basic ones – we can have formulas expressing the result of sequential composition or non-deterministic choice between actions, and in some cases even parallel execution of them. The following works incorporate revising (merging) operations as actions within a modal language.

Dynamic Doxastic Logic (DDL) [47, 48] was introduced *"with the aim of representing the meta-linguistically expressed belief revision operator as an object-linguistic sentence operator in the style of dynamic modal logic"* [39]. The main operations in Belief Revision are, after all, actions, and hence we can use the dynamic logic framework to describe belief change. In DDL, we have (doxastic) actions of the form $+\varphi$ for expansion by φ, $-\varphi$ for contraction and $*\varphi$ for revision. Semantically, the idea is that a belief state should not only represent the agent's beliefs but also how she would respond to new information. Based on the work of Lewis [42] and Grove [29], Segerberg proposed that a belief state can be represented with a non-empty set of theories and a doxastic action can be represented with a binary relation between belief states. Then, the effect of a doxastic action in a belief state is described as a change to another belief state following the corresponding relation.

There is another major manifestation of the "dynamic turn" in logic: *Dynamic Epistemic Logic* (DEL) is the combination of two traditions: *Epistemic Logic* (EL) and *Dynamic semantics*. While *Epistemic Logic* is concerned with reasoning about knowledge, the main idea in *Dynamic Semantics* is that *"the meaning of a syntactic unit is best described as the change it brings about in the state of a human being or a computer"* [25]. In DEL languages, an EL language is extended with operators that describe information-changing actions. On the semantics side, such operators differ from the usual modal ones in that they are not interpreted as relations between worlds, but as operations that *modify* the whole model. In Public Announcement Logic (PAL, [24, 25, 46]), for example, the public announcement operation removes those worlds of the model where the announced formula is false.

In [6], the authors extended the PAL framework by using Kripke structures to represent not only the epistemic state but also *epistemic actions*: actions about which the agents may be incompletely informed. The epistemic state after the execution of an action is obtained by what is called *product update*, reflecting the idea that the uncertainty of an agent about a situation after an action takes place is the result of her uncertainty about the situation *before* the action and her uncertainty *about the action* ([56] provides a nice presentation of product update and other dynamic epistemic logic topics). Further extensions can be found in [7, 8], where the notion of belief has been incorporated, allowing us to describe agents with knowledge and beliefs about the situations and also about the executed actions. With this notion of *doxastic action*, it is possible to deal with both static and dynamic belief revision and also *"implement various belief-revision policies in a unified framework"* [9].

One of the main conceptual contributions of [6] was to put the description of the static situation and that of the actions at the same level. In [52], van Benthem extends this symmetry by noting that product update (updating of the epistemic state through an epistemic action) is actually an aggregation of the (epistemic) relations of the static model and those of the action model. Aggregation is usually conceived as merging of different relations over *the same* domain; product update generalizes it by merging different relations over *different* domains. When we perform product update, two new items are built: the new domain (as the cartesian product of the previous ones) and a relation over it (based on the previous relations). In the paper, the author introduces a static modal language with modalities for the weak and strict versions of the ordering relation, whose logic is similar to the one presented in [53]; on top of which he adds the dynamic operators working as product updates following a *priority update rule*, reflecting the general idea of the Andreka et. al. approach, described below.

In [2], Andreka et. al. present an algebraic treatment for combining relations (which, in particular, can represent a plausibility order over possible situations). They define the concept of a *priority operator*: an operator that, given a family of relations with priority among them, returns a single relation representing their lexicographic combination. It is shown that priority operators are the only way of combining relations with different priorities to get a result that satisfy certain natural conditions, similar to those proposed by Arrow [4] in the context of *Social Aggregation*. Moreover, it is shown that any finitary priority operator can be expressed by the binary operators "||" (*"on the other hand"* operator, indicating the aggregation of relations with the same priority) and "/" (*"but"* operator, indicating the aggregation of relations with different priority). It should be noted how the construction of the aggregation relation is then given by a sequence of operations that defines the priority among the aggregated individual relations.

In Chapter 6 of [28], Girard presents a modal logic for order aggregation based on these priority operators. Following approaches for preference logic like [55], he presents a language that allows us to express not only individual preferences but also their aggregation as the result of operations between the corresponding relations.

5.2.2.2 Connectionist Approaches

As we mentioned earlier, connectionist approaches consider revising and merging of beliefs as a result of dynamics in interconnected networks (i.e., graphs) made up of simple units. At any point of time, each unit in the network has a value representing some aspect. This value is either given by an external input (which remains constant throughout the process) or else by a combination of values of other units. At each step, values are re-calculated, and hence the effect of the external inputs spreads to all other units in the network over time. Some of the units are considered output units and, whenever their values become stable, they are considered as the outcome of the process. There are many forms of connectionism, but the most common forms use neural network models.

An *artificial neural network* (NN) is a mathematical model based on a group of simple processing elements (neurons) whose behaviour is determined by their interconnections and their individual parameters. The NN model has its origin in biological neural networks and is typically an input–output model. They provide an interesting framework to model *reasoning*: if we consider the inputs as incoming information, the outputs can be considered as the result of a reasoning process. Then, an NN can represent situations with multiple sources of information (in general any finite number of inputs) and, moreover, represents their non-uniformity (by specifying different ways in which the different inputs will affect the output). This makes neural networks an attractive tool for modelling merging processes.

Analyzing logical concepts from a connectionist point of view is another stimulating area of study. Interesting examples are Gaifmann Pointer semantics [22, 23] and the revision theory of truth developed by Herzeberger, Gupta and Belnap [30, 33]. The latter provides a semantics for sets of sentences with self-reference by looking at patterns of truth-values that are stable under subsequent revisions. One of the main features of this concept is the *backward propagation* of truth values along the edges of the graph representing the set of sentences. A more belief-related approach is that of *Bayesian Belief Nets* (see [45, 57]), based on the idea of assigning probabilities to the units of the network and using the Bayesian rule for the propagation of values.

A related work is the *Assertion Networks* of [26]. Sources of information (considered as agents) and facts are uniformly represented as units of the network, and their interconnection represents opinions of the agents about relevant facts and also about themselves. The approach uses the graph framework of *Bayesian Belief Nets* (though the numerals used are interpreted in a more deterministic way), and perhaps more crucially, considers the notion of *stability of revision sequences* to decide the outcome of the merging process.

The novelty of this setting is that we can now talk about opinions that agents have about other agents. This is not about the precedence among the multiple information sources; the formalism allows each one of them to have a ranking of the other sources. This also involves common concepts in security, like *Trust* and *Obligation* [17, 43].

Going back to our discussion on neural networks, one of their main drawbacks is the incapacity to provide an explanation for the underlying reasoning mechanism, that is, the incapacity to provide a *logical* description of the merging process. Several attempts have been made to provide some relation between inputs and outputs in connectionist models.

In [12], the authors discussed some of the main problems in the knowledge extraction methods of neural network models and proposed a way for these drawbacks to be amended. The basic ideas are to incorporate a partial order on the input vectors of the net and to use a number of pruning and simplification rules. For a class of networks, the *regular* ones, the algorithm is sound and complete. For the *non-regular* networks, it is shown that they contain regularities in their subnetworks, and the previous method can be applied in a decompositional fashion.

In [38], the author described *interpreted dynamical systems*. A dynamical system is in essence a neural network: we have states representing the neurons and a next-state function representing the connections between them. With an interpretation mapping, we identify formulas of the propositional language with states.

The internal dynamics of the system is given by stepwise iteration of the next-state function. External information modifies the internal flow, affecting the next state of the one to which it is plugged. The resultant dynamics of the system are described in terms of qualitative laws for which a satisfaction clause is defined. Moreover, it is shown that the resulting descriptions are related to systems of non-monotonic logic. The relation between non-monotonic logic systems and the symbolic representation of the *reasoning process* performed by neural networks is of particular relevance, as shown not only in [38], but also by [5] and [11], among others.

5.3 Merging Opinions, Preferences and Beliefs

Real life communication situations are possibly the best exemplification of this merging of information. Various approaches to model it, both from the logical and the connectionist point of view, have been discussed in the previous section. As evident from the existing literature, the logical framework provides a much more global approach regarding "what can be achieved" or "what is the final outcome", but does not give an account of the "micro-structure" of mutual influences and nuances of the agents involved, which has been brought to the fore by the connectionist approaches. In other words, though logical approaches do talk about change, they look at the big steps (the outcome of a revision) and not at the small ones (the "discussion" that takes place while revising).

The primary goal of this work is to provide a logical framework that describes the of the communication networks and capture a reasonable way of amalgamating all the information coming from different sources so as to bring out the global behavior of the system.

Our proposal is based on the approach of [26], where an *Assertion Network* takes the perspective of an external observer that collects opinions of relevant agents about particular facts. Both agents and facts are represented as nodes of a graph, and opinions of the agents about themselves and the facts are represented by labelled edges. We emphasize the fact that agents are not represented by their preferences, beliefs or knowledge, but as tangible objects in the model.

The starting situation is given by some initial assignment, indicating the degree of belief (numerical values) the observer has about each agent and fact (each node). Then, just as connectionist networks, the process through which the observer *merges* the different agents' opinions and her own initial beliefs is represented by the iterative updating of such values until a stable value is reached (which in general is not the case, e.g. *Liar* sentence).

We consider a semantic model close to the described one. It consists of a graph with two kinds of nodes, one representing agents and the other representing possible situations. We give a logical perspective here, and as such, develop a language for describing such a model. Since agents are represented by nodes, it is natural to consider a language that allows us to *name* such nodes. Our work is based on the hybrid logic approach [3] which provides a way to talk about individual states. In this section, we formally define the languages as well as the semantic models, and provide sound and complete axiom systems for them. The logics that we discuss here are devoid of any dynamic component, which we leave for future investigations. A short discussion of the interesting issues that arise from the notion of dynamics can be found in Section 5.4.

Before proceeding further, it should be noted that the distinction made between *"belief"* and *"opinions"* is that, we understand *"opinion"* as a *first-order* concept while *"belief"* is a concept that can be extended to higher orders. In the logic of opinions (*LO*) proposed below, we can express agent's opinions about facts and about each other; opinions about opinions are meaningless for the framework. On the other hand, the logic of opinions and beliefs (LOB^-), an extension of *LO*, allows us to express not only agent's beliefs about facts and about each other, but also belief about beliefs, and so on. We also introduce formulas to express agents' preferences over other agents, which is an useful tool in representing communication situations, but aggregation of preferences so as to model the effect of group preferences is not discussed here.

5.3.1 Logic of Opinions

The logic of opinions basically represents situations comprising of agents and facts, together with opinions of agents about these facts and also about other agents. Facts are represented by propositional variables, world-names and their boolean combinations, whereas agents are expressed by agent-names. The world-names (nominals), besides giving us uniformity by allowing us to name *all* nodes in our graph, allows to talk about agent's opinions not only about facts that may be true in several possible worlds, but also about any such complete world's descriptions. The language of this logic (*LO*) is given as follows:

Definition 5.1. Let *PROP* be a set of atomic propositions, *AG* be a set of agent-names and *NOM* be a set of world-names (nominals). Formulas φ of *LO* are inductively defined by:

$$\varphi := p \mid \bot \mid i \mid \neg\varphi \mid \varphi \wedge \psi \mid \boxplus_a \varphi \mid \boxminus_a \varphi \mid \oplus_{a:b} \mid \ominus_{a:b} \mid @_i \varphi$$

where $p \in PROP$, $i \in NOM$ and $a, b \in AG$. We assume a restricted language in the sense that nesting of opinion modalities are not allowed. In formulas of the form $\boxplus_a \varphi$ and $\boxminus_a \varphi$, φ's are restricted to atomic propositions, nominals and their boolean and $@_i$ combinations.

Other connectives ($\vee$, $\rightarrow$ and $\leftrightarrow$) are defined as usual. The *diamond* version of opinion formulas $\langle + \rangle_a \varphi$ and $\langle - \rangle_a \varphi$ are defined as the duals of their *box* counterparts: $\langle + \rangle_a \varphi \leftrightarrow \neg \boxplus_a \neg \varphi$ and $\langle - \rangle_a \varphi \leftrightarrow \neg \boxminus_a \neg \varphi$, respectively.

The intended meaning of formulas of the form $\boxplus_a \varphi$ ($\boxminus_a \varphi$) is *"agent a has positive (negative) opinion about* φ*"*. Similarly, formulas of the form $\oplus_{a:b}$ ($\ominus_{a:b}$) are read as *"agent a has positive (negative) opinion about agent b"*. It should be noted that, in the language of *LO*, agent-names just appear in opinion formulas.

For the semantic model, we consider graphs with two kinds of nodes: agent-nodes representing agents and world-nodes representing possible situations. Relations between such nodes indicate the agents' opinions about possible situations and other agents. The basic link between the semantic model and the language is given by a couple of functions: a standard hybrid valuation indicating the world-nodes where elements of *PROP* and *NOM* are true (with each $i \in NOM$ being true at one and only one world-node) and a naming function assigning a different agent-node to each element of *AG*. Formally, we have the following.

Definition 5.2. An opinion model is a structure of the form

$$\mathcal{M} = \langle W, A, R^+, R^-, O^+, O^-, V, N \rangle$$

where

- W is the set of world-nodes,
- A is the set of agent-nodes (with A disjoint from W),
- $R^+ \subseteq A \times W$ is a serial binary relation from agent-nodes to world-nodes,
- $R^- \subseteq W \times W$ is a serial binary relation from agent-nodes to world-nodes,
- $O^+ \subseteq A \times A$ is a binary relation from agent-nodes to agent-nodes,
- $O^- \subseteq A \times A$ is a binary relation from agent-nodes to agent-nodes,
- $V : PROP \cup NOM \rightarrow 2^W$ is a standard hybrid valuation (that is, for each $i \in NOM$, $V(i)$ is a singleton), indicating the world-nodes where atomic propositions and nominals are true, and,
- $N : AG \rightarrow A$ is an injection assigning a different agent-node to each agent-name.

The opinion model $\mathcal{M}$ is *named* if every world-node in the model is the denotation of some nominal, that is, for each $w \in W$, there is a nominal $i \in NOM$, such that $V(i) = \{w\}$.

We emphasize the fact that the set of world-nodes and that of agent-nodes are disjoint. Although $\mathcal{M}$ is a graph consisting of both these kind of nodes, atomic propositions and nominals get truth values only in world-nodes, and therefore formulas of the language are evaluated just in them. While the valuation V allows us to give truth value to atomic propositions and nominals in the standard way, the naming function N and the opinion relations (R^+ and R^- for opinions about facts, O^+

and O^- for opinions about agents) allow us to take care of opinion formulas by just looking at the outgoing arrows from the agent-nodes named after the agent-names. Negations, conjunctions and the $@_i$ operator are defined in the usual way.

Definition 5.3. Let $\mathcal{M} = \langle W, A, R^+, R^-, O^+, O^-, V, N \rangle$ be an named opinion model. The truth definition for formulas φ in $\mathcal{M}$ at a world w is given below.

$$
\begin{array}{lll}
\mathcal{M}, w \models p & \text{iff} & w \in V(p) \\
\mathcal{M}, w \models i & \text{iff} & \{w\} = V(i) \\
\mathcal{M}, w \models \neg\varphi & \text{iff} & \mathcal{M}, w \not\models \varphi \\
\mathcal{M}, w \models \varphi \wedge \psi & \text{iff} & \mathcal{M}, w \models \varphi \text{ and } \mathcal{M}, w \models \psi \\
\mathcal{M}, w \models \boxplus_a \varphi & \text{iff} & \text{for all } u \in W \text{ such that } N(a)R^+u,\ \mathcal{M}, u \models \varphi \\
\mathcal{M}, w \models \boxminus_a \varphi & \text{iff} & \text{for all } u \in W \text{ such that } N(a)R^-u,\ \mathcal{M}, u \models \neg\varphi \\
\mathcal{M}, w \models \oplus_{a:b} & \text{iff} & N(a)O^+N(b) \\
\mathcal{M}, w \models \ominus_{a:b} & \text{iff} & N(a)O^-N(b) \\
\mathcal{M}, w \models @_i\varphi & \text{iff} & \mathcal{M}, u \models \varphi, \text{ where } V(i) = \{u\}
\end{array}
$$

where $p \in PROP$, $i \in NOM$ and $a, b \in AG$.

It follows that $\mathcal{M}, w \models \langle + \rangle_a \varphi$ iff there exists $u \in W$ such that $N(a)R^+u$ and $\mathcal{M}, u \models \varphi$, and $\mathcal{M}, w \models \langle - \rangle_a \varphi$ iff there exists $u \in W$ such that $N(a)R^-u$ and $\mathcal{M}, u \models \neg\varphi$. Note that, by the above definition, the opinion formulas are either true or false in the whole model, that is, if an opinion formula is true (false) in some world in $\mathcal{M}$, then it is true (false) in *every* world in $\mathcal{M}$.

A communication situation representing agents' opinion about events and about each other can be described by finite conjunction of the modal formulas introduced above. For example, the formula $\neg \boxplus_a \varphi \wedge \neg \boxminus_a \varphi$ is read as *"the agent does not have any opinion about φ"*, whereas $\boxplus_a \varphi \wedge \boxminus_a \varphi$ corresponds to a being undecided about φ. In terms of epistemic attitudes of an agent, there is a difference between *having no opinion* and *being undecided* about a certain event. One can be undecided whether to take an umbrella or not while she is going out, but she may have no opinion about who should win the next US presidential elections, as she is simply not interested in the issue. In terms of opinions concerning other agents, these attitudes are typically indistinguishable. Moreover, because of the way we have interpreted the modal formulas which is independent of the states, the intuitively inconsistent formulas like $\boxplus_a \varphi \wedge \boxplus_a \neg\varphi$ as well as $\boxminus_a \varphi \wedge \boxminus_a \neg\varphi$ are not satisfiable.

Consider the following simple example.

> Suppose Professor Calculus wants to know how good a singer Bianca Castafiore is. In come Thomson and Thompson, and convey the following.
>
> `Thomson`: "*She is a very good singer.*"
> `Thompson`: "*Aha! I do not think so. I really dislike her singing*"

This network of opinions can be represented by the following formula:

$$\boxplus_a \varphi \wedge \boxminus_b \varphi$$

with a representing Thomson, b Thompson, and φ representing the fact that "Bianca Castafiore is a good singer". The obvious question here is what would Professor Calculus infer in this situation of conflicting opinions. We will come back to this example in Section 5.3.2.

We now provide a sound and complete axiomatization for *LO*.

Theorem 5.1. *LO is sound and complete with respect to countable named opinion models and its validities are axiomatized by,*

(a) all propositional tautologies and inference rules,

(b) axioms and rules for $@_i$:

$\vdash @_i(p \to q) \to (@_i p \to @_i q)$

$\vdash @_i p \leftrightarrow \neg @_i \neg p$

$\vdash i \wedge p \to @_i p$

$\vdash @_i i$

$\vdash @_i j \leftrightarrow @_j i$

$\vdash @_i j \wedge @_j p \to @_i p$

$\vdash @_j @_i p \leftrightarrow @_i p$

if $\vdash \varphi$, *then* $\vdash @_i \varphi$ *for every* $i \in NOM$

(c) positive opinion axioms and rules:

$\vdash \boxplus_a(\varphi \to \psi) \to (\boxplus_a \varphi \to \boxplus_a \psi)$ *(+normal)*

$\vdash \boxplus_a \varphi \to \langle + \rangle_a \varphi$ *(+ser)*

$\vdash \langle + \rangle_a i \wedge @_i \varphi \to \langle + \rangle_a \varphi$ *(+translation)*

$\vdash \langle + \rangle_a @_i \varphi \to @_i \varphi$ *(+back)*

if $\vdash \varphi$, *then* $\vdash \boxplus_a \varphi$, *for every* $a \in AG$ *(+gen)*

(d) negative opinion axioms and rules:

$\vdash \boxminus_a(\neg\varphi \wedge \psi) \to (\boxminus_a \varphi \to \boxminus_a \psi)$ *(-normal)*

$\vdash \boxminus_a \varphi \to \langle + \rangle_a \varphi$ *(–ser)*

$\vdash \langle - \rangle_a i \wedge @_i \neg\varphi \to \langle - \rangle_a \varphi$ *(–translation)*

$\vdash \langle - \rangle_a @_i \varphi \to \neg @_i \varphi$ *(-back)*

if $\vdash \neg\varphi$, *then* $\vdash \boxminus_a \varphi$, *for every* $a \in AG$ *(–gen)*

(e) agreement axioms:

$\vdash \langle + \rangle_a \varphi \leftrightarrow @_i \langle + \rangle_a \varphi$

$\vdash \langle - \rangle_a \varphi \leftrightarrow @_i \langle - \rangle_a \varphi$

$\vdash \oplus_{a:b} \leftrightarrow @_i \oplus_{a:b}$

$\vdash \ominus_{a:b} \leftrightarrow @_i \ominus_{a:b}$

(f) name, paste and substitution rules:

if $\vdash i \to \varphi$, *then* $\vdash \varphi$ *for* i *not occurring in* φ.

if $\vdash (@_i \langle + \rangle_a j \wedge @_j \varphi) \to \psi$,
then $\vdash @_i \langle + \rangle_a \varphi \to \psi$ *for* $i \neq j$ *and* j *not occurring in* φ *or* ψ.

if $\vdash (@_i \langle - \rangle_a j \wedge @_j \neg\varphi) \to \psi$,
then $\vdash @_i \langle - \rangle_a \varphi \to \psi$ *for* $i \neq j$ *and* j *not occurring in* φ *or* ψ.

if $\vdash \varphi$, *then* $\vdash \varphi\sigma$ *where* σ *is a substitution that uniformly replaces atomic propositions by formulas, agent-names by agent-names and nominals by nominals.*

Proof. Soundness of the axioms and rules is straightforward. The completeness proof is based on that of hybrid logic as presented in [10]; see appendix 5.5 for details.

5.3.2 Logic of Opinions and Beliefs

Let us now go back to the example we mentioned earlier, and analyze it further. Evidently, if Calculus believes Thomson more than Thompson, then he would believe that "Bianca is a good singer", otherwise he would believe that "she is not". In the retrospect, to express such decision making in presence of conflict in opinions we want to have a language where we can say that if our belief in a is more than our belief in b then we may believe in φ, and if our belief in b is more than our belief in a, we may believe in $\neg\varphi$. To incorporate the belief strength part, we add operators B_a (belief of a) and $\succeq_a$ (a's preferences among agents) for each agent a to the language of *LO*, to form the language of the logic of opinions and beliefs. We first present a very simple logic with a straightforward axiomatization which does not take into account the interaction between belief, opinions and preferences. We name it LOB^-, whose language is given in the following,

$$\varphi := p \mid \bot \mid i \mid \neg\varphi \mid \varphi \wedge \psi \mid \langle + \rangle_a \varphi \mid \langle - \rangle_a \varphi \mid \oplus_{a:b} \mid \ominus_{a:b} \mid @_i \varphi \mid B_a \varphi \mid a \succeq_a a,$$

with the following restriction that φ occurring in $B_a\varphi$ should only be propositions, nominals and their boolean combinations. We use $\langle B_a \rangle \varphi$ as an abbreviation for $\neg B_a \neg \varphi$, and $b \succ_a c$ for $(b \succeq_a c) \wedge \neg(c \succeq_a b)$.

The model becomes $\mathcal{M} = \langle W, A, R^+, R^-, O^+, O^-, \{R_a : a \in A\}, \{\geq_a : a \in A\}, V, N \rangle$, where for each a, R_a is a serial, transitive and Euclidean relation on W and $\geq_a$ is a reflexive, transitive and connected relation over A. We name the model as *bop model*. The truth definition is given by,

$$\begin{array}{lll} \mathcal{M}, w \models b \succeq_a c & \text{iff} & N(b) \geq_{N(a)} N(c) \\ \mathcal{M}, w \models B_a \varphi & \text{iff} & \text{for all } w' \in W \text{ such that } wR_{N(a)}w',\ \mathcal{M}, w' \models \varphi \end{array}$$

If Professor Calculus is expressed as c, then the earlier situations can now be expressed as:

$$((a \succ_c b) \wedge \boxplus_a \varphi \wedge \boxminus_b \varphi) \to \langle B_c \rangle \varphi$$
$$((b \succ_c a) \wedge \boxplus_a \varphi \wedge \boxminus_b \varphi) \to \langle B_c \rangle \neg\varphi$$

Let us now provide a complete axiomatization for LOB^-.

Theorem 5.2. *LOB is sound and complete with respect to countable named bop models and its validities are axiomatized by,*

(a) LO axioms and rules,
(b) if $\vdash \varphi$ *then* $\vdash B_a \varphi$,

(c) *belief axioms:*
$\vdash B_a(p \to q) \to (B_a p \to B_a q)$
$\vdash \langle B_a \rangle @_i p \to @_i p$
$\vdash B_a i \to \langle B_a \rangle i$
$\vdash B_a i \to B_a B_a i$
$\vdash \neg B_a i \to B_a \neg B_a i$

(d) *paste rule:*
if $\vdash @_i \langle B_a \rangle j \wedge @_j \psi \to \varphi$, *then* $\vdash @_i \langle B_a \rangle \psi \to \varphi$,
where, $i \neq j$ *and* j *not occurring in* φ *or* ψ

(e) *belief order axioms:*
$\vdash b \succeq_a b$
$\vdash (b \succeq_a c) \wedge (c \succeq_a d) \to (b \succeq_a d)$
$\vdash (b \succeq_a c) \vee (c \succeq_a b)$
$\vdash (b \succeq_a c) \leftrightarrow @_i(b \succeq_a c)$

Proof. Soundness of the axioms and rules is once again straightforward. For the completeness proof, which is an extension of that of LO, see Section 5.6.

It is evident that the logic LOB^- is suitable for expressing communication situations, but to reason in these situations we definitely need some axioms or rules that brings out the correspondence between opinions, preferences and beliefs, which this logic does not provide for.

In what follows we make some preliminary ventures into providing bop models with different properties depicting some such intuitive correspondences and finding out the logical validities that follow from such properties of the models.

Consider $\mathcal{M} = \langle W, A, R^+, R^-, O^+, O^-, \{R_a : a \in A\}, \{\geq_a : a \in A\}, V, N \rangle$ to be a *bop model*. Let us now list some intuitive interactive properties in this model, with the validities they ensure.

- Positive and negative opinions may lead to preferences: $b \in Ran_{O^+}(a)$ and $c \in Ran_{O^-}(a)$ imply $b \geq_a c$.
  Validity: $\oplus_{a:b} \wedge \ominus_{a:c} \to b \succeq_a c$.
- Preferences help in decision making under conflicting opinions: $a \geq_c b$ implies for each $w \in W$, $\exists w' \in (Ran_{R^+}(a) \setminus Ran_{R^-}(b))$, such that $wR_c w'$, and $\exists w'' \in (Ran_{R^-}(a) \setminus Ran_{R^+}(b))$, such that $wR_c w''$.
  Validities: $(a \succ_c b \wedge \boxplus_a \varphi \wedge \boxminus_b \varphi) \to \langle B_c \rangle \varphi$,
  $(b \succ_c a \wedge \boxplus_a \varphi \wedge \boxminus_b \varphi) \to \langle B_c \rangle \neg \varphi$.

It can be easily seen that we can model the communication situation expressed in the example in Section 5.3.1, and reason about it in *bop models* satisfying one of the conditions given earlier. But this is a very simple case of interaction between opinions, belief and preferences. As is evident, a careful and systematic study remains to be done to bring out all such plausible interactive properties. This also gives rise to the following interesting issue:

Question What would be a complete axiomatization of the full interactive logic of beliefs, opinions and preferences?

To give the readers an idea about what we intend to model, let us refer back to the example in the previous section, and make it a little more complicated:

> While Professor Calculus was mulling over whether to believe Thomson more than Thompson, Captain Haddock enters the scene and observes: `Haddock`: "*Thomson and Thompson are a pair of jokers. Do not rely on any of them*"

Obviously, if Calculus had a high opinion of Haddock, then he will not be able to come to any conclusion regarding the singing quality of Bianca, whatever Thomson and Thompson say. The model should be able to reason about these detailed intricacies also.

5.4 Conclusion

The paper brings out the "micro-structure" of the communication networks describing the mutual influences of the agents over themselves as well as the events involved, together with their effect on the different epistemic attitudes of the agents, viz. beliefs, opinions, preferences, in a logical framework. Thus, a first attempt is made towards capturing the hitherto unheralded territory of the mutual enhancing and dampening effects of communication between agents from the logical point of view, which have been aptly described in the connectionist approaches.

It is natural, for example, to ask for opinions to lead to preferences, that is, if an agent a has a positive opinion about another, say b but a negative opinion about some other c, then she should prefer b over c. Such properties are reasonable when considering the outcome of the merging, and they can occur anywhere in the merging process, be as an initial information, or an intermediate step or as a final outcome. As it stands, LOB^{-} describes a very general (logical) framework with which we can express *static* opinions, preferences and beliefs of agents, without particular restrictions on the relations between them. What we need to do now is to find a *complete* list of desired interactive properties between beliefs, opinions and preferences (belief strengths) that the bop model should satisfy, and then add a *dynamic* component in the existing framework to take care of the merging process. This dynamic component should modify the model (i.e., should modify opinions, preferences and beliefs) with reasonable operations in order to get all the desired properties mentioned above. As in the works of [26, 27] (in connectionist systems, as well), the dynamicity does not need to be just one action, but a collective iteration of operations; that will end whenever we have reached a model with the desired properties, that is, an iteration of some model-changing operations that end in a *stable* situation.

Clearly enough, this is not the only kind of dynamics that can be studied. Consistent opinions, beliefs and preferences may also change due to new incoming information. One of the many possibilities is to introduce *suggestions*; announcements about beliefs made by some agent: if Professor Calculus has a good opinion about Captain Haddock, then any suggestion of the latter will modify the beliefs of the former. Some further avenues of investigation are mentioned below.

Aggregation of preferences. Together with individual preferences, the notion of group preferences as introduced in [28], which is an emerging area of study can well be incorporated into the framework provided and will add in the expressiveness of the logic. A detailed study of the various notions of preferences as well as their properties and their effects in communication with a focus on the interplay of preferences and beliefs in such situations is also a project worth pursuing.

Dynamic notions. We have already talked about introducing the notion of dynamicity to the framework in the form of "suggestions" which affect the opinions of agents regarding certain events. A host of existing notions of dynamicity can be added to the language of LOB^- so as to study their effects. To name a few, the notions of soft beliefs [51], preference upgrade [54], lexicographic upgrade [51], elite change [51], agent promotion [28] and others. One can also study the effects of the change in opinions about agents, change in opinion about events on that of agents and vice versa which will bring this study closer to that in [27].

5.5 Completeness for *LO*

The completeness proof follows the idea of that appearing in Section 7.3 [10]. Let Σ be a consistent set of formulas of *LO*, with *PROP*, *NOM* and *AG* the sets of atomic propositions, nominals and agent-names, respectively. We extend Σ to a named and pasted maximal consistent set Σ^+. We define the sets of formulas Δ_i (for $i \in NOM$) and Δ_a (for $a \in AG$) as follows:

$$\begin{aligned}\Delta_i &:= \{\varphi \mid @_i\varphi \in \Sigma^+\}\\ \Delta_a &:= \{\langle + \rangle_a\varphi \mid \langle + \rangle_a\varphi \in \Sigma^+\} \cup \{\langle - \rangle_a\varphi \mid \langle + \rangle_a\varphi \in \Sigma^+\} \cup \\ &\quad\ \{\oplus_{a:b} \mid \oplus_{a:b} \in \Sigma^+\} \cup \{\ominus_{a:b} \mid \ominus_{a:b} \in \Sigma^+\}\end{aligned}$$

We call Δ_i a *named world yielded by* Σ^+, and Δ_a a *named agent yielded by* Σ^+. Note that for each $a \in AG$, Δ_a is non-empty, because of the *gen* and *ser* axioms. So for different agent-names a and b, Δ_a and Δ_b are different sets.

From Σ^+, we build the model $\mathscr{M} = \langle W, A, R^+, R^-, O^+, O^-, V, N\rangle$ as follows:

- $W := \{\Delta_i \mid \Delta_i$ is a named world yielded by $\Sigma^+\}$,
- $A := \{\Delta_a \mid \Delta_a$ is a named agent yielded by $\Sigma^+\}$,
- $\Delta_a R^+ \Delta_i$ iff for all non-opinion formulas φ, $\varphi \in \Delta_i$ implies $\langle + \rangle_a\varphi \in \Delta_a$,
- $\Delta_a R^- \Delta_i$ iff for all non-opinion formulas φ, $\neg\varphi \in \Delta_i$ implies $\langle - \rangle_a\varphi \in \Delta_a$,
- $\Delta_a O^+ \Delta_b$ iff $\oplus_{a:b} \in \Delta_a$,
- $\Delta_a O^- \Delta_b$ iff $\ominus_{a:b} \in \Delta_a$,
- $V(x) := \{\Delta_i \in W \mid x \in \Delta_i\}$ for $x \in (PROP \cup NOM)$,
- $N(a) := \Delta_a$ for $a \in AG$.

Note that W and A are disjoint. Also, because of the axioms, V is an hybrid valuation assigning a singleton to every nominal i (see definition 7.26 of [10]); moreover, N is an injection (in fact, a bijection in this case). We first show that R^+ and R^- are serial. Then we move onto proving existence lemmas for these relations and finally the truth lemma.

R^+ is serial: Consider any Δ_a. Since it is non-empty, we have that, for some formula φ, $@_i\langle+\rangle_a\varphi \in \Sigma^+$ (by *agreement* axiom and definition of Δ_a). So, as Σ^+ is *pasted*, $@_i\langle+\rangle_a j \wedge @_j\varphi \in \Sigma^+$ for some j. Then, $\langle+\rangle_a j \in \Sigma^+$. Now consider any propositional variable, nominal or their boolean combination, ψ in Δ_j. Then $@_j\psi \in \Sigma^+$. This implies that $\langle+\rangle_a\psi \in \Sigma^+$, (by *+translation* axiom) and so, $\langle+\rangle_a\psi \in \Delta_a$. So, $\Delta_a R^+ \Delta_j$.

R^- is serial: Consider any Δ_a. Since it is non-empty, we have that, for some formula φ, $@_i\langle-\rangle_a\varphi \in \Sigma^+$ (by *agreement* axiom and definition of Δ_a). So, as Σ^+ is *pasted*, $@_i\langle-\rangle_a j \wedge @_j\varphi \in \Sigma^+$ for some j. Then, $\langle-\rangle_a j \in \Sigma^+$. Now consider any propositional variable, nominal or their boolean combination, ψ such that $\neg\psi \in \Delta_j$. Then $@_j\neg\psi \in \Sigma^+$. This implies that $\langle-\rangle_a\psi \in \Sigma^+$, (by *–translation* axiom) and so, $\langle-\rangle_a\psi \in \Delta_a$. So, $\Delta_a R^- \Delta_j$.

The proofs of the existence lemmas for both R^+ and R^- follow similarly with the additional fact to note that, if $@_j\varphi \in \Sigma^+$, then $\varphi \in \Delta_j$. We now prove the truth lemma.

The clauses for atomic propositions, nominals, negation, conjunctions and the satisfaction operator @ are standard. For opinion formulas, we have

$\mathcal{M},\Delta_i \models \langle+\rangle_a\varphi$	iff	$\exists\, \Delta_j \in W$ s.t. $N(a)R^+\Delta_j$ and $\mathcal{M},\Delta_j \models \varphi$	def. of $\models$
	iff	$\exists\, \Delta_j \in W$ s.t. $\Delta_a R^+\Delta_j$ and $\varphi \in \Delta_j$	def. of N and I.H.
	iff	$\langle+\rangle_a\varphi \in \Delta_a$	def. of R^+ and *existence lemma*
	iff	$\langle+\rangle_a\varphi \in \Sigma^+$	def. of Δ_a
	iff	$@_i\langle+\rangle_a\varphi \in \Sigma^+$	agree axiom
	iff	$\langle+\rangle_a\varphi \in \Delta_i$	def. of Δ_i
$\mathcal{M},\Delta_i \models \langle-\rangle_a\varphi$	iff	$\exists\, \Delta_j \in W$ s.t. $N(a)R^-\Delta_j$ and $\mathcal{M},\Delta_j \models \neg\varphi$	def. of $\models$
	iff	$\exists\, \Delta_j \in W$ s.t. $\Delta_a R^-\Delta_j$ and $\neg\varphi \in \Delta_j$	def. of N and I.H.
	iff	$\langle-\rangle_a\varphi \in \Delta_a$	def. of R^- and *existence lemma*
	iff	$\langle-\rangle_a\varphi \in \Sigma^+$	def. of Δ_a
	iff	$@_i\langle-\rangle_a\varphi \in \Sigma^+$	agree axiom
	iff	$\langle-\rangle_a\varphi \in \Delta_i$	def. of Δ_i
$\mathcal{M},\Delta_i \models \oplus_{a:b}$	iff	$N(a)O^+N(b)$	def. of $\models$
	iff	$\Delta_a O^+ \Delta_b$	def. of N
	iff	$\oplus_{a:b} \in \Delta_a$	def. of O^+
	iff	$\oplus_{a:b} \in \Sigma^+$	def. of Δ_a
	iff	$@_i\oplus_{a:b} \in \Sigma^+$	agree. axiom
	iff	$\oplus_{a:b} \in \Delta_i$	def. of Δ_i
$\mathcal{M},\Delta_i \models \ominus_{a:b}$	iff	$N(a)O^-N(b)$	def. of $\models$
	iff	$\Delta_a O^- \Delta_b$	def. of N
	iff	$\ominus_{a:b} \in \Delta_a$	def. of O^-
	iff	$\ominus_{a:b} \in \Sigma^+$	def. of Δ_a
	iff	$@_i\ominus_{a:b} \in \Sigma^+$	agree. axiom
	iff	$\ominus_{a:b} \in \Delta_i$	def. of Δ_i

Because of the name and paste rules as well as the seriality of R^+ and R^-, the model $\mathcal{M}$ is a named opinion model. Now, we also have that $\Sigma^+ = \Delta_k$ for some nominal k. Hence, by truth lemma, $\mathcal{M}, \Delta_k \models \Sigma$, so our original consistent set of formulas is satisfiable.

5.6 Completeness for *LOB*$^-$

As before, let Σ be a consistent set of formulas of *LOB*, with *PROP*, *NOM* and *AG* the sets of atomic propositions, nominals and agent-names, respectively. We extend Σ to a named and pasted maximal consistent set Σ^+. We define the sets of formulas Δ_i as before, while Δ_a (for $a \in AG$) is defined as follows:

$$\begin{aligned}\Delta_a := & \{\langle + \rangle_a \varphi \mid \langle + \rangle_a \varphi \in \Sigma^+\} \cup \{\langle - \rangle_a \varphi \mid \langle - \rangle_a \varphi \in \Sigma^+\} \cup \\ & \{\oplus_{a:b} \mid \oplus_{a:b} \in \Sigma^+\} \cup \{\ominus_{a:b} \mid \ominus_{a:b} \in \Sigma^+\} \cup \\ & \{b \succeq_a c \mid b \succeq_a c \in \Sigma^+\}\end{aligned}$$

From Σ^+, we build the model $\mathcal{M} = \langle W, A, R^+, R^-, O^+, O^-, \{R_{\Delta_a} : \Delta_a \in A\}, \{\geq_{\Delta_a} : \Delta_a \in A\}, V, N \rangle$, as earlier with R_{Δ_a}, and $\geq_{\Delta_a}$ (for each Δ_a) defined as follows:

- $\Delta_i R_{\Delta_a} \Delta_j$ iff for all formulas φ, $\varphi \in \Delta_j$ implies $\langle B_a \rangle \varphi \in \Delta_i$,
- $\Delta_b \geq_{\Delta_a} \Delta_c$ iff $b \succeq_a c \in \Delta_a$.

The proof of existence lemma for the R_{Δ_a}'s is similar to that in Lemma 7.27 in [10]. The reflexivity, transitivity and connectedness of the relations $\geq_{\Delta_a}$ follow from the definition of the corresponding Δ_a's. Let us now focus on the remaining part of the truth lemma.

$\mathcal{M}, \Delta_i \models \langle B_a \rangle \varphi$	iff $\exists\, \Delta_j \in W$ s.t. $\Delta_i R_{N(a)} \Delta_j$ and $\mathcal{M}, \Delta_j \models \varphi$	def. of $\models$
	iff $\exists\, \Delta_j \in W$ s.t. $\Delta_i R_{\Delta_a} \Delta_j$ and $\varphi \in \Delta_j$	def. of N and I.H.
	iff $\langle B_a \rangle \varphi \in \Delta_i$	def. of R_{Δ_a} and *existence lemma*
$\mathcal{M}, \Delta_i \models b \succeq_a c$	iff $N(b) \geq_{N(a)} N(c)$	def. of $\models$
	iff $\Delta_b \geq_{\Delta_a} \Delta_c$	def. of N
	iff $b \succeq_a c \in \Delta_a$	def. of $\geq_{\Delta_a}$
	iff $b \succeq_a c \in \Sigma^+$	def. of Δ_a
	iff $@_i(b \succeq_a c) \in \Sigma^+$	axiom
	iff $b \succeq_a c \in \Delta_i$	def. of Δ_i

Completeness follows as in the case *LO*, noting the fact that, any set of pure formulas Σ (i.e. formulas without propositional variables), when added to an extension of $\mathcal{K}^+_{\mathcal{H}(@)}$ [49], becomes complete with respect to the frames definable by Σ.

References

1. C.E. Alchourròn, P. Gardenfors, and D. Makinson. On the logic of theory change: Partial meet contraction and revision functions. *The Journal of Symbolic Logic*, 50(2):510–530, 1985.
2. H. Andréka, M.D. Ryan, and P.-Y. Schobbens. Operators and laws for combining preference relations. *Journal of Logic and Computation*, 12(1):13–53, 2002.
3. C. Areces and B. ten Cate. Hybrid logics. In P. Blackburn, J. van Benthem, and F. Wolter, editors, *Handbook of Modal Logic,* volume 3 of *Studies in Logic and Practical reasoning*. Elsevier, Amsterdam, The Netherlands, 2005.
4. K.J. Arrow. A difficulty in the concept of social welfare. *Journal of Political Economy*, 58:328–346, 1950.
5. C. Balkenius and P. Gärdenfors. Nonmonotonic inferences in neural networks. In J.A. Allen, R. Fikes, and E. Sandewall, editors, *Principles of knowledge representation and reasoning*, pages 32–39. Morgan Kaufmann, San Mateo, CA, 1991.
6. A. Baltag, L.S. Moss, and S. Solecki. *The logic of public announcements, common knowledge and private suspicious*. Technical Report SEN-R9922, CWI, Amsterdam, 1999.
7. A. Baltag and S. Smets. Conditional doxastic models: A qualitative approach to dynamic belief revision. *Electronic Notes in Theoretical Computer Science*, 165:5–21, 2006.
8. A. Baltag and S. Smets. Dynamic belief revision over multi-agent plausibility models. In W. van der Hoek and M. Wooldridge, editors, *Proceedings of LOFT 2006,* pages 11–24. University of Liverpool, Liverpool, 2006. Available at http://www.vub.ac.be/CLWF/SS/loft.pdf
9. A. Baltag and S. Smets. A qualitative theory of dynamic interactive belief revision. In G. Bonanno, W. van der Hoek, and M. Wooldridge, editors, *Logic and the Foundation of Game and Decision Theory (LOFT7),* volume 3 *of Texts in Logic and Games,* pages 13–60. Amsterdam University Press, Amsterdam, The Netherlands, 2008.
10. P. Blackburn, M. de Rijke, and Y. Venema. *Modal Logic*. Cambridge University Press, Cambridge, 2001.
11. R. Blutner and P.D. Doherty. Nonmonotonic logic and neural networks. Available at http://citeseer.ist.psu.edu/160749.html
12. A. d'Avila Garcez, K. Broda, and D. Gabbay. Symbolic knowledge extraction from trained neural networks: A sound approach. *Artificial Intelligence,* 125(1–2):155–207, January 2001.
13. J. de Kleer. An assumption-based truth maintenance system. *Artificial Intelligence,* 28(2):127–162, 1986.
14. J. de Kleer. A perspective on assumption-based truth maintenance. *Artificial Intelligence*, 59(1–2):63–67, 1993.
15. M. de Rijke. *Extending Modal Logic*. PhD thesis, Institute for Logic, Language and Computation (Universiteit van Amsterdam), 1993. ILLC Dissertation Series DS-1993–04.
16. M. de Rijke. Meeting some neighbours. In J. van Eijk and A. Viser, editors, *Logic and Information Flow*, pages 170–196. The MIT Press, Cambridge, MA, 1994.
17. E. Lorini and R. Demolombe. From binary trust to graded trust in information sources: a logical perspective. In *Trust in Agent Societies, Lecture Notes in Computer Science*, volume 5396, pages 205–225. Springer, Berlin, 2008.
18. J. Doyle. A truth maintenance system. *Artificial Intelligence,* 12(3):231–272, 1979.
19. D. Eckert and G. Pigozzi. Belief merging, judgment aggregation and some links with social choice theory. In J.P. Delgrande, J. Lang, H. Rott, and J.-M. Tallon, editors, *Belief Change in Rational Agents,* volume 05321 of *Dagstuhl Seminar Proceedings*. Internationales Begegnungs- und Forschungszentrum für Informatik (IBFI), Schloss Dagstuhl, Germany, 2005.
20. A. Fuhrmann. Reflective modalities and theory change. *Synthese*, 81(1):115–134, October, 1989.
21. D.M. Gabbay, G. Pigozzi, and O. Rodrigues. Common foundations for belief revision, belief merging and voting. In G. Bonanno, J.P. Delgrande, J. Lang, and H. Rott, editors, *Formal*

Models of Belief Change in Rational Agents, volume 07351 of *Dagstuhl Seminar Proceedings*. Internationales Begegnungs- und Forschungszentrum fuer Informatik (IBFI), Schloss Dagstuhl, Germany, 2007.
22. H. Gaifman. Operational pointer semantics: solution to self-referential puzzles i. In TARK '88: *Proceedings of the 2nd Conference on Theoretical Aspects of Reasoning About Knowledge*, pages 43–59. Morgan Kaufmann Publishers Inc., San Francisco, CA, 1988.
23. H. Gaifman. Pointers to truth. *Journal of Philosophy*, 5(89):223–261, 1992.
24. J. Gerbrandy. *Bisimulations on Planet Kripke*. PhD thesis, Institute for Logic, Language and Computation (Universiteit van Amsterdam), 1999. ILLC Dissertation Series DS-1999–01.
25. J. Gerbrandy. Dynamic epistemic logic. In L.S. Moss, J. Ginzburg, and M. de Rijke, editors, *Logic, Language and Computation*, volume 2, pages 67–84. Center for the Study of Language and Information, Stanford, CA, 1999.
26. S. Ghosh, B. Löwe, and E. Scorelle. Belief flow in assertion networks. In U. Priss, S. Polovina, and R. Hill, editors, *Proceedings of the 15th International Conference on Conceptual Structures (ICCS 2007), Sheffield, UK*, volume 4604 of *LNAI*, pages 401–414. Springer, Heidelberg, July 2007.
27. S. Ghosh and F.R. Veläzquez-Quesada. Expressing belief flow in assertion networks. In P. Bosch, D. Gabelaia, and J. Lang, editors, *TbiLLC*, volume 5422 of *Lecture Notes in Computer Science*, pages 124–138. Springer, Berlin, Germany, 2007.
28. P. Girard. *Modal Logic for Belief and Preference Change*. PhD thesis, Department of Philosophy (Stanford University), Stanford, CA, February 2008. ILLC Dissertation Series DS-2008–04.
29. A. Grove. Two modellings for theory change. *Journal of Philosophical Logic*, 17(2):157–170, May 1988.
30. A. Gupta and N.D. Belnap. *The Revision Theory of Truth*. The MIT Press, Cambridge, MA, 1993.
31. S. Ove Hansson. A survey of non-prioritized belief revision. *Erkenntnis*, 50(2–3):413–427, 1999.
32. W.L. Harper. Rational conceptual change. In *Proceedings of the Meeting of the Philosophy of Science Association (PSA 1976)*, volume 2, pages 462–494. Philosophy of Science Association, The University of Chicago Press, Chicago, IL, 1977.
33. H.G. Herzberger. Naive semantics and the liar paradox. *The Journal of Philosophy*, 79:479–497, 1982.
34. S. Konieczny. *Sur la logique du changement: Révision et fusion de bases de connaissance*. PhD thesis, Laboratoire d'Informatique Fondamentale de Lille, Université de Lille 1, Lille, France, November 1999.
35. S. Konieczny and R. Pino-Pérez. On the logic of merging. In *Sixth International Conference on Principles of Knowledge Representation and Reasoning (KR'98)*, pages 488–498. Morgan Kaufmann, Trento, June 2–5, 1998.
36. S. Konieczny and R. Pino-Pérez. Merging information under constraints: A logical framework. *Journal of Logic and Computation*, 12(5):773–808, 2002.
37. S. Konieczny and R. Pino-Pérez. Propositional belief base merging or how to merge beliefs/goals coming from several sources and some links with social choice theory. *European Journal of Operational Research*, 160(3):785–802, 2005.
38. H. Leitgeb. Interpreted dynamical systems and qualitative laws: From neural networks to evolutionary systems. *Synthese*, 146(1–2):189–202, August 2005.
39. H. Leitgeb and K. Segerberg. Dynamic doxastic logic: Why, how, and where to? *Synthese*, 155(2):167–190, 2007.
40. I. Levi. Subjunctives, dispositions, and chances. *Synthese*, 34:423–455, 1977.
41. I. Levi. *The Enterprise of Knowledge*. MIT Press, Cambridge, MA, 1980.
42. D.K. Lewis. *Counterfactuals*. Blackwell, Cambridge, MA, 1973.
43. E. Lorini and R. Demolombe, Trust and norms in the context of computer security: a logical formalization. In van der Meyden, Ron and van der Torre, Leendert editors, *Deontic Logic in Computer Science*, LNCS 5076, pages 50–64, Springer, Berlin, 2008.

44. T. Meyer A.K. Ghose and S. Chopra. Non-prioritized ranked belief change. *Journal of Philosophical Logic*, 32(4):117–143, 2003.
45. J. Pearl. *Probabilistic Reasoning in Intelligent Systems: Networks of Plausible Inference*. Morgan Kaufmann, Santa Mateo, CA, September 1988.
46. J. Plaza. Logics of public communications. In M.L. Emrich, M.S. Pfeifer, M. Hadzikadic, Z. Ras, editors, *Proceedings of the 4th International Symposium on Methodologies for Intelligent Systems*, pages 201–216. Charlotte, NC, 1989.
47. K. Segerberg. Belief revision from the point of view of doxastic logic. *Logic Journal of IGPL*, 3(4):535–553, 1995.
48. K. Segerberg. Two traditions in the logic of belief: Bringing them together. In H.J. Ohlbach and U. Reyle, editors, *Logic, Language and Reasoning. Essays in Honour of Dov Gabbay*, pages 135–147. Kluwer Academic Publishers, Dordrecht, 1999.
49. B. ten Cate. *Model theory for extended modal languages*. PhD thesis, Institute for Logic, Language and Computation (Universiteit van Amsterdam), 2005. ILLC Dissertation Series DS-2005–01.
50. J. van Benthem. Semantic parallels in natural language and computation. In *Logic Colloquium '87*, pages 331–375. Elsevier Science Publishers, Amsterdam, 1989.
51. J. van Benthem. Dynamic logic for belief revision. *Journal of Applied Non-Classical Logics*, 17(2):129–155, 2007.
52. J. van Benthem. Priority product update as social choice. Available at http://dare.uva.nl/document/93918, 2007.
53. J. van Benthem, P. Girard, and O. Roy. Everything else being equal: A modal logic approach for ceteris paribus preferences. *Journal of Philosophical Logic*, 38:83–125, 2009.
54. J. van Benthem and F. Liu. Dynamic logic of preference upgrade. *Journal of Applied Non-Classical Logics*, 17(2):157–182, 2007.
55. J. van Benthem, S. van Otterloo, and O. Roy. Preference logic, conditionals and solution concepts in games. In *Festschrift for Krister Segerberg*. University of Uppsala, Uppsala, 2005.
56. H. van Ditmarsch, W. van der Hoek, and B. Kooi. *Dynamic Epistemic Logic*, volume 337 of *Synthese Library Series*. Springer, Berlin, Germany, 2007.
57. J. Williamson. *Bayesian Nets and Causality: Philosophical and Computational Foundations*. Oxford University Press, Oxford, 2005.

Chapter 6
Modal Logic for Lexicographic Preference Aggregation

Patrick Girard

Bad news abounds when it comes to theories of aggregation. The most famous result is Arrow's theorem (cf., [3]), stating the impossibility of a non-dictatorial aggregation procedure satisfying desirable properties. More recently, research in judgment aggregation has produced a plethora of analogous results pertaining to judgments (cf., [5, 9, 10]). One may get the impression, from this series of results, that it is hopeless to look for a systematic and satisfying aggregation procedure.

Yet we do succeed in acting as groups in such a way that individual standpoints are reflected in groups' actions. We trust the committees we have at various levels of our societies (e.g., we gave a Nobel prize to the *Intergovernmental Panel on Climate Change)* and we still cherish our aggregation procedures for democratic societies - we do not have dictators.

How do we succeed?

Impossibility results tell us that we cannot aggregate systematically while obeying a set of prescribed aggregation conditions, but they are impervious to common achievements of groups. They do not state the metaphysical impossibility of aggregation procedures, but that we cannot use a unique one in any given circumstance.

In this paper, I adopt an optimistic standpoint and show that modal logic for preference aggregation is possible, provided that we change our expectations about what is to be achieved. Of course, because of Arrow's theorem, something has to give. One way to achieve this is by shifting the scope of the analysis. Instead of looking for general results that apply uniformly, one looks for procedures applying to simple groups, such as committees or even simpler, a group of friends ordering food in a restaurant together.

This is the quest for possibility results.

My main goal here is to show that lexicographic preference reordering can be treated inside Modal Logic. I base most of my results on [1], where a possibility result is proved for *lexicographic reordering*. In this setting, groups of agents are taken

Patrick Girard
Department of Philosophy, University of Auckland, Auckland, New Zealand,
e-mail: p.girard@auckland.ac.nz

J. van Benthem et al. (eds.), *Games, Norms and Reasons*, Synthese Library 353,
DOI 10.1007/978-94-007-0714-6_6,

to be ordered and aggregation proceeds in a compensatory way: if every member of the group prefers x, then the group also prefers x, otherwise the group's preferences follows that of the most influential agents. To achieve this, I take the logic of preference exposed in [14] as basic and show that expending the language with so-called nominals is sufficient to modalize lexicographic reordering. I call the resulting logic *Group Preference Logic* and denote it GPL.

The paper is divided as follows. In Section 6.1, I outline the basic Preference Logic. In Section 6.2, I collect the results from [1] over which I base the rest of the paper. In Section 6.3, I define GPL and provide its complete axiomatization. In Section 6.4, I investigate various applications of GPL. I show how the complete relational algebra of [1] can be derived and how to lift comparisons over states to preferences over sets of states. Finally, in Section 6.5, I introduce dynamics in the logic. This comes in two stages: (1) public announcement and preference upgrade, and (2) *agent promotion*. Agent promotion is the action of changing the hierarchy of the group by putting one agent on top. My approach is advantageous in two ways: (1) it provides a simple modal logic for preference aggregation and (2) it yields a straightforward dynamification of group preferences with a new kind of action, over the hierarchy of the group.

6.1 Basic Preference Logic

Preference Logic has been at the core of different systems under various guises. The version I use in this paper is based on the work of a few authors in selected papers, originating in Boutilier [4] and van Benthem, van Otterloo and Roy [16], but fully developed with a *ceteris paribus* rider in [14].

Language and Semantics

Let PROP be a set of propositional letters. The *Preference Language*, denoted $\mathscr{L}_{\mathscr{P}}$, is inductively defined by the following rules:

$$\mathscr{L}_{\mathscr{P}} := p \mid \varphi \vee \psi \mid \neg\varphi \mid \Diamond^{\leq}\varphi \mid \Diamond^{<}\varphi \mid E\varphi$$

The class of formulas of $\mathscr{L}_{\mathscr{P}}$ is denoted "FORM". The intended reading of $\Diamond^{\leq}\varphi$ is "φ is true in a state that is considered to be at least as good as the current state", whereas that of $\Diamond^{<}\varphi$ is "φ is true in a state that is considered to be strictly better than the current state". $E\varphi$ is interpreted as "there is a state where φ is true".

I write $\Box^{\leq}\varphi$ to abbreviate $\neg\Diamond^{\leq}\neg\varphi$, and use $\Box^{<}\varphi$ and $U\varphi$ for the duals of $\Diamond^{<}\varphi$ and $E\varphi$ respectively.

Definition 6.1 (Models). A *preference model* $\mathfrak{M}$ is a triple $\mathfrak{M} = \langle W, \preceq, V\rangle$ where W is a set of states, $\preceq$ is a reflexive and transitive relation (a *preorder*) and V is a standard propositional valuation. The strict subrelation $\prec$ is defined in terms of $\preceq$: $u \prec v := u \preceq \& v \not\preceq u$. Finally, a *pointed order model* is a pair $\mathfrak{M}, u$ where $u \in W$.

Definition 6.2 (Truth definition). Formulas of $\mathcal{L}_{\mathcal{P}}$ are interpreted in pointed order models.

$$\begin{array}{ll}
\mathfrak{M}, u \models p & \text{iff } u \in V(p) \\
\mathfrak{M}, u \models \neg\varphi & \text{iff } \mathfrak{M}, u \not\models \varphi \\
\mathfrak{M}, u \models \varphi \vee \psi & \text{iff } \mathfrak{M}, u \models \varphi \text{ or } \mathfrak{M}, u \models \psi \\
\mathfrak{M}, u \models \Diamond^{\leq}\varphi & \text{iff } \exists v \text{ s.t. } u \preceq v \,\&\, \mathfrak{M}, v \models \varphi \\
\mathfrak{M}, u \models \Diamond^{<}\varphi & \text{iff } \exists v \text{ s.t. } u \prec v \,\&\, \mathfrak{M}, v \models \varphi \\
\mathfrak{M}, u \models E\varphi & \text{iff } \exists v \text{ s.t. } \mathfrak{M}, v \models \varphi
\end{array}$$

Definition 6.3. A formula φ is said to be *satisfiable* in a model $\mathfrak{M}$ if there is a state u such that $\mathfrak{M}, u \models \varphi$ and *valid* if it is true at every state in every model.

For a complete axiomatization $\Lambda^{\mathcal{L}_{\mathcal{P}}}$ of Preference Logic, the reader should consult [14]. Of interest is the realization that the strict subrelation $\prec$ of $\preceq$ is not definable in $\mathcal{L}_{\mathcal{P}}$, even though the completeness result, using Segerberg's bulldozing technique (cf. [11]), guarantees that the canonical model can be transformed into an *adequate* preference model, i.e., one where $u \prec v$ iff $u \preceq v \,\&\, v \not\preceq u$. This is taken care of by the following axioms of $\Lambda^{\mathcal{L}_{\mathcal{P}}}$:

$$\Diamond^{<}\varphi \to \Diamond^{\leq}\varphi \tag{6.1}$$

$$\Diamond^{\leq}\Diamond^{<}\varphi \to \Diamond^{<}\varphi \tag{6.2}$$

$$\Diamond^{<}\Diamond^{\leq}\varphi \to \Diamond^{<}\varphi \tag{6.3}$$

$$\varphi \wedge \Diamond^{\leq}\psi \to (\Diamond^{<}\psi \vee \Diamond^{\leq}(\psi \wedge \Diamond^{\leq}\varphi)) \tag{6.4}$$

Deriving Binary Preferences

The reader might be worried by my choice of nomenclature for the basic language. Admittedly, calling $\mathcal{L}_{\mathcal{P}}$ a "preference language" is abusive in that its modalities do not express genuine preference statements.[1] $\Diamond^{\leq}\varphi$ says that φ is true in a better state, but preferences are typically expressed as statements comparing (at least) two things: I prefer beer over wine, blue over red and vampires over zombies.

Such preference statements can be understood in various ways. Saying that I prefer blue over red might mean that I prefer every blue object to every red one, or that for every red object, there is a blue one which I prefer, and so on. There is flexibility in defining preferences and this flexibility is reflected in $\mathcal{L}_{\mathcal{P}}$. I illustrate this by considering two examples, the $\leq_{\exists\exists}$ and the $\leq_{\forall\exists}$ binary preference operators, with semantics given by:

[1] Accordingly, in [7], I have called the basic logic "Order Logic".

Definition 6.4 (Binary preference statements).

$$\mathfrak{M},u \models \varphi \leq_{\exists\exists} \psi \text{ iff } \exists s, \exists t : \mathfrak{M},s \models \varphi \,\&\, \mathfrak{M},t \models \psi \,\&\, s \preceq t \tag{6.5}$$

$$\mathfrak{M},u \models \varphi \leq_{\forall\exists} \psi \text{ iff } \forall s, \exists t : \mathfrak{M},s \models \varphi \Rightarrow \mathfrak{M},t \models \psi \,\&\, s \preceq t \tag{6.6}$$

Fact 6.1. The preference operators of Definition 6.4 can be defined in $\mathscr{L}_{\mathscr{P}}$.

$$\varphi \leq_{\exists\exists} \psi := E(\varphi \wedge \Diamond^{\leq} \psi) \tag{6.7}$$

$$\varphi \leq_{\forall\exists} \psi := U(\varphi \rightarrow \Diamond^{\leq} \psi) \tag{6.8}$$

The choice of the basic logic is more important than the name given to it. What matters here is that $\mathscr{L}_{\mathscr{P}}$ offers a unifying framework to deal with various notions of preferences. One of its great advantages is that it is a normal multi-modal logic, hence meta-theoretical results such as completeness are easily treated. Furthermore, and this is the main contribution of the present paper, a simple expansion of $\mathscr{L}_{\mathscr{P}}$ yields a logic appropriate for lexicographic preference aggregation. Before I show how this is carried out, I describe what lexicographic reordering is.

6.2 Lexicographic Reordering

The lexicographic reordering treatment for preference aggregation is fully developed in an algebraic setting in [1]; I call this system "ARS". The modal logic for preference aggregation presented in Section 6.3 relies heavily on some of the results contained therein. For a complete presentations of ARS, the reader should consult [1]. It is outside the scope of this paper to present the details of this system, which I take for granted here, simply listing the results needed for Group Preference Logic.

The main motivation for adopting ARS as the aggregation policy is the *possibility* result proved in [1], namely that priority operators are the only ones satisfying the following conditions: (1) independence of irrelevant alternatives, (2) based on preferences only, (3) unanimous with abstention, (4) preserving transitivity and (5) non-dictatorial. The fifth property is derivable from the four others and is where the divergence with Arrow's theorem becomes most important; lexicographic operators are non-dictatorial![2] Of course, this is not in contradiction with Arrow's result, since his conditions are more general, but it provides a useful touchstone to investigate possibility results in the fields of preference and judgment aggregation, as well as to compare them with well-known and abundant impossibility results.

The two tools which play a fundamental role in ARS are *priority graph* and *priority operator*. A priority graph imposes a hierarchy among basic relations and a priority operator maps these relations to a single relation lexicographically, following the hierarchy provided by the graph. Although the results hold for the aggregation of arbitrary orders, I confine my discussion to the special case of preorders interpreted as preference orders. No generality is lost by this choice. I thus take a

[2] For a proof of this important theorem, the reader should consult [1].

priority graph as a hierarchy imposed on a group of agents and a priority operator as the aggregation of their preferences.[3]

Let W be a set containing at least two elements, standing for the set of objects to be compared and over which agents give their preference orders $\preceq_i$. For the remainder of this paper, I identify agents with the order they give on W, but I keep the notation $i, j, \ldots$ to refer to agents instead of $\preceq_i, \preceq_j, \ldots$, for the sake of readability. For two objects $u, v \in W$, I say that agent i prefers v over u if $u \preceq_i v$; this is the order given by i between u and v.

To aggregate preferences, a partial strict order $<$ is imposed over the agents, which can occur several times in the order. To help build the intuition, one can think of this order as providing credibility or reliability criteria. Suppose, for example, that you have a committee of scientists (geologists, physicists, chemists etc.) investigating the effect of human societies on global climate change. One would expect that the interpretation of results of each scientist would be given more value in their respective field of expertise. Physicists would be given more credibility in answering questions pertaining to physics, geologists to geological questions, and so on. One can think of the multiple occurrences of agents in the order as representing exactly this: expertise. The hierarchy of agents is represented in a *priority graph*:

Definition 6.5 (Priority graph). Let X be a set of variables. A *priority graph* is a tuple $P = \langle N, <, V \rangle$ where N is a set (of agents), $<$ is a strict partial order on N and V is function from N to X.

Something has to be said about the use of variables in priority graphs. First, as I have noted above, agents are identified with the order they give on W. Two agents giving the same order can be identified in priority graphs, thus eliminating redundancy in graphs. Second, priority graphs allow variables to occur several times, so that assigning agents to variables allows a representation of expertise in graphs; in one occurrence an agent is the expert and in another, it is dominated. Finally, the repetition of variables in priority graphs increases the expressivity, although I do not prove this here.

Example 6.1. Let $N = \{i, j, k\}$, $V(i) = V(j) = y$, $V(k) = x$ and let $j < i, k < i$. The priority graph $g = \langle N, <, V \rangle$ is represented graphically in Figure 6.1. In the figure (and in the remainder of this paper), a variable x occurs above a variable y if $y < x$, i.e., if x is higher in the hierarchy than y.

Next, I define the notion of a priority operator. A priority operator aggregates individual preferences lexicographically according to the priority graph. A lexicographic order $\leq$ between two elements is an order such that:

$$(a,b) \leq (a',b') \text{ iff } a < a' \text{ or } (a = a' \text{ and } b \leq b')$$

[3] I write $i < j$ to express that j is strictly better than i, the opposite notation of that used in [1]. I also draw pictures for priority graphs by putting best agents on top of the graph where [1] puts them at the bottom and I write the ARS "on the other hand" operator x/y with the reverse interpretation. The reader is asked to keep this in mind if reading this paper and [1] in parallel and I apologize for confusion that may ensue; I find my notation more intuitive and easier to work with.

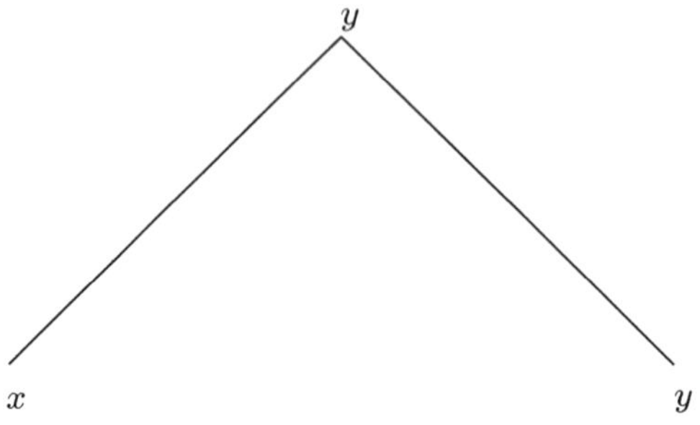

Fig. 6.1 Graphical representation of the priority graph g defined in Example 6.1. In the figure, a variable occurring above another variable is prioritized

This definition generalizes to n-tuples in the following way:

$$(a_1, a_2, ...a_n) \leq (b_1, b_2, ..., b_n) \text{ iff } \exists m > 0, \forall i < m(a_i = b_i \text{ and } a_m <_m b_m)$$

A familiar example of a lexicographic ordering is the alphabetical order used in a dictionary, where the priority is given to letters on the left.

In the case of priority graphs with single occurrences of variables, a priority operator o orders the relations of the graph lexicographically when:

$$a \preceq b \text{ iff } \forall x \in V(a \preceq_x b \text{ or } \exists y \in V(x < y \& a \prec_y b))$$

Since priority graphs allow variables to occur multiple times, a further generalization of the lexicographic rule is needed:

Definition 6.6 (Priority operator). A priority graph g denotes a *priority operator* o if:

$$ao((\preceq_x)_{x \in X} b \Leftrightarrow \forall i \in N(a \preceq_{V(i)} b \vee \exists j \in N(i < j \wedge a \prec_{V(j)} b)) \quad (6.9)$$

Example 6.2. Consider the priority graph given in Example 6.1. Let $a, b \in M$, then according to Definition 6.6:

$$\begin{aligned} ao(\preceq_x, \preceq_y)b \text{ iff } & (a \preceq_{V(i)} b \wedge a \preceq_{V(j)} b \wedge a \preceq_{V(k)} b) \vee a \prec_{V(i)} b \\ \text{iff } & (a \preceq_x b \wedge a \preceq_y b) \vee a \prec_y b \end{aligned}$$

Therefore, the group consisting of $\{i, j, k\}$ considers b better than a if they reach a consensus or if both i and j strictly prefer b to a.

The next two theorems are crucial in modalizing Group Preference Logic. Theorem 6.2 states that every priority operator is equivalent to one built from two fundamental operators given in Definition 6.7 and Theorem 6.3 provides a complete relational algebra in terms of these operators.

Definition 6.7. The two operators $x \parallel y$ and x/y are called the *but* and *on the other hand* operators respectively and are defined by:

$$x \parallel y = x \cap y$$
$$x/y \;\; = (x \cap y) \cup x^{<}$$

The two operators are depicted in Figure 6.2. Here are the two crucial theorems:

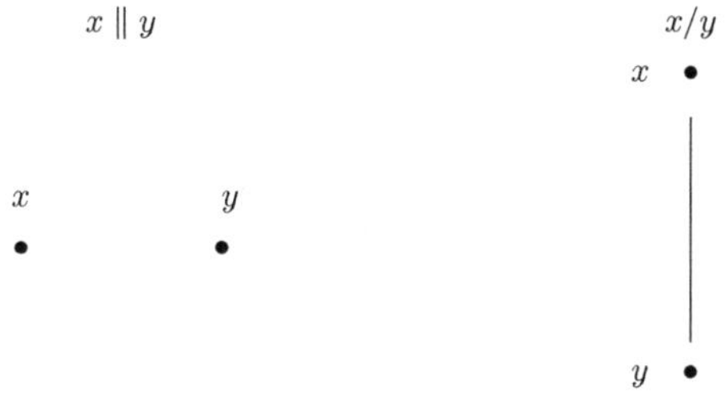

Fig. 6.2 The priority graph of the *but* and *on the other hand* operators

Theorem 6.2. *Any finitary priority operator is denoted by a term built from $/$, $\parallel$ and the variables occurring in the priority graph for the operator.*

Theorem 6.3. *An equation is true in all preferential algebras iff it is derivable from the following axioms:*

$$x \parallel x = x \tag{6.10}$$
$$x \parallel (y \parallel z) = (x \parallel y) \parallel z \tag{6.11}$$
$$x \parallel y = y \parallel x \tag{6.12}$$
$$(x/x) = x \tag{6.13}$$
$$(x/y)/z = x/(y/z) \tag{6.14}$$
$$x/(y \parallel z) = (x/y) \parallel (x/z) \tag{6.15}$$
$$(x/y) \parallel y = x \parallel y \tag{6.16}$$

Theorems 6.2 and 6.3 show that the complete algebra of ARS can be captured by a set of equations involving two fundamental operators. In the next section, I show that relativizing the basic modalities of $\mathscr{L}_{\mathscr{P}}$ with priority operators yield a logic for group preferences and I show in Theorem 6.5 that the complete algebra of Theorem 6.3 can be derived in the complete logic for group preferences. Therefore, Group Preference Logic modalizes the ARS system in a natural way, but also provides a platform to investigate dynamification of the system, as I demonstrate in Section 6.5.

6.3 Modal logic for Preference Aggregation

As I mentioned above, the language for Group Preference Logic is obtained by expanding $\mathscr{L}_{\mathscr{P}}$ with nominals. In contrast to propositional variables, which are typically evaluated at sets of states, nominal are evaluated at single states. On the basis of this standpoint, states in models can be interpreted as simple objects and nominals as names for these objects. Preference Logic can then be seen as reasoning about preference orderings over sets of objects. As I have shown in Section 6.1, the existential modality $E\varphi$ can be used to lift these basic preferences over objects to complex (binary) preferences over sets of objects (or propositions). In the present section, I show how a second kind of lift, this time from individual preferences to group preferences, can be performed with nominals. I thus introduce nominals in the logic to lift individual orders to group orders, using lexicographic reordering in the ARS style. Furthermore, with the existential modality, I can then lift group orders among objects to group preferences over sets of objects, just as I did in the individual case. Group preference logic thus operates on two orthogonal lifts: one from individual to group orders over objects and the other from group orders over objects to group preferences over propositions.

Language and Semantics

Definition 6.8 (Language). Let PROP be a set of propositional variables with $p \in$ PROP, NOM a set of nominals with $s \in$ NOM and N a set of agents with $i \in N$. The Group Preference Language $\mathscr{L}_{\mathscr{G}\mathscr{P}}$ is defined by the following recursive rules:

$$\varphi := s \mid p \mid \neg\varphi \mid \varphi \vee \psi \mid \langle X\rangle^{\leq}\varphi \mid \langle X\rangle^{<}\varphi \mid E\varphi$$
$$X := i \mid X/Y \mid X \parallel Y$$

The intended reading of the modalities is as follows. $\langle i\rangle^{\leq}\varphi$ stands for 'agent i thinks that an accessible state where φ holds is at least as good as the current state' and $\langle i\rangle^{<}\varphi$ stands for "according to agent i, there is an accessible state that is strictly better where φ holds". Complex modalities with capital variables X,Y,Z and their combinations with the operators / and $\parallel$ stand for group orders. Giving an intuitive and succinct reading of modalities for groups of more than two agents is not straightforward, although the intention is quite clear. I illustrate the meaning of these modalities with a simple group of two agents i and j. In the case where there is a hierarchy, say of a master i and a student j, the modality $\langle i/j\rangle^{\leq}\varphi$ is read as 'the group consisting of a master i and a student j considers an accessible state where φ holds to be at least as good as the current state. If the two agents i and j have the same weight in the decision making, the case where $i \parallel j$, the modality $\langle i \parallel j\rangle^{\leq}\varphi$ is read as "the group consisting of two agents of incomparable rank considers an accessible state where φ holds to be at least as good as the current state".[4] The strict modalities receive the obvious analogous readings.

Definition 6.9 (Models). Models are tuples $\mathfrak{M} = \langle W, G, N, \{\preceq_X\}_{X \in G}, V\rangle$, where W is a set of states, G is a set of graphs, N a set of agents, $\{\preceq_X\}_{X\in G}$ is a family of relations induced by priority graphs and $V : \text{PROP} \cup \text{NOM} \to \mathscr{P}(W)$ is valuation assigning singleton sets for members of NOM.

Definition 6.10 (Semantics).

$$\begin{array}{ll}
\mathfrak{M}, u \models p & \text{iff } u \in V(p) \\
\mathfrak{M}, u \models s & \text{iff } \{u\} = V(s) \\
\mathfrak{M}, u \models \neg\varphi & \text{iff } \mathfrak{M}, u \not\models \varphi \\
\mathfrak{M}, u \models \varphi \vee \psi & \text{iff } \mathfrak{M}, u \models \varphi \text{ or } \mathfrak{M}, u \models \psi \\
\mathfrak{M}, u \models \langle X\rangle^{\leq}\varphi & \text{iff } \exists v \in W \text{ s.t. } u \preceq_X v \,\&\, \mathfrak{M}, v \models \varphi \\
\mathfrak{M}, u \models \langle X\rangle^{<}\varphi & \text{iff } \exists v \in W \text{ s.t. } u \prec_X v \,\&\, \mathfrak{M}, v \models \varphi \\
\mathfrak{M}, u \models E\varphi & \text{iff } \exists v \text{ s.t. } \mathfrak{M}, v \models \varphi
\end{array}$$

In case $X = \{i\}$, $\preceq_i$ is the preference relation for a single agent. Complex relations $\preceq_X$ are recursively reduced to individual preference relations using $\preceq_{X/Y} = (\preceq_X \cap \preceq_Y) \cup \prec_Y$ and $\preceq_{X\|Y} = \preceq_X \cap \preceq_Y$. The strict subrelations $\prec_X$ are defined in the standard way, i.e., $u \prec_X v$ iff $u \preceq_X v \,\&\, \neg(v \preceq_X u)$.

Remark 6.1. The hybrid binder $@_i\varphi$ is definable with the existential modality and nominals by $E(i \wedge \varphi)$. Therefore, Group Preference Logic is a superset of the hybrid logic $\mathscr{H}(@)$ (cf., [2]). This will be used in the axiomatization below, appealing to the hybrid logic rules of NAME and PASTE in their existential form (cf., [12]).

Axiomatization

Theorem 6.4. *The following set of axioms is complete for Group Preference Logic. I call the logic* $\Lambda_{\mathscr{L}\mathscr{G}\mathscr{P}}$.

(a) Classical tautologies and normality axioms for each modality.
(b) Axioms for the existential modality:

$$\varphi \to E\varphi \tag{6.17}$$

$$EE\varphi \to E\varphi \tag{6.18}$$

$$\varphi \to UE\varphi \tag{6.19}$$

$$Ei \tag{6.20}$$

$$E(i \wedge \varphi) \to U(i \to \varphi) \tag{6.21}$$

[4] It might help the reader to read the modality $\langle i/j\rangle^{\leq}$ as "diamond weak i over j, φ", $\langle i/j\rangle^{<}$ as "diamond strict i over j, φ", $\langle i \parallel j\rangle^{\leq}\varphi$ as "diamond weak i and j, φ" and $\langle i \parallel j\rangle^{<}\varphi$ as "diamond strict i and j, φ".

(c) Axioms defining properties of $\preceq$ and $\prec$:

$$s \to \langle X\rangle^{\leq} s \quad \textit{Reflexivity of } \leq_X \tag{6.22}$$
$$\langle X\rangle^{\leq}\langle X\rangle^{\leq} s \to \langle X\rangle^{\leq} s \quad \textit{Transitivity of} \leq_X \tag{6.23}$$
$$s \to \neg\langle X\rangle^{<} s \quad \textit{Irreflexivity of } <_X \tag{6.24}$$
$$s \to \neg\langle X\rangle^{<}\langle X\rangle^{<} s \quad \textit{Assymetry of } <_X \tag{6.25}$$
$$\langle X\rangle^{<} s \to \langle X\rangle^{\leq} s \quad \textit{Inclusion} \tag{6.26}$$
$$\langle X\rangle^{\leq}\langle X\rangle^{<} s \to \langle X\rangle^{<} s \quad \textit{Mix 1} \tag{6.27}$$
$$\langle X\rangle^{<}\langle X\rangle^{\leq} s \to \langle X\rangle^{<} s \quad \textit{Mix 2} \tag{6.28}$$
$$s \wedge \langle X\rangle^{\leq} t \to (\langle X\rangle^{<} t \vee \langle X\rangle^{\leq}(t \wedge \langle X\rangle^{\leq} s)) \quad \textit{Mix 3} \tag{6.29}$$

(d) Mixed axioms:

$$\langle X\rangle^{\leq}\varphi \to E\varphi \tag{6.30}$$
$$\langle X\rangle^{<}\varphi \to E\varphi \tag{6.31}$$

(e) Group axioms:

$$\langle X \parallel Y\rangle^{\leq} s \leftrightarrow \langle X\rangle^{\leq} s \wedge \langle Y\rangle^{\leq} s \tag{6.32}$$
$$\langle X/Y\rangle^{\leq} s \leftrightarrow (\langle X\rangle^{\leq} s \wedge \langle Y\rangle^{\leq} s) \vee \langle X\rangle^{<} s \tag{6.33}$$
$$\langle X \parallel Y\rangle^{<} s \leftrightarrow (\langle X\rangle^{<} s \wedge \langle Y\rangle^{\leq} s) \vee (\langle X\rangle^{\leq} s \wedge \langle Y\rangle^{<} s) \tag{6.34}$$
$$\langle X/Y\rangle^{<} s \leftrightarrow (\langle X\rangle^{\leq} s \wedge \langle Y\rangle^{<} s) \vee \langle X\rangle^{<} s \tag{6.35}$$

In addition, Λ_{GP} has the rules of Modus Ponens, Necessitation and the hybrid logic rules NAME *and* PASTE *(with $\Diamond$ instantiated for all modalities $\langle X/Y\rangle^{\leq}, \langle X/Y\rangle^{<}, \langle X \parallel Y\rangle^{\leq}$ and $\langle X \parallel Y\rangle^{<}$) :*

$$\vdash i \to \varphi \Rightarrow \vdash \varphi, \textit{ if } i \notin \varphi \quad \text{NAME} \tag{6.36}$$
$$\vdash E(i \wedge \Diamond j) \to E(j \wedge \varphi) \Rightarrow \vdash E(i \wedge \Box\varphi), \textit{ if } i \neq j \textit{ and } i, j \notin \varphi \quad \text{PASTE} \tag{6.37}$$

Remark 6.2. (a) Transitivity of $\langle X\rangle^{<}\varphi$ is derivable, using Axioms 6.26 and 6.27.
(b) Axioms 6.32 and 6.33 analyze the (weak) modalities $\langle X \parallel Y\rangle^{\leq}$ and $\langle X/Y\rangle^{\leq}$ in terms of $\langle X\rangle^{\leq}$, $\langle X\rangle^{<}$ and $\langle Y\rangle^{<}$. Similarly, the strict modalities $\langle X||Y\rangle^{<}$ and $\langle X/Y\rangle^{<}$ can be analyzed in terms of more basic modalities by Axioms 6.34 and 6.35.

Before proving completeness, a preliminary result is needed, namely that the canonical model constructed is an *adequate* preference model, in the sense that the relation $\prec_X$ is the appropriate subrelation of $\preceq_X$, defined as:

Definition 6.11 ($\prec_X$-adequacy). A model is called *$\prec_X$-adequate* if the following are equivalent:

(a) $u \prec_X v$
(b) a. $u \preceq_X v$ and

b. $v \not\preceq_X u$.

Lemma 6.1. *For every frame $\mathfrak{F}$, Axioms 6.24, 6.26, 6.28 and 6.29 are valid on $\mathfrak{F}$ iff $\mathfrak{F}$ is $\prec_X$-adequate*

Proof. I do not prove the right to left direction. In the other direction, assume that $u \prec_x v$ and take any model $\mathfrak{M}$ with $V(s) = u$ and $V(t) = v$. Then $\mathfrak{M}, u \models \langle X \rangle^{<} t$, so $\mathfrak{M}, u \models \langle X \rangle^{\leq} t$ by Axiom 6.26. Therefore $u \preceq_X v$. Now, assume that $v \preceq_X u$, then $\mathfrak{M}, v \models \langle X \rangle^{\leq} s$, which implies that $\mathfrak{M}, u \models \langle X \rangle^{<} \langle X \rangle^{\leq} s$. Hence, by Axiom 6.28, $\mathfrak{M}, u \models \langle X \rangle^{<} s$, which implies that $\mathfrak{M}, u \models s \wedge \langle X \rangle^{<} s$, contradicting Axiom 6.24. Therefore, $u \prec_X v$ implies that $u \preceq_X v$ and $v \not\preceq_X u$.

Finally, suppose that $u \preceq_X v$ and $v \not\preceq_X u$ and let $\mathfrak{M}$ be a model with the same valuation as above. Thus, $\mathfrak{M}, u \models s \wedge \Diamond^{\leq} t$. By Axiom 6.29, $\mathfrak{M}, u \models \Diamond^{<} t \vee \Diamond^{\leq} (t \wedge \Diamond^{\leq} s)$. Thus, for some w, either $u < w \& w = v$ (i.e., $u < v$) - and we are done – or $u \leq v \leq u$. Therefore, $u \preceq_X v$ and $v \not\preceq_X u$ implies that $u \prec_x v$.

Proof of Theorem 6.4. I only show soundness of Axioms 6.34 and 6.35.

To show the soundness of Axioms 6.34, it is enough to show that:

$$\prec_{X \| Y} = (\prec_X \cap \preceq_Y) \cup (\preceq_X \cap \prec_Y)$$

In the first direction, assume that $u \prec_{X\|Y} v$. By definition, $u \preceq_{X\|Y} v$ and $v \not\preceq_{X\|Y} u$. Thus, $u \preceq_X v$ and $u \preceq_Y v$. It is now enough to show that $u \prec_X v$ or $u \prec_Y v$. Suppose not, then $v \preceq_X u$ and $v \preceq_Y u$, which implies that $v \preceq_{X\|Y} u$, a contradiction.

In the other direction, assume that $(u \prec_X v \& u \preceq_Y v)$ or $(u \preceq_X v \& u \prec_Y v)$. In either case, $u \preceq_{X\|Y} v$. Now, if $u \not\prec_{X\|Y} v$, then it must be that $v \preceq_{X\|Y} u$, i.e., $v \preceq_X u$ and $v \preceq_Y u$. Hence, $u \not\prec_X v$ and $u \not\prec_Y v$, a contradiction.

For Axiom 6.35, I show the following:

$$u \prec_{X/Y} v = (u \preceq_X v \& u \prec_Y v) \text{ or } u \prec_X v \tag{6.38}$$

In the first direction, assume that $u \prec_{X/Y} v$. Then $u \preceq_{X/Y} v$ and $v \not\preceq_{X/Y} u$, which implies, by definition, that $(u \preceq_X v \& u \preceq_Y v)$ or $u \prec_X v$. In the latter case, the result follows, so I show that $(u \preceq_X v \& u \preceq_Y v) \& v \not\preceq_{X/Y} u$ implies the right-hand side of 6.38. Since $u \preceq_X v$, it is enough to show that $u \prec_X v$ or $u \prec_Y v$. Suppose not, then $v \preceq_X u$ and $v \preceq_Y u$, since $u \preceq_X v$ and $u \preceq_Y v$. Thus, $v \preceq_{X/Y} u$, a contradiction.

In the other direction, assume that the right-hand side of (6.38) holds. Since $u \prec_Y v$ implies that $u \preceq_Y v$, we have that $u \preceq_{X/Y} v$. Suppose that $u \not\prec_{X/Y} v$, then we must have that $v \preceq_{X/Y} u$, i.e., $(v \preceq_X u \& v \preceq_Y u)$ or $v \prec_X u$. The first disjunct implies that $u \not\prec_X v$ and $\not{u} \prec_Y v$, whereas the second implies that $u \not\preceq_X v$, and we obtain a contradiction in either case.

For completeness, I use the following well-known result of hybrid logic, stated as Corollary 5.4.1 in [12]:

> Let Σ be any set of pure $\mathcal{H}(E)$-formulas,[5] then $K^{+}_{\mathcal{H}(E)}$ is strongly complete for the class of frames defined by Σ.

[5] A formula is pure if it contains no propositional variables.

From this result, every consistent set Φ is satisfiable in the canonical model (named and pasted). Furthermore, thanks to Lemma 6.1, this model is adequate. Finally, thanks to Axioms 6.32, 6.33, 6.34 and 6.35, the completeness for group modalities is reduced to that of individual modalities. QED

For an alternative (and more economical) axiomatization of GPL, as well as a cut-free Gentzen sequent calculus, see [8].

6.4 Applications

Group Preference Logic vs Equational Algebra

In this section, I show that the equational algebra of Theorem 6.2 can be derived inside Λ_{GP} and that binary preferences statements between propositions can be defined as in basic Preference Logic.

Theorem 6.5. *The complete relational algebra of Section is derivable in group order logic.*

Proof. I first translate the equations given in Theorem 6.3:

$$\langle X \parallel X\rangle^{\leq} s \leftrightarrow \langle X\rangle^{\leq} s \tag{6.39}$$

$$\langle X \parallel (Y \parallel Z)\rangle^{\leq} s \leftrightarrow \langle (X \parallel Y) \parallel Z\rangle^{\leq} s \tag{6.40}$$

$$\langle X \parallel Y\rangle^{\leq} s \leftrightarrow \langle Y \parallel X\rangle^{\leq} s \tag{6.41}$$

$$\langle (X/X)\rangle^{\leq} s \leftrightarrow \langle X\rangle^{\leq} s \tag{6.42}$$

$$\langle (X/Y)/Z)\rangle^{\leq} s \leftrightarrow \langle X/(Y/Z)\rangle^{\leq} s \tag{6.43}$$

$$\langle X/(Y \parallel Z)\rangle^{\leq} s \leftrightarrow \langle (X/Y) \parallel (X/Z)\rangle^{\leq} s \tag{6.44}$$

$$\langle (X/Y) \parallel Y\rangle^{\leq} s \leftrightarrow \langle X \parallel Y\rangle^{\leq} s \tag{6.45}$$

Equations (6.39), (6.40) and (6.41) are easily derivable using Axiom (6.32). I show how to derive the remaining formulas, keeping Equation (6.43) for the last, at it is the most difficult:

(a) Equation (6.42):

$$\begin{aligned} \langle (X/X)\rangle^{\leq} s &\leftrightarrow (\langle X\rangle^{\leq} s \wedge \langle X\rangle^{\leq} s) \vee \langle X\rangle^{<} s && \text{(Axiom 6.33)} \\ &\leftrightarrow \langle X\rangle^{\leq} s && \text{(Logic and Axiom 6.26)} \end{aligned}$$

(b) Equation (6.44):

$$\begin{aligned} \langle X/(Y \parallel Z)\rangle^{\leq} s &\leftrightarrow (\langle X\rangle^{\leq} s \wedge \langle Y \parallel Z\rangle^{\leq} s) \vee \langle X\rangle^{<} s && \text{(Axiom 6.33)} \\ &\leftrightarrow (\langle X\rangle^{\leq} s \wedge \langle Y\rangle^{\leq} s \wedge \langle Z\rangle^{\leq} s) \vee \langle X\rangle^{<} s && \text{(Axiom 6.32)} \\ &\leftrightarrow ((\langle X\rangle^{\leq} s \wedge \langle Y\rangle^{\leq} x) \vee \langle X\rangle^{<} s) \\ &\quad \wedge ((\langle X\rangle^{\leq} s \wedge \langle Z\rangle^{\leq} s) \vee \langle X\rangle^{<} s) && \text{(Logic)} \\ &\leftrightarrow \langle (X/Y) \parallel (X/Z)\rangle^{\leq} s && \text{(Axiom 6.32,6.33)} \end{aligned}$$

(c) Equation (6.45). The right to left direction follows from Axiom 6.32 and Logic. I prove the left to right direction:

$$\begin{aligned}
\langle (X/Y) \parallel Y\rangle^{\leq} s &\rightarrow (((\langle X\rangle^{\leq} s \wedge \langle Y\rangle^{\leq} s) \vee \langle X\rangle^{<} s) \wedge \langle Y\rangle^{\leq} s && \text{(Axiom 6.32, 6.33)}\\
&\rightarrow (\langle X\rangle^{\leq} s \wedge \langle Y\rangle^{\leq} s) \vee (\langle X\rangle^{<} s \wedge \langle Y\rangle^{\leq} s) && \text{(Logic)}\\
&\rightarrow \langle X\rangle^{\leq} s \wedge \langle Y\rangle^{\leq} s && \text{(Logic and Axiom 6.26)}\\
&\rightarrow \langle X \parallel Y\rangle^{\leq} s && \text{(Axiom 6.32)}
\end{aligned}$$

(d) Equation (6.43). I first prove a preliminary lemma that is crucial in the main proof:

Lemma 6.2. $\vdash (\langle X\rangle^{\leq} s \wedge \langle Y\rangle^{\leq} s) \vee \langle X\rangle^{<} s \leftrightarrow \langle X\rangle^{\leq} s \wedge (\langle Y\rangle^{\leq} s \vee \langle X\rangle^{<} s)$*:*

Proof.

$$\begin{aligned}
(\langle X\rangle^{\leq} s \wedge \langle Y\rangle^{\leq} s) \vee \langle X\rangle^{<} s &\leftrightarrow (\langle X\rangle^{\leq} s \vee \langle X\rangle^{<} s) \wedge (\langle Y\rangle^{\leq} s \vee \langle X\rangle^{<} s) && \text{(Logic)}\\
&\leftrightarrow \langle X\rangle^{\leq} s \wedge (\langle Y\rangle^{\leq} s \vee \langle X\rangle^{<} s) && \text{(Axiom 6.26)}
\end{aligned}$$

I use the following abbreviations:

$$\begin{aligned}
\alpha &:= \langle X\rangle^{\leq} s \wedge \langle Y\rangle^{\leq} s \wedge \langle Z\rangle^{\leq} s\\
\beta &:= (\langle X\rangle^{\leq} s \wedge \langle Y\rangle^{<} s) \vee \langle X\rangle^{<} s
\end{aligned}$$

Now for the main proof:

$$\begin{aligned}
\langle (X/Y)/Z)\rangle^{\leq} s &\leftrightarrow (\langle X/Y\rangle^{\leq} s \wedge \langle Z\rangle^{\leq} s) \vee \langle X/Y\rangle^{<} s && \text{(Axiom 6.33)}\\
&\leftrightarrow [((\langle X\rangle^{\leq} s \wedge \langle Y\rangle^{\leq} s) \vee \langle X\rangle^{<} s) \wedge \langle Z\rangle^{\leq} s] \vee \beta && \text{(Axioms 6.33, 6.35)}\\
&\leftrightarrow [\langle X\rangle^{\leq} s \wedge ((\langle Y\rangle^{\leq} s \vee \langle X\rangle^{<} s) \wedge \langle Z\rangle^{\leq} s)] \vee \beta && \text{(Lemma 6.2)}\\
&\leftrightarrow \alpha \vee (\langle X\rangle^{\leq} s \wedge \langle X\rangle^{<} s \wedge \langle Z\rangle^{\leq} s) \vee \beta && \text{(Logic)}\\
&\leftrightarrow \alpha \vee \beta && \text{(Logic!)}\\
&\leftrightarrow [((\langle Y\rangle^{\leq} s \wedge \langle Z\rangle^{\leq} s) \vee \langle Y\rangle^{<} s) \wedge \langle X\rangle^{\leq} s] \vee \langle X\rangle^{<} s && \text{(Logic)}\\
&\leftrightarrow (\langle X\rangle^{\leq} s \wedge \langle Y/Z\rangle^{\leq} s) \vee \langle X\rangle^{<} s && \text{(Axiom 6.33)}\\
&\leftrightarrow \langle X/(Y/Z)\rangle^{\leq} s && \text{(Axiom 6.33)}
\end{aligned}$$

Corollary 6.1. *The equational algebra of Theorem 6.3 is decidable.*

Proof. Notice that in the proof of Theorem 6.5, I only appealed to Axioms 6.26 and the Group Axioms 6.32, 6.33, 6.34 and 6.35. Furthermore, I did not use the existential modality. Hence, the equational algebra of Theorem 6.5 is derivable in the fragment of group order logic that only uses normal modalities and nominals, which is a decidable system.

Theorem 6.5 establishes that GPL subsumes the algebraic treatment in a natural way, just as regular algebra gets subsumed in a straightforward manner by *PDL*. The decidability of ARS is not surprising, as it is translatable into the two-variable fragment of first-order logic (p.c. Hajnal Andréka). It is still an open question whether the full Group Preference Logic is decidable. Indeed, since the basic order relations are transitive, we are no longer in the two-variable fragment of first-order logic. Furthermore, the presence of transitivity may make a system undecidable, as in the case of the guarded fragment with transitivity (cf., [6]).

Binary Group Preferences

In Section 6.1, I reduced binary preference statements among formulas to sentences using unary modalities and the existential modality. Exactly the same can be done for binary group preferences between formulas:

Definition 6.12 (Binary group preference statements).

$$\varphi \leq^X_{\exists\exists} \psi := E(\varphi \wedge \langle X\rangle^{\leq}\psi) \tag{6.46}$$
$$\varphi \leq^X_{\forall\exists} \psi := U(\varphi \rightarrow \langle X\rangle^{\leq}\psi) \tag{6.47}$$

Hence, as I have claimed above, GPL performs two kinds of lifts: (1) from individual orders to group orders and (2) from basic order relations between objects to binary preferences between propositions.

6.5 Dynamics

To introduce dynamics in Group Preference Logic, I use the method known as *Dynamic Logic* (cf., [17]). A typical example of a dynamic action is that of *public announcement*. A public announcement of some information A is the truthful announcement of A. If A is true at a state u and is announced, then all $\neg A$-state are deleted from the model along with accessibility relations from A to $\neg A$-states. Public announcements are represented by modalities of the form $\langle !A\rangle\varphi$ for every A and the modalities are interpreted by:

$$\mathfrak{M}, u \models \langle !A\rangle\varphi \text{ iff } \mathfrak{M}, u \models A \,\&\, \mathfrak{M}|_A, u \models \varphi \tag{6.48}$$

where $\mathfrak{M}|_A$ is the submodel whose domain is given by the set of states that satisfy A ($W|_A$) with a corresponding restriction of the accessibility relation to $W|_A$. The effect of announcing A is depicted in Figure 6.3. The left model is divided into two zones, the A and the $\neg A$-zones. The right model is the result of publicly announcing A, thus eliminating all $\neg A$-states as well as the relations to or from $\neg A$-states.

An action pertaining more specifically to Preference Logic has been investigated in [15]: *preference upgrade*. Preference upgrade can be seen as a relation change describing a public suggestion to make A better than $\neg A$, i.e., the change in the model is such that every links from $\neg A$-state to A-state is deleted while keeping the

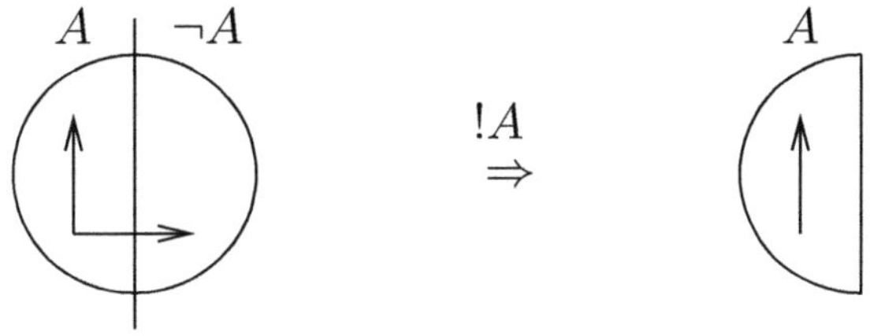

Fig. 6.3 The effect of publicly announcing A

relation unchanged in the two respective zones. Preference upgrade is denoted by $\#A$ and its action on models can can defined by the following:

Definition 6.13. Given a model $\mathfrak{M} = \langle W, G, N, \{\preceq_X\}_{X\in G}, V\rangle$, the *upgraded model* with A upgraded for X is given by $\mathfrak{M}_X^{\#A} = \langle W, G, N, \preceq_X^{\#A}, V\rangle$, where

$$\preceq_X^{\#A} = \preceq_X - \{(u,v) : \mathfrak{M}, u \models A \,\&\, \mathfrak{M}, v \models \neg A\} \tag{6.49}$$

Preference upgrade is depicted in Figure 6.4. As above, to get an Group Preference Logic with preference upgrade, one augments $\mathscr{L}_{\mathscr{GP}}$ with a modality $\langle \#A, X\rangle\varphi$ with semantics given by:

$$\mathfrak{M}, u \models \langle \#A, X\rangle\varphi \text{ iff } \mathfrak{M}_X^{\#A}, u \models \varphi \tag{6.50}$$

Notice that this definition of preference upgrade for groups is a (natural) generalization of the single agent case studied in [15].

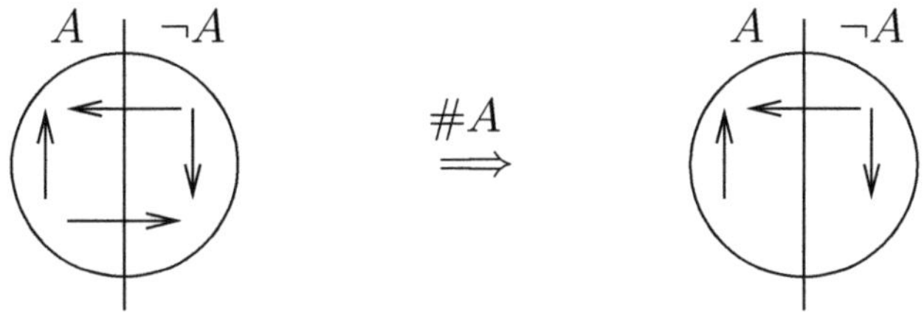

Fig. 6.4 Illustration of preference upgrade

Axiomatization and Completeness

A great tool used in dynamic logic is the method of *compositional analysis via reduction axioms*. Reduction axioms analyze the effect of actions inside the basic language, thus reducing the completeness of the extended logic to that of the basic one. Reduction axioms have a twofold advantage: (1) they provide an explicit analysis of actions on models and (2) they yield free completeness results. For instance, a typical principle analyzing epistemic effect of public announcement is the following reduction axiom:

$$\mathsf{PA}A\Diamond\varphi \leftrightarrow A \wedge \Diamond\mathsf{PA}A\varphi \tag{6.51}$$

Axiom 6.51 can be read as stating that a φ-state is accessible after the public announcement of A if and only if A is true in the current world – therefore can be announced – and there is an accessible state which becomes a φ-state after the announcement of A.

An important feature of Axiom 6.51 is that, on the left-hand side, the action modality $\mathsf{PA}A$ is outside the scope of $\Diamond$, whereas on the right-hand side, it is inside it. Since there are reduction axioms for every component of the basis language, one can push the action modalities all the way to propositional letters, where they do not act any further and can be fully eliminated.

Preference upgrade can be similarly analyzed compositionally with the following reduction principle:

$$\begin{aligned}\langle\#A\rangle\Diamond\varphi \leftrightarrow\; & A \wedge \Diamond(A \wedge \langle\#A\rangle\varphi)\\ & \vee\; \neg A \wedge \Diamond\langle\#A\rangle\varphi\end{aligned} \tag{6.52}$$

The reduction principles 6.51 and 6.52 can be imported in Group Preference Logic, yielding the following completeness theorem:

Theorem 6.6. *The complete logic of Group Preference Logic with public announcement and preference upgrade is axiomatized by: (1) the logic $\Lambda_{\mathscr{L}\mathscr{G}\mathscr{P}}$ together with (2) the reduction axioms of public announcement and preference upgrade:*

$$\mathsf{PA}Ap \leftrightarrow A \wedge p \tag{6.53}$$

$$\mathsf{PA}A\neg\varphi \leftrightarrow A \wedge \neg\mathsf{PA}A\varphi \tag{6.54}$$

$$\mathsf{PA}A(\varphi \vee \psi) \leftrightarrow \mathsf{PA}A\varphi \vee \mathsf{PA}A\psi \tag{6.55}$$

$$\mathsf{PA}A\langle X\rangle^{\leq}\varphi \leftrightarrow A \wedge \langle X\rangle^{\leq}\mathsf{PA}A\varphi \tag{6.56}$$

$$\mathsf{PA}A\langle X\rangle^{<}\varphi \leftrightarrow A \wedge \langle X\rangle^{<}\mathsf{PA}A\varphi \tag{6.57}$$

$$\mathsf{PA}AE\varphi \leftrightarrow E\mathsf{PA}A\varphi \tag{6.58}$$

$$\langle\#A\rangle p \leftrightarrow p \tag{6.59}$$

$$\langle\#A\rangle\neg\varphi \leftrightarrow \neg\langle\#A\rangle\varphi \tag{6.60}$$

$$\langle\#A\rangle(\varphi \vee \psi) \leftrightarrow \langle\#A\rangle\varphi \vee \langle\#A\rangle\psi \tag{6.61}$$

$$\begin{aligned}\langle\#A,X\rangle\langle Y\rangle^{\leq}\varphi \leftrightarrow\; & A \wedge \langle Y\rangle^{\leq}(A \wedge \langle\#A,X\rangle\varphi)\\ & \vee\; \neg A \wedge \langle Y\rangle^{\leq}\langle\#A,X\rangle\varphi\end{aligned} \tag{6.62}$$

$$\begin{aligned}\langle\#A,X\rangle\langle Y\rangle^{<}\varphi \leftrightarrow\; & A \wedge \langle Y\rangle^{<}(A \wedge \langle\#A,X\rangle\varphi)\\ & \vee\; \neg A \wedge \langle Y\rangle^{<}\langle\#A,X\rangle\varphi\end{aligned} \tag{6.63}$$

$$\langle\#A\rangle E\varphi \leftrightarrow E\langle\#A\rangle\varphi \tag{6.64}$$

Proof. Notice first that no special work has to be done for the completeness part, since the axioms reduce the analysis of an arbitrary formula of the extended language to that of $\mathscr{L}_{\mathscr{G}\mathscr{P}}$ and the corresponding complete logic $\Lambda^{\mathscr{L}_{\mathscr{G}\mathscr{P}}}$. To see this, consider an arbitrary formula φ. Working inside-out, consider (one of) the innermost occurrence of an action modality. By applying successively the relevant axioms listed above until only propositional letters are in the scope of that modality, its occurrence can be eliminated using either Axiom 6.53 or 6.59. This procedure can be iterated until φ is transformed into and equivalent formula φ' containing no action modalities. The completeness of the extended logic is therefore reduced to that of $\Lambda^{\mathscr{L}_{\mathscr{P}}}$. Thus, unlike in most cases of completeness proofs, the interesting part for dynamic logic is the soundness of the axioms! This is left as an exercise for the reader.

In the remainder of this paper, I show that GPL motivates new topics in dynamic logic.

Agent Promotion

Given that I base aggregation of preferences on a given hierarchy between agents, it seems natural to inquire what happens when the ranks of agents change in the hierarchy. Several reordering of group hierarchy are conceivable, but I focus my attention on an obvious first choice: putting an agent on top of the group. In the present section, I investigate a different and new kind of dynamics for group of agents, this time where the hierarchy inside the group is changed by upgrading an agent to become the master of the group. I call this action *agent promotion*, namely when an agent in a (sub)group is promoted to a higher rank.

I introduce some preliminary notations. The promotion of an an gent i in group X, simply written "i/X", is given by the graph X' whose hierarchy is the same as in X with $j < i$ added for every $j \in X$. If X does not contain i, then $X' = X$.[6] An illustration of a promotion is provided in Figure 6.5.

The action of promoting $i \in X$ in a model $\mathfrak{M} = \langle W, G, N, \{\preceq_X\}_{X \in G}, V\rangle$, denoted $\mathfrak{M} \uparrow i, X$ is given by the model $\mathfrak{M}' = \langle W, G', N, \{\preceq_X\}_{X \in G'}\rangle V$, where each graph $Y \in G$ that has non-empty intersection with X is replaced with the graph $Y' \in G'$ where i has been promoted in $X \cap Y$, as described above.

To talk about agent promotion in subgroups, I expand the language with a modality $\langle \uparrow i, X\rangle\varphi$, which should be read as "after promoting agent i in (sub)group X, φ is the case." The semantics of this new modality is given by:

$$\mathfrak{M}, u \models \langle \uparrow i, X\rangle\varphi \text{ iff } (\mathfrak{M} \uparrow i, X), u \models \varphi \tag{6.65}$$

[6] The notation just introduced is somewhat abusive, as I should write "$V(i)/X$", but I keep the original one for the sake of readability. I also illustrate graphs with nodes labeled with individual constants instead of variables.

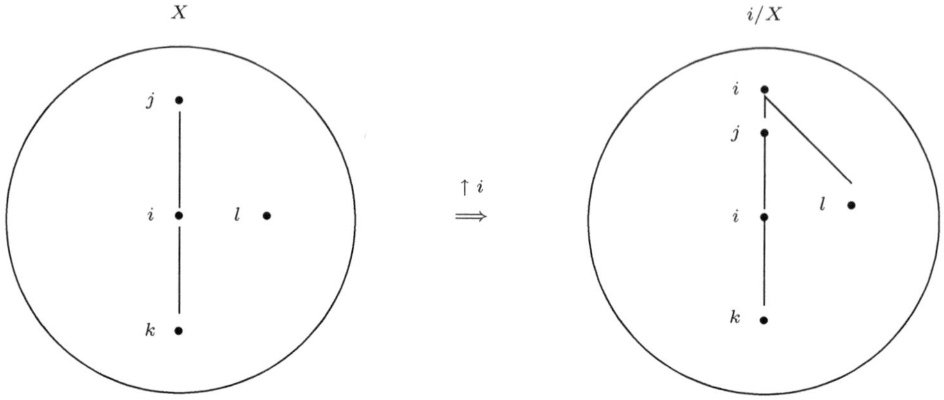

Fig. 6.5 Illustration of the promotion of an agent *i* inside a group *X*

For the axiomatization of agent promotion, I use reduction axioms viewed as syntactic relativizations. As van Benthem and Liu note in [15], "the reduction axioms for public announcement merely express the inductive facts about the modal assertion $\langle !\varphi\rangle\psi$ referring to the left-hand side, relating these on the right to relativization instructions creating $(\psi)^{\varphi}$" (p.171). On the basis of this standpoint, a reduction axiom may be seen as a syntactic relativization expressed in the principle:

$$\langle R := def(R)\rangle\langle R\rangle\varphi \leftrightarrow \langle def(R)\rangle\langle R := def(R)\rangle\varphi \tag{6.66}$$

In the case of agent promotion, I denote $def(R)$ by "$\uparrow i,X : Y$", standing for the substitution of the priority graph $i/(X \cap Y)$ for every occurrence of $X \cap Y$ in Y. Notice that $\uparrow i,X : Y$ is defined over the intersection of X and Y. There are thus 4 cases that may arise: (1) $X \subseteq Y$, (2) $Y \subset X$, (3) $X \cap Y \neq \varnothing$ and $X \cap Y = \varnothing$. The first case is depicted in Figure 6.6. The second case, $Y \subset X$ implies that $\uparrow i,X : Y = \uparrow i,Y : Y = \uparrow i,Y$ is the same as in Figure 6.5. The third case is depicted in Figure 6.7. The fourth case is obvious: if $X \cap Y = \varnothing$, promoting i in X has no effect on Y. The next definition provides a recursive construction of $\uparrow i,X : Y$[7]:

Definition 6.14.

$$\uparrow i,X : j = \begin{cases} i/j & \text{if } j \in X \\ j & \text{if } j \notin X \end{cases} \tag{6.67}$$

$$\uparrow i,X : (Y \parallel Z) = (\uparrow i,X : Y) \parallel (\uparrow i,X : Z) \tag{6.68}$$

$$\uparrow i,X : (Y/Z) = (\uparrow i,X : Y)/(\uparrow i,X : Z) \tag{6.69}$$

The following theorem provides a compositional analysis of this modality in the base group preference language.

[7] Thanks to Alexandru Baltag for suggesting Definition 6.14.

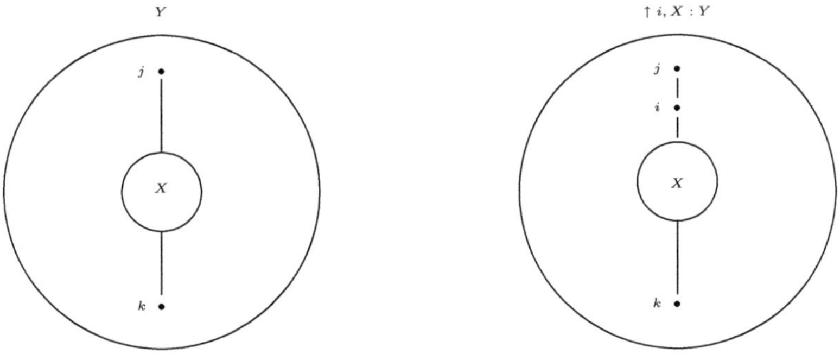

Fig. 6.6 Illustration of the promotion of an agent i inside a subgroup X of Y

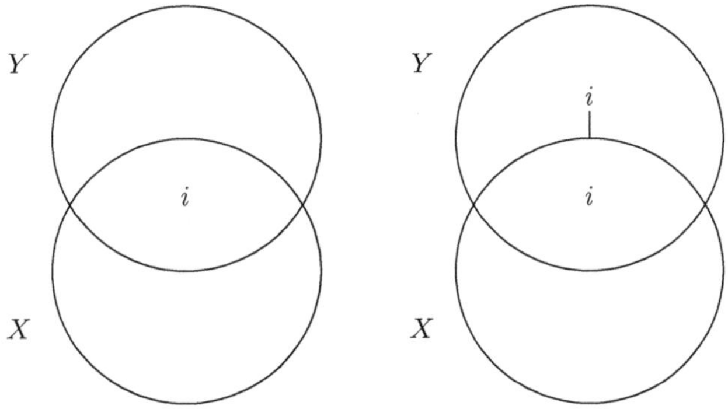

Fig. 6.7 Illustration of the promotion of an agent i inside X when $X \cap Y$ but neither $X \subseteq Y$ nor $Y \subseteq X$

Theorem 6.7. *The logic of group preference with public announcement, preference upgrade and agent promotion is given by (1) Λ_{GP}, (2) the reduction axioms of public announcement and preference upgrade of Theorem 6.6 and (3) the following reduction principles:*

$$\langle \uparrow i, X \rangle s \leftrightarrow s \tag{6.70}$$

$$\langle \uparrow i, X \rangle p \leftrightarrow p \tag{6.71}$$

$$\langle \uparrow i, X \rangle \neg \varphi \leftrightarrow \neg \langle \uparrow i, X \rangle \varphi \tag{6.72}$$

$$\langle \uparrow i, X \rangle (\varphi \vee \psi) \leftrightarrow \langle \uparrow i, X \rangle \varphi \vee \langle \uparrow i, X \rangle \varphi \tag{6.73}$$

$$\langle \uparrow i, X \rangle E \varphi \leftrightarrow E \langle \uparrow i, X \rangle \varphi \tag{6.74}$$

$$\langle \uparrow i, X \rangle \langle Y \rangle^{\leq} \varphi \leftrightarrow \langle \uparrow i, X : Y \rangle^{\leq} \langle \uparrow i, X \rangle \varphi \tag{6.75}$$

$$\langle \uparrow i, X \rangle \langle Y \rangle^{<} \varphi \leftrightarrow \langle \uparrow i, X : Y \rangle^{<} \langle \uparrow i, X \rangle \varphi \tag{6.76}$$

Proof. The soundness is immediate since the definition of promotion in models and Definition 6.14 are in a perfect match.

6.6 Conclusion

In this paper, I have shown how to extend Preference Logic to Group Preference Logic. For this, all that was required was to incorporate nominals into $\mathscr{L}_{\mathscr{P}}$. This addition to the language allowed two kinds of lifts: (1) from individual orders to group orders, using lexicographic upgrade and (2) from orders between states to preferences between propositions. I have also provided a complete axiomatization and shown how standard dynamic actions can be included in the logic via compositional analysis. Finally, I have investigated a new kind of action, this time acting on the group hierarchy, which I have called *promotion*. The innovations in this paper were to modalize the algebraic setting of [1], thus getting a modal logic for aggregating individual orders into group orders. This modalization of the algebra has also enabled the introduction of dynamics into the system as well as suggesting a new kind of dynamics, namely *promotion*.

As was mentioned in Section 6.2, one motivation for choosing ARS as the model of aggregation for Group Preference Logic is that it is non-dictatorial, yet satisfies interesting aggregation properties. From [1]'s results, however, lexicographic reordering is the *only* operator satisfying the conditions listed above. This then opens up questions as to the reliability of the lexicographic reordering and the quest for possibility results. Is giving more weight to certain agents in a society justifiable from a democratic point of view? What kind of voting procedures and responses to the opinions of the majority would obtain? Are we better to stay in a society where the aggregation of preferences is known to be non-uniform (because of Arrow's theorem)? But also, can we find other sets of conditions that are represented by operators which still yield possibility results?

I have kept the focus of my analysis in this paper to Preference Logic, but ARS could also be introduced in other settings, such as belief revision, especially with the dynamic logic approach of the latter as developed in [13]. Finally, an even more general standpoint on modalizing ARS could be taken so as to get a logic of graph manipulations. For a starter, the fundamental priority operators $X \parallel Y$ and X/Y can be seen as the operations on graphs of disjoint unions and *sequential composition*. But many more actions on graphs could be defined, and it would be interesting to see how this could be treated in a fashion similar to the one adopted in this paper for preference aggregation. I leave these issues for future research.

References

1. H. Andréka, M. Ryan, and P.-Y. Schobbens. Operators and laws for combining preference relations. *Journal of Logic and Computation*, 12(1):13–53, 2002.
2. C. Areces and B. ten Cate. Hybrid logic. In P. Blackburn, J. van Benthem, and F. Wolter, editors, *Handbook of Modal Logic*, volume 3 of *Studies in Logic and Practical Reasoning*, chapter 14, pages 821–868. Elsevier Science Inc., New York, NY, 2007.
3. K. Arrow. *Social Choice and Individual Value*. Wiley, New York, NY, 1951.

4. C. Boutilier. Toward a logic for qualitative decision theory. In J. Doyle, E. Sandewall, and P. Torasso, editors, *Principles of Knowledge Representation and Reasoning*, pages 75–86, Morgan Kaufmann, Bonn, May 24–27, 1994.
5. F. Dietrich and C. List. Arrow's theorem in judgment aggregation. *Social Choice and Welfare*, 29(1):19–33, July 2007.
6. H. Ganzinger, C. Meyer, and M. Veanes. The two-variable guarded fragment with transitive relations. *Logic in Computer Science, 1999. Proceedings. 14th Symposium on*, pages 24–34, 1999.
7. P. Girard. *Modal Logic for Belief and Preference Change*. PhD thesis, Stanford University, 2008.
8. P. Girard and J. Seligman. An analytic logic of aggregation. In *Logic and Its Applications*, volume 5378 of *Lecture Notes in Computer Science*, pages 146–161, 2009.
9. C. List and P. Pettit. Aggregating sets of judgments: an impossibility result. *Economics and Philosophy*, 18:89–110, April 2002.
10. M. Pauly and M. van Hees. Logical constraints on judgment aggregation. *Journal of Philosophical Logic*, 35:569–585, 2006.
11. K. Segerberg. *An Essay in Classical Modal Logic*, volume 13 of *Filosofiska Studier*. Filosofiska föreningen och Filosofiska institutionen vid Uppsala universitet, Uppsala, 1971.
12. B. ten Cate. *Model Theory for Extended Modal Languages*. PhD thesis, University of Amsterdam, ILLC Dissertation Series DS-2005-01, 2005.
13. J. van Benthem. Dynamic logic for belief revision. *Journal of Applied Non-Classical Logic*, 17(2):129–155, 2007.
14. J. van Benthem, P. Girard, and O. Roy. Everything else being equal: A modal logic for *ceteris paribus* preferences. *Journal of Philosophical Logic*, 38(1):83–125, February 2009.
15. J. van Benthem and F. Liu. The dynamics of preference upgrade. *Journal of Applied Non-Classical Logics*, 17(2):157–182, 2007.
16. J. van Benthem, S. van Otterloo, and O. Roy. Preference logic, conditionals, and solution concepts in games. In H. Lagerlund, S. Lindström, and R. Sliwinski, editors, *Modality Matters: Twenty-Five Essays in Honour of Krister Segerberg*, Uppsala Philosophical Studies, Uppsala, 2006.
17. H. van Ditmarsch, B. Kooi, and W. van der Hoek. *Dynamic Epistemic Logic*, volume 337 of *Synthese Library*. Springer, Heidelberg, 2007.

Chapter 7
No-Φ-Regret: A Connection between Computational Learning Theory and Game Theory

Amy Greenwald, Amir Jafari, and Casey Marks

7.1 Introduction

We analyze learning among a group of agents[1] playing an infinitely-repeated matrix game. At each stage, each agent chooses among a set of actions. The outcome, which is jointly determined by all the agents' choices, assigns a reward to each agent. A learning algorithm is a mapping from a history of past actions, outcomes, and rewards to a current choice of action. Our goal is to characterize the dynamics of multiple agents playing according to no-regret learning algorithms.

In the no-regret framework, the efficacy of learning is determined by comparing the performance of a learning algorithm to the performance of an alternative set of strategies. At each time t, we compare the action (i.e., pure strategy) a_t dictated by the learning algorithm with an alternative mixed strategy $\varphi(a_t)$. The function φ is called an action transformation. The agent's regret is the difference between the rewards obtained by playing action a_t and the rewards it would have expected to obtain had it instead played the transformed action $\varphi(a_t)$. Given a set Φ of action transformations, the Φ-regret vector (at time t) is the vector of regrets the agent experiences for not having played according to each $\varphi \in \Phi$. By definition, *no-Φ-regret* learning algorithms have the property that the time-averaged Φ-regret vector approaches the negative orthant.

For example, consider the set of all constant strategies. (A constant strategy always plays action a, for some action a.) Learning algorithms that perform at least

Amy Greenwald
Department of Computer Science, Brown University, Providence, RI 02912, USA,
e-mail: amy@cs.brown.edu
Amir Jafari
Mathematics Department, Duke University, Durham, NC 27708, USA,
e-mail: amir@math.duke.edu
Casey Marks
Five Hut Consulting, Providence, RI 02906, USA, e-mail: casey@fivehut.com

[1] In this paper, we use the terms "agent" and "player" interchangeably.

J. van Benthem et al. (eds.), *Games, Norms and Reasons*, Synthese Library 353,
DOI 10.1007/978-94-007-0714-6_7, © Springer Science+Business Media B.V. 2011

as well as this strategy set are said to exhibit *no external regret* [18]. As another example, consider a strategy that is identical to the strategy dictated by the learning algorithm, except that every play of action a suggested by the learning algorithm is replaced by action a', for some a and a'. Learning algorithms that perform at least as well as all such strategies are said to exhibit *no internal regret* [12]. The following results are well-known (see, for example, Hart and Mas-Colell [19, 20]): In two-player, zero-sum, repeated games, if each player plays using a no-external-regret learning algorithm, then each player's empirical distribution of play converges to his set of minimax strategies.[2] In general-sum, repeated games, if each player plays using a no-internal-regret learning algorithm, then the empirical distribution of play converges to the set of correlated equilibria.

In this paper, we define a general class of no-regret learning algorithms, called no-Φ-regret learning algorithms, which spans the spectrum from no external regret to no internal regret, and beyond. The set Φ describes a set of strategies into which the play of a given learning algorithm is transformed. Such a learning algorithm satisfies no-Φ-regret if no regret is experienced for playing as the algorithm prescribes, rather than playing according to any of the transformations of the algorithm's play prescribed by the elements of Φ. The existence of no-Φ-regret learning algorithms is established here (and elsewhere [3, 15, 23]), for all finite Φ.

Analogously, we define a class of game-theoretic equilibria, called $\vec{\Phi}$-equilibria, for $\vec{\Phi} = (\Phi_i)_{1 \le i \le n}$. An important contribution of this paper is to show that the empirical distribution of play of no-Φ_i-regret algorithms converges to the set of $\vec{\Phi}$-equilibria. We obtain as corollaries of our theorem the aforementioned results on convergence of no-external-regret learning (no-internal-regret learning) to the set of minimax equilibria (correlated equilibria) in zero-sum (general-sum) games. Furthermore, we establish a necessary condition for convergence to the set of $\vec{\Phi}$-equilibria, namely that the time-averaged Φ_i-regret experienced by each agent i approaches the negative orthant.

This work was originally motivated by an attempt to design a no-regret learning scheme that would converge to a tighter solution concept than the set of correlated equilibria (e.g., the convex hull of the set of Nash equilibria). We imagined that by comparing an agent's play to a larger set of alternative strategies, we could perhaps design a more powerful algorithm than no-internal-regret learning. Perhaps surprisingly, we find that the strongest form of no-Φ-regret algorithms are no-internal-regret algorithms. Consequently, the tightest game-theoretic solution concept to which no-Φ-regret algorithms converge is correlated equilibrium.

This paper is organized as follows. In the next section, we formally define no-Φ-regret learning and we show that no external regret and no internal regret are special cases of no Φ-regret. We prove our existence theorem in Section 7.3. In Section 7.4, we define the notion of $\vec{\Phi}$-equilibrium; then, in Section 7.5, we establish necessary

[2] In fact, Hart and Mas-Colell [20] establish this convergence result for what they call "better-play" algorithms (see Section 7.7), but their proof extends immediately to the class of no-external-regret learning algorithms.

and sufficient conditions for convergence to the set of $\overrightarrow{\Phi}$-equilibria. In Section 7.6, we show that no internal regret is the strongest form of no Φ-regret.

7.2 Φ-Regret

Our goal in this section is to define no-Φ-regret learning in the framework of Blackwell's approachability theory. We start by reviewing Blackwell's framework, and stating the variant of Blackwell's theorem on which our existence theorem is based (see Greenwald et al. [16] for details). Next, we introduce the notion of an action transformation: a mapping from a pure strategy to a mixed strategy. Finally, for finite sets Φ of action transformations, we formally define no-Φ-regret learning.

7.2.1 Blackwell's Approachability Theory

Consider an agent with a finite set of actions A playing a game against a set of opponents who play actions in the (arbitrary) joint action space A'. (The opponents' joint action space can be interpreted as the product of independent action sets.) Associated with each possible outcome is a vector given by the function $\rho : A \times A' \to V$, where V is a vector space over $\mathbb{R}$ with an inner product $\cdot$ and a distance metric d defined by the inner product in the standard manner: i.e., $d(x,y) = \|x-y\|_2$, for all $x,y \in V$.

A *vector-valued game* is a 4-tuple $\Gamma = (A, A', V, \rho)$. We study *infinitely-repeated* vector-valued games Γ^∞ in which the agent interacts with its opponents repeatedly and indefinitely. Recall that the agent's action set A is assumed to be finite. We denote by $\Delta(A)$ the set of probability distributions over the set A, and we allow agents to play *mixed strategies*, which means that rather than selecting an action $a \in A$ to play at each round, the agent selects a probability distribution $q \in \Delta(A)$. More specifically, an arbitrary round t (for $t \geq 1$) proceeds as follows:

(a) the agent selects a mixed strategy $q_t \in \Delta(A)$,
(b) the agent plays an action $a_t \in A$ (which is sampled according to the distribution q_t);
simultaneously, the opponents play action a'_t
(c) the agent observes reward vector $\rho(a_t, a'_t) \in V$

Given an infinitely-repeated vector-valued game Γ^∞ the *set of action histories of length t*, for $t \geq 0$, is denoted by H_t. For $t \geq 1$, H_t is given by $(A \times A')^t$: e.g., $h = \{a_\tau, a'_\tau\}_{\tau=1}^t \in H_t$. The set H_0 is defined to be a singleton. Given an infinitely-repeated vector-valued game Γ^∞, a *learning algorithm* is a sequence of functions $\mathscr{L} = \{L_t\}_{t=1}^\infty$, where $L_t : H_{t-1} \to \Delta(A)$.

Salient examples of learning algorithms include the best-reply heuristic [8] and fictitious play [4, 27]. At time t, the former returns an element of $\Delta(A)$ that maximizes the agent's rewards with respect to only a'_{t-1}, while the latter returns an element of $\Delta(A)$ that maximizes the agent's rewards with respect to the empirical distribution of play through time $t-1$.

We are interested in the properties of learning algorithms employed by an agent playing an infinitely-repeated vector-valued game Γ^∞. Given a learning algorithm $\mathscr{L} = \{L_t\}_{t=1}^\infty$ together with an opposing algorithm $\mathscr{L}' = \{L'_t\}_{t=1}^\infty$, we define a probability space whose universe consists of sequences of joint actions and whose measure can be defined inductively:

$$P[a_t = \alpha \mid a_\tau = \alpha_\tau, \forall \tau = 1,\ldots,t-1] = L_t((\alpha_1,a'_1),\ldots,(\alpha_{t-1},a'_{t-1}))(\alpha) \quad (7.1)$$

for all $\alpha \in A$. (The probabilities on opposing actions are defined analogously.)

In this probability space, we define two sequences of random variables: cumulative rewards $R_t = \sum_{\tau=1}^t \rho(a_\tau, a'_\tau)$ and average rewards $\overline{\rho}_t = \frac{R_t}{t}$.

Now, following Blackwell, we define the notion of approachability as follows:

Definition 7.1 (Approachability). Given an infinitely-repeated vector-valued game Γ^∞, a set $U \subseteq V$, and a learning algorithm $\mathscr{L}$, the set U is said to be *approachable* by $\mathscr{L}$, if for all $\varepsilon > 0$, there exists t_0 such that for any opposing learning algorithm $\mathscr{L}'$, $P[\exists t \geq t_0 \text{ s.t. } d(U, \overline{\rho}_t) \geq \varepsilon] < \varepsilon$.

Hence, if a learning algorithm $\mathscr{L}$ approaches a set $U \subseteq V$, then $d(U, \overline{\rho}_t) \to 0$ almost surely.

The following theorem [16, 22] gives a sufficient condition for the negative orthant, that is, the set $\mathbb{R}^d_- = \{x \in \mathbb{R}^d \mid x_i \leq 0, \text{ for all } 1 \leq i \leq d\} \subseteq \mathbb{R}^d$, to be approachable by a learning algorithm $\mathscr{L}$ in an infinitely-repeated vector-valued game $(A, A', \mathbb{R}^d, \rho)^\infty$ where $d \in \mathbb{N}$ and $\rho(A \times A')$ is bounded. For $x \in \mathbb{R}^d$, define x^+ by $(x^+)_i = \max\{x_i, 0\}$, for all $1 \leq i \leq d$.

Theorem 7.1 (Jafari [22]). *Let $(A, A', \mathbb{R}^d, \rho)^\infty$ be an infinitely-repeated vector-valued game with $d \in \mathbb{N}$ and $\rho(A \times A')$ bounded and a learning algorithm $\mathscr{L} = \{L_t\}_{t=1}^\infty$, the negative orthant $\mathbb{R}^d_- \subseteq \mathbb{R}^d$ is approachable by $\mathscr{L}$ if there exists a constant $c \in \mathbb{R}$ such that for all times $t \geq 1$, for all action histories $h \in H_{t-1}$ of length $t-1$, and for all opposing actions a',*

$$(R_{t-1}(h))^+ \cdot \rho(L_t(h), a') \leq c \quad (7.2)$$

where $R_t(h) \equiv \sum_{\tau=1}^t \rho(a_\tau, a'_\tau)$ and $\rho(q, a') \equiv \mathrm{e}_{a \sim q}[\rho(a, a')]$.

Blackwell's seminal approachability theorem provides a sufficient condition to ensure that, in a vector-valued repeated game, a learner's average rewards approach any closed set $U \subseteq \mathbb{R}^n$ [2, 20]. To prove existence of no-Φ-regret algorithms, we rely on Theorem 7.1, a close cousin of Blackwell's theorem. On the one hand, our theorem specializes Blackwell's theorem: it provides a sufficient condition for the negative orthant $\mathbb{R}^n_- \subseteq \mathbb{R}^n$ to be approachable, rather than an arbitrary closed subset of Euclidean space. On the other hand, our sufficient condition (Equation (7.2)) is weaker than Blackwell's original condition: our condition need only hold for some

$c \in \mathbb{R}$, rather than precisely for $c = 0$. Moreover, in our framework, the opponents (i.e., not the learner) have at their disposal an arbitrary, rather than merely a finite, set of actions.

7.2.2 Action Transformations

An *action transformation* is a function $\varphi : A \to \Delta(A)$.[3] Let $\Phi_{\text{ALL}}(A)$ denote the set of all action transformations over the set A. Following Blum and Mansour [3], we let $\Phi_{\text{SWAP}}(A) \subseteq \Phi_{\text{ALL}}(A)$ denote the set of all action transformations that map actions to distributions with all their weight on a single action (i.e., pure strategies).

There are two well-studied subsets of $\Phi_{\text{SWAP}}(A)$, namely, external and internal action transformations. Let $\delta_a \in \Delta(A)$ denote the distribution with all its weight on action a. An *external* action transformation is simply a constant transformation, so for $a \in A$,

$$\varphi_{\text{EXT}}^{(a)} : x \mapsto \delta_a, \quad \text{for all } x \in A \tag{7.3}$$

An *internal* action transformation behaves like the identity, except on one particular input, so for $a, b \in A$

$$\varphi_{\text{INT}}^{(a,b)} : x \mapsto \begin{cases} \delta_b & \text{if } x = a \\ \delta_x & \text{otherwise} \end{cases} \tag{7.4}$$

The set of external and internal action transformations are denoted by $\Phi_{\text{EXT}}(A)$ and $\Phi_{\text{INT}}(A)$, respectively. Observe that $|\Phi_{\text{INT}}(A)| = |A|^2 - |A| + 1$ and $|\Phi_{\text{EXT}}(A)| = |A|$.

We can represent an action transformation by a stochastic matrix. Given $\varphi \in \Phi_{\text{ALL}}(A)$ and an enumeration of A, we define its matrix representation $[\varphi]$ as:

$$[\varphi]_{ij} = \varphi(a_i)(a_j) \tag{7.5}$$

where a_k is the kth action in the enumeration. For example, for $A = \{1,2,3,4\}$, the action transformations $\varphi_{\text{EXT}}^{(2)}$ and $\varphi_{\text{INT}}^{(23)}$ can be represented as:

$$[\varphi_{\text{EXT}}^{(2)}] = \begin{bmatrix} 0&1&0&0 \\ 0&1&0&0 \\ 0&1&0&0 \\ 0&1&0&0 \end{bmatrix} \qquad [\varphi_{\text{INT}}^{(23)}] = \begin{bmatrix} 1&0&0&0 \\ 0&0&1&0 \\ 0&0&1&0 \\ 0&0&0&1 \end{bmatrix}.$$

[3] Some researchers consider transformations to be functions $\varphi : \Delta(A) \to \Delta(A)$. This alternative definition is applicable in the study of *distribution* regret. The distinction between distribution and *action* regret is discussed in Section 7.7.

7.2.3 No Φ-Regret Learning

A *real-valued* game (A, A', R, r) is a vector-valued game with $R \subseteq \mathbb{R}$. Given a real-valued game $\Gamma = (A, A', R, r)$ and a set of action transformations $\Phi \subseteq \Phi_{\text{ALL}}(A)$, we define the (vector-valued) *Φ-regret game* as $\Gamma^{\Phi} = (A, A', \mathbb{R}^{\Phi}, \rho^{\Phi})$, with the vector-valued function $\rho^{\Phi} : A \times A' \to \mathbb{R}^{\Phi}$ given by[4]:

$$\rho^{\Phi}(a, a') \equiv \left(\rho^{\varphi}(a, a')\right)_{\varphi \in \Phi} \tag{7.6}$$

where

$$\rho^{\varphi}(a, a') = r(\varphi(a), a') - r(a, a') \tag{7.7}$$

and $r(q, a') \equiv \mathrm{e}_{a \sim q}[r(a, a')]$, for all $q \in \Delta(A)$. In words, the φth entry in the regret vector $\rho^{\Phi}(a, a')$ describes the difference between the rewards the agent obtains by playing action a and the rewards the agent would have expected to obtain by playing the mixed strategy $\varphi(a)$ instead, given opposing action a'. We now define no-Φ-regret learning in Blackwell's approachability framework:

Definition 7.2 (No-Φ-Regret Learning). Given a real-valued game $\Gamma = (A, A', R, r)$ and a finite set of action transformations $\Phi \subseteq \Phi_{\text{ALL}}(A)$, a *no-$\Phi$-regret* learning algorithm $\mathscr{L}$ is one that approaches the negative orthant $\mathbb{R}^{\Phi}_{-} \subseteq \mathbb{R}^{s}$ in the infinitely-repeated Φ-regret game $\left(\Gamma^{\Phi}\right)^{\infty}$: i.e., for all $\varepsilon > 0$, there exists t_0 such for any opposing learning algorithm $\mathscr{L}'$,

$$P\left[\exists t \geq t_0 \ s.t. \ d\left(\mathbb{R}^{\Phi}_{-}, \overline{\rho}^{\Phi}_{t}\right) \geq \varepsilon\right] < \varepsilon \tag{7.8}$$

In words, if an agent plays an infinitely-repeated game Γ^{∞} as prescribed by a no-Φ-regret learning algorithm, then the time-averaged Φ-regret experienced by the agent uniformly converges to the negative orthant with probability 1, regardless of the opposing learning algorithm.

7.3 Existence of No-Φ-Regret Learning Algorithms

Indeed, no-Φ-regret learning algorithms exists. In particular, no-external-regret algorithms pervade the literature. The earliest date back to Blackwell [2] and Hannan [18]; but, more recently, Littlestone and Warmuth [24], Freund and Schapire [13], Herbster and Warmuth [21], Fudenberg and Levine [14], Foster and Vohra [10], Hart and Mas-Colell [20], and others have studied such algorithms. Foster and Vohra [12] were the first to design an algorithm exhibits no-internal-regret.

Our first theorem establishes the existence of no-Φ-regret learning algorithms, for all finite Φ.

[4] Given two set X and Y, the notation X^{Y} denotes the set of functions $\{f : Y \to X\}$. Note that, if Y is finite, then $\mathbb{R}^{Y}$ is isomorphic to $\mathbb{R}^{|Y|}$.

Theorem 7.2 (Existence). *Given a real-valued game $\Gamma = (A, A', R, r)$ with $R \subseteq \mathbb{R}$ bounded, for all finite sets of action transformations $\Phi \subseteq \Phi_{ALL}(A)$, there exists a no-$\Phi$-regret learning algorithm: i.e., one that approaches $\mathbb{R}_-^{\Phi}$ in the infinitely-repeated Φ-regret game Γ_Φ^∞.*

Proof. By Theorem 7.1, it suffices to show that there exists a learning algorithm $\mathscr{L} = \{L_t\}_{t=1}^\infty$ and a constant $c \in \mathbb{R}$ such that for all times $t \geq 1$, for all action histories $h \in H^{t-1}$ of length $t-1$, and for all opposing actions a',

$$(R_{t-1}^{\Phi}(h))^+ \cdot \rho^{\Phi}(L_t(h), a') \leq c \tag{7.9}$$

where $R_t^{\Phi}(h) = \sum_{\tau=1}^{t} \rho^{\Phi}(a_\tau, a'_\tau)$ and $\rho^{\Phi}(q, a') = \mathrm{e}_{a\sim q}[\rho^{\Phi}(a, a')]$.

Case 1. $R_{t-1}^{\Phi}(h) \in \mathbb{R}_-^{\Phi}$: If $R_{t-1}^{\Phi}(h) \in \mathbb{R}_-^{\Phi}$, so that $(R_{t-1}^{\Phi}(h))^+ = 0$, Equation (7.9) holds for $c = 0$.

Case 2. $R_{t-1}^{\Phi}(h) \notin \mathbb{R}_-^{\Phi}$: We show that for all $x^{\Phi} \notin \mathbb{R}_-^{\Phi}$ there exists $q \equiv q(x^{\Phi}) \in \Delta(A)$ such that for all $a' \in A'$, $\left(x^{\Phi}\right)^+ \cdot \rho^{\Phi}(q, a') = 0$. Then, letting $L_t(h) = q(R_{t-1}^{\Phi}(h))$, Equation (7.9) holds for $c = 0$.

Let q be the (row) vector representation of a mixed strategy.

$$\begin{aligned}
0 &= \left(x^{\Phi}\right)^+ \cdot \rho^{\Phi}(q, a') && (7.10)\\
&= \sum_{\varphi\in\Phi} (x^{\varphi})^+ \left(r(q[\varphi], a') - r(q, a')\right) && (7.11)\\
&= \sum_{\varphi\in\Phi} (x^{\varphi})^+ \left(\sum_{a\in A} r(a, a')(q[\varphi])_a - \sum_{a\in A} r(a, a') q_a\right) && (7.12)\\
&= \sum_{a\in A} r(a, a') \left[\left(q \sum_{\varphi\in\Phi} (x^{\varphi})^+ [\varphi]\right)_a - \left(q \sum_{\varphi\in\Phi} (x^{\varphi})^+\right)_a\right] && (7.13)
\end{aligned}$$

Equation (7.11) follows from the definitions of the inner product and ρ^{Φ}. Equation (7.12) follows from the definition of expectation. Equation (7.13) follows via algebra.

Now it suffices to show the following:

$$q \sum_{\varphi\in\Phi} (x^{\varphi})^+ [\varphi] = q \sum_{\varphi\in\Phi} (x^{\varphi})^+ \tag{7.14}$$

Define the matrix M as follows:

$$M = \frac{\sum_{\varphi\in\Phi} (x^{\varphi})^+ [\varphi]}{\sum_{\varphi\in\Phi} (x^{\varphi})^+} \tag{7.15}$$

Since M is a convex combination of stochastic matrices, M itself is a stochastic matrix with at least one fixed point with non-negative entries that sum to 1. Any algorithm for computing such a fixed point of M gives rise to a no-Φ-regret learning algorithm.

Algorithm 1 No-Regret Learning Algorithm $((A,A',R,r),\Phi \subseteq \Phi_{\text{ALL}}(A))$

1: initialize $x_0 = 0$ and q_0 to be an arbitrary distribution over A
2: **for** $t = 1,2,\ldots,$ **do**
3:   sample pure action $a \sim q_t$
4:   choose opposing actions $a'_t \in A'$
5:   observe reward vector $r_t = r(\cdot, a'_t) \in R^A$
6:   **for all** $\varphi \in \Phi$ **do**
7:     compute instantaneous regret $y_t^\varphi = r_t \cdot e_a[\varphi] - r_t \cdot e_a$
8:     update cumulative regret vector $x_t^\varphi = x_{t-1}^\varphi + y_t^\varphi$
9:   **end for**
10:   **if** $(x_t^\Phi)^+ = 0$ **then**
11:     set $q_{t+1} \in \Delta(A)$ arbitrarily
12:   **else**
13:     let $M_t = \sum_{\varphi \in \Phi} (x_t^\varphi)^+ [\varphi] / \sum_{\varphi \in \Phi} (x_t^\varphi)^+$
14:     solve for a fixed point $q_{t+1} = q_{t+1} M_t$
15:   **end if**
16: **end for**

Algorithm 1 lists the steps in the no-Φ-regret learning algorithm derived in the proof of the existence theorem. At time t, the agent plays the mixed strategy q_t by sampling a pure action a according to the distribution q_t, after which it observes an $|A|$-dimensional reward vector r_t, where $(r_t)_a = r(a, a'_t)$, assuming a'_t is the opponents' pure action vector at time t. Given this reward vector, the agent computes its instantaneous regret in all dimensions $\varphi \in \Phi$: specifically, $\rho^\varphi(a_t, a'_t) = r(\varphi(a_t), a'_t) - r(a_t, a'_t)$, which, since $r(q, a')$ is an expectation, we compute via dot products in Step 7. The cumulative regret vector is then updated accordingly, after which its positive part is extracted. If this quantity is zero, then the algorithm outputs an arbitrary mixed strategy. Otherwise, a fixed point of the stochastic matrix M derived in Equation (7.15) is returned.

Complexity. Each iteration of Algorithm 1 has time complexity $O(\max\{|\Phi||A|^2, |A|^3\})$. Updating the cumulative regret vector in steps 6–9 takes time $O(|\Phi||A|)$, since computing instantaneous regret for each $\varphi \in \Phi$ (step 7) is an $O(|A|)$ operation. Computing the stochastic matrix M in step 13 takes time $O(|\Phi||A|^2)$, since each matrix $[\varphi]$ has dimensions $|A| \times |A|$. Finding the fixed point of an $n \times n$ stochastic matrix (step 14) takes $O(n^3)$ time.

If, however, $\Phi \subseteq \Phi_{\text{SWAP}}(A)$, then the time complexity reduces to $O(\max\{|\Phi||A|, |A|^3\})$, since in this case, (i) computing instantaneous regret for each $\varphi \in \Phi$ (Step 7) takes constant time so that updating the cumulative regret vector takes time $O(|\Phi|)$; and (ii) computing the stochastic matrix M in Step 13 is only an $O(|\Phi||A|)$ operation, since there are only $|A|$ nonzero entries in each $\varphi \in \Phi$. In particular, if $\Phi = \Phi_{\text{INT}}(A)$, then the time complexity reduces to $O(|A|^3)$, because $|\Phi_{\text{INT}}(A)| = O(|A|^2)$. Moreover, if $\Phi = \Phi_{\text{EXT}}(A)$, then the time complexity reduces even further to $O(|A|)$, because matrix manipulation is not required in the special case of no-external-regret learning. The rows of M are constant in this case: each is a copy of the (normalized) cumulative regret vector, which is precisely the fixed point of M.

The space complexity of Algorithm 1 is $O(|\Phi||A|^2) = O(\max\{|\Phi||A|^2, |A|^2\})$ because it is necessary to store $|\Phi|$ matrices, each with dimensions $|A| \times |A|$, and computing the fixed point of an $|A| \times |A|$ stochastic matrix (via Gaussian elimination) requires $O(|A|^2)$ space. If, however, $\Phi \subseteq \Phi_{\text{SWAP}}(A)$, then the space complexity reduces to $O(\max\{|\Phi||A|, |A|^2\})$, since, in this case, there are only $|A|$ nonzero entries in each $\varphi \in \Phi$. In particular, if $\Phi = \Phi_{\text{INT}}(A)$ then the space complexity reduces to $O(|A|^2)$, since it suffices to store cumulative regrets in a matrix of size $|A| \times |A|$. Similarly, if $\Phi = \Phi_{\text{EXT}}(A)$, then the space complexity reduces to $O(|A|)$, since it suffices to store cumulative regrets in a vector of size $|A|$. The above discussion of the time and space complexity of Algorithm 1 is summarized in Table 7.1.

	Time	Space
$\Phi \subseteq \Phi_{\text{ALL}}$	$O(\max\{|\Phi||A|^2, |A|^3\})$	$O(|\Phi||A|^2)$
$\Phi \subseteq \Phi_{\text{SWAP}}$	$O(\max\{|\Phi||A|, |A|^3\})$	$O(\max\{|\Phi||A|, |A|^2\})$
$\Phi = \Phi_{\text{INT}}$	$O(|A|^3)$	$O(|A|^2)$
$\Phi = \Phi_{\text{EXT}}$	$O(|A|)$	$O(|A|)$

Table 7.1 Complexity of no-Φ-regret learning

7.4 $\overrightarrow{\Phi}$-Equilibria

In this section, we define the notion of $\overrightarrow{\Phi}$-equilibria, of which correlated, Nash, and minimax equilibria are all special cases. We show that the set of $\overrightarrow{\Phi}$-equilibria is convex, for all $\overrightarrow{\Phi}$.

In a (real-valued) *n-player game* $\Gamma_n = \langle (A_i, r_i)_{1 \leq i \leq n} \rangle$, each player i chooses an action from the finite set A_i, and the rewards earned by player i are determined by the function $r_i : A_1 \times \ldots \times A_n \to \mathbb{R}$. We abbreviate action profile $(a_1, \ldots, a_n)$ by $(a_i, a_{-i}) \in A_i \times \prod_{j \neq i} A_j$, or simply by $a \in \prod A_j$.

Definition 7.3 ($\overrightarrow{\Phi}$-Equilibrium). Given an n-player game $\Gamma_n = \langle (A_i, r_i)_{1 \leq i \leq n} \rangle$ and a vector $\overrightarrow{\Phi} = \langle \Phi_i \rangle_{1 \leq i \leq n}$ such that $\Phi_i \subseteq \Phi_{\text{ALL}}(A_i)$ for $1 \leq i \leq n$, an element $q \in \Delta(A_1 \times \ldots \times A_n)$ is called a $\overrightarrow{\Phi}$-equilibrium iff $\mathrm{e}_{a \sim q}[\rho^{\varphi}(a)] \leq 0$, for all players i and for all $\varphi \in \Phi_i$.

If for all players i, each Φ_i is of the same type, e.g., $\Phi_i = \Phi_{\text{EXT}}(A_i)$, then we refer to the $\overrightarrow{\Phi}$-equilibrium accordingly, e.g., Φ_{EXT}-equilibrium.

7.4.1 Examples of $\overrightarrow{\Phi}$-Equilibria

Correlated, Nash, and minimax equilibria are all special cases of $\overrightarrow{\Phi}$-equilibria.

Correlated Equilibrium. Given an n-player game $\Gamma_n = \langle (A_i, r_i)_{1\le i\le n}\rangle$, let $\Phi_i = \Phi_{\text{INT}}(A_i)$ for all players i. The joint distribution $q \in \Delta(A_1 \times \ldots \times A_n)$ is a Φ_{INT}-equilibrium if and only if for all players i and for all $\alpha, \beta \in A_i$,

$$0 \ge \mathrm{e}_{a\sim q}\left[r\left(\varphi_{\text{INT}}^{(\alpha,\beta)}(a_i), a_{-i}\right) - r(a)\right] \tag{7.16}$$

$$= \sum_{a\in\prod A_i} q(a)\left(r\left(\varphi_{\text{INT}}^{(\alpha,\beta)}(a_i), a_{-i}\right) - r(a)\right) \tag{7.17}$$

$$= \sum_{a_{-i}\in A_{-i}} \left[q(\alpha, a_{-i})\left(r(\beta, a_{-i}) - r(a)\right) + \sum_{a_i\neq\alpha} q(a_i, a_{-i})\left(r(a_i, a_{-i}) - r(a)\right)\right] \tag{7.18}$$

$$= \sum_{a_{-i}\in A_{-i}} q(\alpha, a_{-i})\left(r(\beta, a_{-i}) - r(a)\right) \tag{7.19}$$

which is precisely the definition of correlated equilibrium [1].

Coarse Correlated Equilibrium. Given an n-player game $\Gamma_n = \langle (A_i, r_i)_{1\le i\le n}\rangle$, let $\Phi_i = \Phi_{\text{EXT}}(A_i)$ for all players i. The joint distribution $q \in \Delta(A_1 \times \ldots \times A_n)$ is a Φ_{EXT}-equilibrium if and only if for all players i and for all $\alpha \in A_i$,

$$0 \ge \mathrm{e}_{a\sim q}\left[r\left(\varphi_{\text{EXT}}^{(\alpha)}(a_i), a_{-i}\right) - r(a)\right] \tag{7.20}$$

$$= \mathrm{e}_{a\sim q}\left[r(\alpha, a_{-i}) - r(a)\right] \tag{7.21}$$

$$= \mathrm{e}_{a\sim q}\left[r(\alpha, a_{-i})\right] - r(q) \tag{7.22}$$

which is the definition of coarse correlated equilibrium (also called weak correlated equilibrium) [25].

Zero-Sum Games. A coarse correlated equilibrium need not be a correlated equilibrium. This observation is intuitive for general-sum games, but perhaps less so for zero-sum games.

For example, in the following zero-sum game, with row as maximizer and column as minimizer, the joint distribution with half its weight on (T,L) and the other half on (B,M) is a coarse correlated equilibrium, but not a correlated equilibrium. It is a coarse correlated equilibrium because row has no incentive to deviate from its marginal distribution (half its weight on T and half on B), and column has no incentive to deviate from its marginal distribution (half its weight on L and half on M). If column were to deviate to R, it would expect to lose $\frac{1}{2}$ instead of 0. It is not, however, a correlated equilibrium: if column is advised to play L, then row is playing T, in which case column actually prefers to play R, where it would win 1 instead of 0.

In the case of two-player, zero-sum games, we obtain the following result for coarse correlated equilibria (and consequently correlated equilibria), which is related to the result in Forges [9]:

Proposition 7.1. *Given a two-player, zero-sum game Γ with reward function r and value v. If q is a coarse correlated equilibrium, then (i) $r(q) = v$ and (ii) each*

	L	M	R
T	0	0	−1
B	0	0	2

Fig. 7.1 Sample zero-sum game

player's marginal distribution is an optimal strategy (i.e., optimal for the maximizing player means: guarantees he wins at least v; optimal for the minimizing player means: guarantees he loses at most v).

Proof. Let q_1 and q_2 denote the marginal distributions of the maximizer and the minimizer in q, respectively. First, $r(q) \geq \max_{\alpha \in A_1} r(\alpha, q_2) \geq v$ since q is a coarse correlated equilibrium and v is the value of the game. Symmetrically, $r(q) \leq \max_{\beta \in A_2} r(q_1, \beta) \leq v$. Hence, $r(q) = v$.

Second, applying the definition of coarse correlated equilibrium again together with the above result, $v = r(q) \geq \max_{\alpha \in A_1} r(\alpha, q_2)$, so by playing q_2, player 2 loses at most v. Symmetrically, $v = r(q) \leq \max_{\beta \in A_2} r(q_1, \beta)$, so by playing q_1, player 1 wins at least v.

Note that the sets of coarse correlated equilibria and minimax equilibria need not coincide, since the former allows for correlations while the latter does not. For this reason, we refer to coarse correlated equilibria in two-player, zero-sum games as *generalized* minimax equilibria.

Nash Equilibrium. Given an n-player game $\Gamma_n = \langle (A_i, r_i)_{1 \leq i \leq n} \rangle$, a Nash equilibrium [26] is an independent element $q \in \Delta(A_1 \times \ldots \times A_n)$ such that $r(q) \geq r(q_1, \ldots, a_i, \ldots, q_n)$, for all players i and for all actions $a_i \in A_i$. An element $q \in \Delta(A_1 \times \ldots \times A_n)$ is called *independent* if it can be written as the product of n independent elements $q_i \in \Delta(A_i)$: i.e., $q = q_1 \times \ldots \times q_n$. Thus, by definition, a Nash equilibrium is an independent coarse correlated equilibrium. However, a Nash equilibrium is also an independent correlated equilibrium. Therefore, the set of independent coarse correlated equilibria and independent correlated equilibria coincide.

7.4.2 Properties of $\overrightarrow{\Phi}$-Equilibrium

Next we discuss two convexity properties of the set of $\overrightarrow{\Phi}$-equilibria.

Proposition 7.2. *Given an n-player game* $\Gamma_n = \langle (A_i, r_i)_{1 \leq i \leq n} \rangle$ *and a vector* $\overrightarrow{\Phi} = \langle \Phi_i \rangle_{1 \leq i \leq n}$ *such that* $\Phi_i \subseteq \Phi_{\text{ALL}}(A_i)$ *for* $1 \leq i \leq n$, *the set of* $\overrightarrow{\Phi}$*-equilibria is convex.*

Proof. Let $q, q' \in \Delta(A_1 \times \ldots \times A_n)$ be $\overrightarrow{\Phi}$-equilibria. For arbitrary $\lambda \in [0,1]$, let $q^* = \lambda q + (1-\lambda) q'$. Then, for all players i and for all $\varphi \in \Phi_i$,

$$\begin{aligned}
\mathrm{e}_{a\sim q^*}\left[\rho^{\varphi}(a)\right] &= \sum_{a\in A} q^*(a)\rho^{\varphi}(a) && (7.23)\\
&= \sum_{a\in A}\left(\lambda q(a)+(1-\lambda)q'(a)\right)\rho^{\varphi}(a) && (7.24)\\
&= \lambda\sum_{a\in A} q(a)\rho^{\varphi}(a)+(1-\lambda)\sum_{a\in A} q'(a)\rho^{\varphi}(a) && (7.25)\\
&= \lambda\mathrm{e}_{a\sim q}\left[\rho^{\varphi}(a)\right]+(1-\lambda)\mathrm{e}_{a\sim q'}\left[\rho^{\varphi}(a)\right] && (7.26)\\
&\leq 0 && (7.27)
\end{aligned}$$

Thus, q^* is a $\vec{\Phi}$-equilibrium.

Given a set of actions A, let I denote the identity map: i.e., $I(a)=\delta_a$ for all $a\in A$.

Definition 7.4. Given a set of actions A and a set of action transformations $\Phi \subseteq \Phi_{\text{ALL}}(A)$, we define the *super convex hull* of Φ, denoted $\text{SCH}(\Phi)$, as follows:

$$\text{SCH}(\Phi)=\left\{\left(\sum_{j=1}^{k}\alpha_j\varphi_j\right)+\beta I \;\middle|\; k\in\mathbb{N},\ \varphi_j\in\Phi,\ \alpha_j\geq 0,\ \beta\in\mathbb{R},\text{and}\sum_{j=1}^{k}\alpha_j+\beta=1\right\} \qquad (7.28)$$

Proposition 7.3. *Given an n-player game* $\Gamma_n = \langle (A_i, r_i)_{1\leq i\leq n}\rangle$ *and a vector* $\vec{\Phi} = \langle \Phi_i\rangle_{1\leq i\leq n}$ *such that* $\Phi_i \subseteq \Phi_{\text{ALL}}(A_i)$ *for* $1\leq i\leq n$, *if q is a* $\vec{\Phi}$*-equilibrium, then q is also a* $\vec{\Phi}'$*-equilibrium, where* $\vec{\Phi}' = (\text{SCH}(\Phi_i))_{1\leq i\leq n}$.

Proof. Let q be a $\vec{\Phi}$-equilibrium. Let i be an arbitrary player and let φ^* be an arbitrary element of $\text{SCH}(\Phi_i)$. Choose $k\in\mathbb{N}$, $\varphi_j\in\Phi_i$ and $\alpha_j\geq 0$ for all $1\leq j\leq k$, and $\beta\in\mathbb{R}$ such that $\varphi^* = \left(\sum_{j=1}^{k}\alpha_j\varphi_j\right)+\beta I$ and $\sum_{j=1}^{k}\alpha_j+\beta=1$.

$$\begin{aligned}
r_i(q) &= \sum_{j=1}^{k}\alpha_j r_i(q)+\beta r_i(q) && (7.29)\\
&\geq \sum_{j=1}^{k}\alpha_j \mathrm{e}_{a\sim q}\left[r_i(\varphi_j(a_i),a_{-i})\right]+\beta r_i(q) && (7.30)\\
&= \mathrm{e}_{a\sim q}\left[\sum_{j=1}^{k}\alpha_j r_i(\varphi_j(a_i),a_{-i})+\beta r_i(a)\right] && (7.31)\\
&= \mathrm{e}_{a\sim q}\left[r_i\left(\sum_{j=1}^{k}\alpha_j\varphi_j(a_i)+\beta\delta_{a_i},a_{-i}\right)\right] && (7.32)\\
&= \mathrm{e}_{a\sim q}\left[r_i(\varphi^*(a_i),a_{-i})\right] && (7.33)
\end{aligned}$$

Line (7.30) follows because $\varphi_j\in\Phi_i$ and q is a $\vec{\Phi}$-equilibrium. Line (7.31) follows from the linearity of expectations. Line (7.32) follows because r_i is linear in its ith argument. Finally, line (7.33) follows from the definition of φ^*.

7.5 Convergence of No-$\overrightarrow{\Phi}$-Regret Learning Algorithms

In this section, we establish a fundamental relationship between no-regret learning algorithms and game-theoretic equilibria. We prove that learning algorithms that satisfy no-$\overrightarrow{\Phi}$-regret converge to the set of $\overrightarrow{\Phi}$-equilibria. We derive as corollaries of this theorem the following two specific results: no-Φ_{EXT}-regret algorithms (i.e., no-external-regret algorithms) converge to the set of Φ_{EXT}-equilibria, which correspond to generalized minimax equilibria in zero-sum games; and no-Φ_{INT}-regret algorithms (i.e., no-internal-regret algorithms) converge to the set of Φ_{INT}-equilibria, which correspond to correlated equilibria in general-sum games. This latter result is well-known [19]. By Proposition 7.1, we arrive at another known result, namely, in two-player, zero-sum games, if each player plays using a no-external-regret learning algorithm, then each player's empirical distribution of joint play converges to his set of minimax strategies [20].

In addition to giving sufficient conditions for convergence to the set of $\overrightarrow{\Phi}$-equilibria, we also give *necessary* conditions. We show that multiagent learning converges to the set of $\overrightarrow{\Phi}$-equilibria only if the time-averaged Φ_i-regret experienced by each player i converges to the negative orthant.

Given an infinitely-repeated n-player game Γ_n^∞, a *run* of the game is a sequence of action vectors $\{\mathbf{a}_\tau\}_{\tau=1}^\infty$ with each $\mathbf{a}_\tau \in A_1 \times \ldots \times A_n$. Given a run $\{\mathbf{a}_\tau\}_{\tau=1}^\infty$ of Γ_n^∞, the *empirical distribution of joitn play through time t*, denoted z_t, is the element of $\Delta(A_1 \times \ldots \times A_n)$ given by:

$$z_t(\mathbf{b}) = \frac{1}{t}\sum_{\tau=1}^{t} \mathbf{1}_{\mathbf{a}_\tau=\mathbf{b}} \tag{7.34}$$

where $\mathbf{1}_{x=y}$ denotes the indicator function, which equals 1 whenever $x = y$, and 0 otherwise.

The results in this section rely on a technical lemma, the statement and proof of which appear in Section 7.9. We apply this lemma via the following corollary, which relates the empirical distribution of joint play at equilibrium to the players' rewards at equilibrium.

Corollary 7.1. *Given an n-player game Γ_n and a vector of sets of action transformations $\overrightarrow{\Phi} = (\Phi_i)_{1\le i\le n}$ such that $\Phi_i \subseteq \Phi_{\mathrm{ALL}}(A_i)$ for $1 \le i \le n$. If Z is the set of $\overrightarrow{\Phi}$-equilibria of Γ_n, then $d(z_t, Z) \to 0$ as $t \to \infty$ if and only if $\mathrm{e}_{a\sim z_t}[\rho^\varphi(a)] \to \mathbb{R}_-$ as $t \to \infty$, for all players i and for all action transformations $\varphi_i \in \Phi_i$.*

Proof. For all players i and action transformations $\varphi_i \in \Phi_i$, let $f_i^{\varphi_i}(q) = \mathrm{e}_{a\sim q}[\rho^\varphi(a)]$ and $Z_i^{\varphi_i} = \{q \in \Delta(A_1 \times \ldots \times A_n) \mid f_i^{\varphi_i}(q) \le 0\}$, for all $q \in \Delta(A_1 \times \ldots \times A_n)$. The set of $\overrightarrow{\Phi}$-equilibria is thus $Z = \bigcap_{1\le i\le n} \bigcap_{\varphi_i\in\Phi_i} Z_i^{\varphi_i}$. For each i and φ_i, apply Lemma 7.4 to $f_i^{\varphi_i}$ and $Z_i^{\varphi_i}$ so that $d(z_t, Z_i^{\varphi_i}) \to 0$ as $t \to \infty$ if and only if $\mathrm{e}_{a\sim z_t}[\rho^\varphi(a)] \to \mathbb{R}_-$ as $t \to \infty$.

In words, Corollary 7.1 states that the empirical distribution of joint play converges to the set of $\overrightarrow{\Phi}$-equilibria if and only if the rewards each player i obtains exceed the rewards player i could have expected to obtain by playing according to any of the action transformations $\varphi_i \in \Phi_i$ of component i of the empirical distribution of joint play.

Theorem 7.3. *Given an n-player game Γ_n and a vector of sets of action transformations $\vec{\Phi} = (\Phi_i)_{1 \leq i \leq n}$ such that $\Phi_i \subseteq \Phi_{\text{ALL}}(A_i)$ is finite for $1 \leq i \leq n$. As $t \to \infty$, the average Φ_i-regret experienced by each player i through time t converges to the negative orthant if and only if the empirical distribution of joint play converges to the set of $\vec{\Phi}$-equilibria of Γ_n.*

Proof. By Corollary 7.1, it suffices to show that, as $t \to \infty$, the average Φ_i-regret through time t experienced by each player i converges to the negative orthant if and only if for all players i and for all $\varphi_i \in \Phi_i$, $\mathrm{e}_{a \sim z_t}[\rho^{\varphi}(a)] \to \mathbb{R}_-$ as $t \to \infty$. But for arbitrary player i and for arbitrary $\varphi_i \in \Phi_i$,

$$\mathrm{e}_{a \sim z_t}[\rho^{\varphi}(a)] = \frac{1}{t} \sum_{\tau=1}^{t} \rho^{\varphi}(a_\tau) \tag{7.35}$$

From this equivalence, the conclusion follows immediately.

By Theorem 7.3, if the time-averaged Φ_i-regret experienced by each player i converges to the negative orthant with probability 1, then empirical distribution of joint play converges to the set of $\vec{\Phi}$-equilibria with probability 1. But if each player i plays according to a no-Φ_i-regret learning algorithm, then the time-averaged Φ_i-regret experienced by each player i converges to the negative orthant with probability 1, regardless of the opposing algorithm:: i.e., on any run of the game. From this discussion, we draw the following general conclusion:

Theorem 7.4. *Given an n-player game Γ_n and a vector of sets of action transformations $\vec{\Phi} = (\Phi_i)_{1 \leq i \leq n}$ such that $\Phi_i \subseteq \Phi_{\text{ALL}}(A_i)$ is finite for $1 \leq i \leq n$. If all players i play no-Φ_i-regret learning algorithms, then the empirical distribution of joint play converges to the set of $\vec{\Phi}$-equilibria of Γ_n with probability 1.*

Thus, we see that if all players abide by no-internal-regret algorithms, then the distribution of play converges to the set of correlated equilibria. Moreover, in two-player, zero-sum games if all players abide by no-external-regret algorithms, then the distribution of play converges to the set of generalized minimax equilibria, that is, the set of minimax-valued joint distributions. Again, by Proposition 7.1, this latter result implies that each player's empirical distribution of joint play converges to his set of minimax strategies, under the stated assumptions.

7.6 The Power of No Internal Regret

Perhaps surprisingly, no internal regret is the strongest form of no Φ-regret, for finite Φ. It follows from the results in Section 7.5 that the tightest game-theoretic solution concept to which no-Φ-regret learning algorithms converge is correlated equilibrium. In particular, Nash equilibrium is not a necessary outcome of learning via no-Φ-regret algorithms.

Theorem 7.5. *Given a real-valued game (A, A', R, r), if a learning algorithm $\mathcal{L}$ satisfies no internal regret, then $\mathcal{L}$ also satisfies no Φ-regret for all finite sets $\Phi \subseteq \Phi_{ALL}(A)$.*

The proof of this theorem follows immediately from the following lemmas.

Lemma 7.1. *Given a real-valued game (A, A', R, r), for all $\Phi \subseteq \Phi_{ALL}(A)$ and $\Phi' \subseteq \text{SCH}(\Phi)$, there exists a constant $c > 0$ such that $d\left(\mathbb{R}_-^{\Phi'}, \overline{\rho}_t^{\Phi'}\right) \leq c\, d\left(\mathbb{R}_-^{\Phi}, \overline{\rho}_t^{\Phi}\right)$, for all t.*

Proof. For each $\varphi' \in \Phi' \subseteq \text{SCH}(\Phi)$ there exist $k \in \mathbb{N}$, $\varphi_j \in \Phi$ and $\alpha_j \geq 0$ for all $1 \leq j \leq k$, and $\beta \in \mathbb{R}$ such that $\varphi' = \left(\sum_{j=1}^k \alpha_j \varphi_j\right) + \beta I$ and $\sum_{j=1}^k \alpha_j + \beta = 1$. Now

$$\rho^{\varphi'}(a, a') = r(\varphi'(a), a') - r(a, a') \tag{7.36}$$

$$= r\left(\left(\sum_{j=1}^k \alpha_j \varphi_j + \beta I\right)(a), a'\right) - r(a, a') \tag{7.37}$$

$$= \sum_{j=1}^k \alpha_j r(\varphi_j(a), a') + (\beta - 1) r(a, a') \tag{7.38}$$

$$= \sum_{j=1}^k \alpha_j (r(\varphi_j(a), a') - r(a, a')) \tag{7.39}$$

$$= \sum_{j=1}^k \alpha_j \rho^{\varphi_j}(a, a') \tag{7.40}$$

Line 7.39 follows because $\sum_{j=1}^k \alpha_j = 1 - \beta$.

Thus, we can define a linear transformation $F : \mathbb{R}^{\Phi} \to \mathbb{R}^{\Phi'}$ such that $F(\rho^{\Phi}(a, a')) = \rho^{\Phi'}(a, a')$, for all a, a'. Because the α_i are all non-negative, F exhibits the following property:

$$\rho^{\Phi}(a, a') \in \mathbb{R}_-^{\Phi} \Rightarrow F(\rho^{\Phi}(a, a')) = \rho^{\Phi'}(a, a') \in \mathbb{R}_-^{\Phi'} \tag{7.41}$$

i.e., $F(\mathbb{R}_-^{\Phi}) \subseteq \mathbb{R}_-^{\Phi'}$. Further, because F is linear,

$$d\left(\mathbb{R}_-^{\Phi'}, \overline{\rho}_t^{\Phi'}\right) = d\left(\mathbb{R}_-^{\Phi'}, F\left(\overline{\rho}_t^{\Phi}\right)\right) \tag{7.42}$$

$$\leq d\left(F\left(\mathbb{R}_-^{\Phi}\right), F\left(\overline{\rho}_t^{\Phi}\right)\right) \tag{7.43}$$

$$\leq c\, d\left(\mathbb{R}_-^{\Phi}, \overline{\rho}_t^{\Phi}\right) \tag{7.44}$$

where $c > 0$ is the operator norm of F.

Lemma 7.2. *Given a real-valued game (A, A', R, r), if a learning algorithm $\mathscr{L}$ satisfies no Φ-regret for some finite set $\Phi \subseteq \Phi_{ALL}(A)$, then $\mathscr{L}$ also satisfies no Φ'-regret, for all finite sets $\Phi' \subseteq \text{SCH}(\Phi)$.*

Proof. By Lemma 7.1, there exists a constant $c > 0$ such that $d\left(\mathbb{R}_-^{\Phi'}, \overline{\rho}_t^{\Phi'}\right) \leq c\, d\left(\mathbb{R}_-^{\Phi}, \overline{\rho}_t^{\Phi}\right)$, for all t. Now for any $\varepsilon > 0$, let $\delta = \min\left\{\frac{\varepsilon}{c}, \varepsilon\right\}$. Since $\delta \leq \frac{\varepsilon}{c}$,

$$d(\mathbb{R}_-^{\Phi'}, \overline{\rho}_t^{\Phi'}) \geq \varepsilon \Rightarrow c\, d(\mathbb{R}_-^{\Phi}, \overline{\rho}_t^{\Phi}) \geq \varepsilon \tag{7.45}$$

$$\Rightarrow d(\mathbb{R}_-^{\Phi}, \overline{\rho}_t^{\Phi}) \geq \delta \tag{7.46}$$

Because $\mathscr{L}$ satisfies no Φ-regret, we can choose t_0 such that for any $a_1', a_2', \ldots$,

$$P\left[\exists t \geq t_0 \text{ s.t. } d\left(\mathbb{R}_-^{\Phi}, \overline{\rho}_t^{\Phi}\right) \geq \delta\right] < \delta \tag{7.47}$$

But then, by Equation (7.46), and since $\delta \leq \varepsilon$,

$$P\left[\exists t \geq t_0 \text{ s.t. } d\left(\mathbb{R}_-^{\Phi'}, \overline{\rho}_t^{\Phi'}\right) \geq \varepsilon\right] < \varepsilon \tag{7.48}$$

Therefore, no Φ-regret implies no Φ'-regret.

Lemma 7.3. *For any (finite) set of actions, A, the super convex hull of $\Phi_{INT}(A)$ is $\Phi_{ALL}(A)$.*

Proof. Let φ^* be an arbitrary element of $\Phi_{\text{ALL}}(A)$. Define $\widehat{\varphi} \in \text{SCH}(\Phi_{\text{INT}}(A))$ by

$$\widehat{\varphi} = \sum_{a,b \in A} \varphi^*(a)(b)\varphi_{\text{INT}}^{(ab)} + (1 - |A|)I \tag{7.49}$$

For any $x \in A$,

$$\widehat{\varphi}(x) = \sum_{a,b \in A} \varphi^*(a)(b)\varphi_{\text{INT}}^{(ab)}(x) + (1 - |A|)\delta_x \tag{7.50}$$

$$= \sum_{a,b \in A} \varphi^*(a)(b) \left\{ \begin{array}{l} \delta_b \text{ if } x = a \\ \delta_x \text{ otherwise} \end{array} \right\} + (1 - |A|)\delta_x \tag{7.51}$$

$$= \sum_{a,b \in A} \varphi^*(a)(b) \left(\sum_{x=a} \delta_b + \sum_{x \neq a} \delta_x \right) + (1 - |A|)\delta_x \tag{7.52}$$

$$= \sum_{b \in A} \varphi^*(x)(b)\delta_b + \sum_{x \neq a} \sum_{b \in A} \varphi^*(x)(b)\delta_x + (1 - |A|)\delta_x \tag{7.53}$$

$$= \sum_{b \in A} \varphi^*(x)(b)\delta_b + (|A| - 1)\delta_x + (1 - |A|)\delta_x \tag{7.54}$$

$$= \sum_{b \in A} \varphi^*(x)(b)\delta_b \tag{7.55}$$

Further, for any $y \in A$,

$$\widehat{\varphi}(x)(y) = \sum_{b \in A} \varphi^*(x)(b)\delta_b(y) \tag{7.56}$$

$$= \varphi^*(x)(y) \tag{7.57}$$

Therefore, $\varphi^* = \widehat{\varphi} \in \text{SCH}(\Phi_{\text{INT}}(A))$.

7.7 Related Work

In this section, we relate our basic algorithm and our main theorems to results published elsewhere.

7.7.1 On the Existence of No-Regret Algorithms

The algorithm presented here is, in the terminology of Greenwald et al. [17], an *action*-regret-based learning algorithm, which means that regret at time t is computed with respect to the action a_t as in Equations (7.6) and (7.7), as opposed a *distribution*-regret-based learning algorithm, in which regret at time t is calculated with respect to the distribution q_t as follows:

$$\rho^{\Phi}(q,a') \equiv \left(\rho^{\varphi}(q,a')\right)_{\varphi \in \Phi} \tag{7.58}$$

where

$$\rho^{\varphi}(q,a') \equiv \mathrm{e}_{a \sim q}[r(\varphi(a),a') - r(a,a')] \tag{7.59}$$

Many well-known no-regret learning algorithms arise as instances, or close cousins, of action- or distribution-regret-based variants of Algorithm 1:

(a) The no-external-regret algorithm of Hart and Mas-Colell [19] (Theorem B) is the special case of Algorithm 1 (action-regret-based) when $\Phi = \Phi_{\text{EXT}}(A)$.
(b) The no-internal-regret algorithm of Foster and Vohra [12] is closely related to Algorithm 1 (distribution-regret-based) when $\Phi = \Phi_{\text{INT}}(A)$. Foster and Vohra calculate the fixed points of a stochastic matrix that is derived from the internal regret vector. Their matrix is identical to M (computed in terms of distribution-based regrets) up to normalization. Consequently, both their matrix and M have the same set of fixed points.[5]

[5] Let $R_t^{(ij)}$ denote the cumulative $\varphi_{\text{INT}}^{(ij)}$ regret at time t. Define the matrix Q_t by $(Q_t)_{ii} = -\sum_j \left(R_t^{(ij)}\right)^+$ and $(Q_t)_{ij} = \left(R_t^{(ij)}\right)^+$ for $i \neq j$. Our algorithm plays the fixed point of a matrix A which can be written as $A = I + \frac{1}{\sum_i |(Q_t)_{ii}|} Q_t$. Foster and Vohra's algorithm plays the fixed point of a matrix A' which can be written as $A' = I + \frac{1}{\max_i |(Q_t)_{ii}|} Q_t$. It can be shown that A and A' have the same set of fixed points: If q is a fixed point of A, then $qA = q \Rightarrow q + \frac{1}{\sum_i |(Q_t)_{ii}|} qQ = q \Rightarrow qQ = 0 \Rightarrow qA' = qI = q$. And similarly in the other direction.

(c) By replacing $+$ operation (i.e., $(x_t^\varphi)^+$) in steps 10 and 13 of Algorithm 1 (distribution-regret-based) with $e^{x_t^\varphi}$, we arrive at Freund and Schapire's Hedge algorithm [13] when $\Phi = \Phi_{\text{EXT}}(A)$, and an instance of an algorithm discussed by Cesa-Bianchi and Lugosi [6] when $\Phi = \Phi_{\text{INT}}(A)$.

Lehrer [23] derives a "wide range no-regret theorem" analogous to our existence result.[6] Lehrer's approach combines "replacing schemes," functions from $\mathscr{H} \times A$ to A, with "activeness functions" from $\mathscr{H} \times A$ to $\{0,1\}$. Given a replacing scheme g and an activeness function I, Lehrer's framework compares the agent's rewards to the rewards that could have been obtained by playing action $g(h_t, a_t)$, but only if $I(h_t, a_t) = 1$, yielding a general form of action regret. Lehrer establishes the existence of *action*-regret-based no-regret algorithms whose regret with respect to any countable set of pairs of replacing schemes and activeness functions, averaged over the number of times each pair is "active," approaches the negative orthant.

Because Lehrer deals with *countable* sets of replacing schemes, whereas we restrict attention to *finite* sets of transformations, it may seem that Lehrer's theorem immediately subsumes ours. However, in Lehrer's framework, the co-domain of each transformation is A, not $\Delta(A)$, as it is in ours. In other words, in our framework, actions are transformed into mixed, rather than pure, strategies. This turns turns out to yield no additional power however. By Theorem 7.5, it suffices to consider the set of internal action transformations, each element of which can be expressed simply as a function from $A \to A$. Hence, in light of Theorem 7.5, Lehrer's theorem can in fact be viewed as subsuming our existence theorem.

Blum and Mansour's [3] framework is similar to Lehrer's, but yields results about *distribution* regret. Their "modification rules" are the same as replacing schemes, but instead of activeness functions, they pair modification rules with "time selection functions," which are functions from $\mathbb{N}$ to the interval $[0,1]$. The rewards an agent could have obtained under each modification rule are weighted according to how "awake" the rule is, as indicated by the corresponding time selection function. They present a method that, given a collection of algorithms whose external distribution regret is bounded above by $f(t)$ at time t, generates an algorithm whose swap distribution regret (and hence, internal distribution regret) is bounded above by $|A|f(t)$.

Cesa-Bianchi and Lugosi [6] develop a framework of "generalized" *distribution* regret. They rely on a notion of "experts," which they define as functions from $\mathbb{N}$ to A, and following Lehrer, they pair experts $f_1, \ldots, f_N$ with activation functions $I_i : A \times \mathbb{N} \to \{0,1\}$. At time t, for each i, if $I_i(a_t, t) = 1$, they compare the agent's rewards to the rewards the agent could have obtained by playing $f_i(t)$. This approach is more general than our action-transformation framework in that alternatives may depend on time. At the same time, it is more limited in that it does not naturally represent swap regret (but this is not necessarily a shortcoming, in light of Theorem 7.5). Cesa-Bianchi and Lugosi's calculations yield bounds on generalized distribution regret.

From the bounds derived by Blum and Mansour and Cesa-Bianchi and Lugosi, one can infer the existence of *distribution*-regret-based no-regret learning algo-

---

[6] Our initial findings [15] were obtained simultaneously and independently.

rithms, by applying the Hoeffding-Azuma lemma (see, for example, the appendix of Cesa-Bianchi and Lugosi [7]).

Finally, it has been observed that swap regret is bounded above by $|A|$ times internal regret. Hence, no-internal-regret implies no-swap-regret,[7] which in turn implies no-Φ-regret, for all $\Phi \subseteq \Phi_{\text{ALL}}$, because the elements of any Φ can be constructed as a convex combination of the elements of Φ_{SWAP}. It follows from this observation that every no-internal-regret algorithm – such as the algorithms of Foster and Vohra [12], Hart and Mas-Colell [20], Cesa-Bianchi and Lugosi [6], and Young [29] – is a no-Φ-regret algorithm, for all $\Phi \subseteq \Phi_{\text{ALL}}$. In other words, the existence of (action- or distribution-based) no-internal-regret algorithms implies the existence of (action- or distribution-based) no-Φ-regret algorithms, for all $\Phi \subseteq \Phi_{\text{ALL}}$.

7.7.2 On the Connection Between Learning and Games

Several authors before us have explored the connection between no-regret (and related) learning algorithms and game-theoretic equilibria. This literature originates with Foster and Vohra [11], who present a (calibrated) learning algorithm such that if all players play according to it, the empirical distribution of joitn play converges to the set of correlated equilibria.

Hart and Mas-Colell [19] exhibit a simple adaptive procedure such that if all players follow this procedure, then the time-averaged internal regret vector of each player converges to zero almost surely, and the empirical distribution of joint play converges to the set of correlated equilibrium almost surely. Their algorithm does not exhibit no-internal-regret against an adaptive adversary, however.[8] More fundamentally, they show that a necessary and sufficient condition for the empirical distribution of joint play to converge to the set of correlated equilibria is that all players' internal regrets converge to zero.[9] In Theorem 7.3, we generalize this result beyond Φ_{INT}.

Hart and Mas-Colell [20] present a class of no-external-regret learning algorithms. They show that in repeated two-player zero-sum games, if both players play according to algorithms in this class, then each player's empirical distribution of joint play converges to his set of minimax strategies and the players' average rewards converge to the minimax value of the game. They also discuss a class of algorithms which are no-internal-regret. They argue that if each player plays according to an algorithm in this class, then the empirical distribution of joint play converges to the set of correlated equilibria almost surely, which is an immediate consequence of their earlier result [19].

---

[7] This observation also follows from Theorem 7.5. Choose $\Phi = \Phi_{\text{SWAP}}$.

[8] Their algorithm does exhibit no-internal-regret against an oblivious opponent [5]. An *adaptive* adversary is able to respond the learning algorithm's choices of actions; an *oblivious* adversary is not.

[9] Stoltz and Lugosi [28] generalize this result to the case where the set of actions is a compact and convex subset of a normed space and the reward function is continuous.

7.8 Summary

In this article, we defined a general class of no-regret learning algorithms, called no-Φ-regret learning algorithms, which spans the spectrum from no-external-regret learning to no-internal-regret learning and beyond. Analogously, we defined a general class of game-theoretic equilibria, called $\vec{\Phi}$-equilibria, and we showed that the empirical distribution of play of no-Φ_i-regret algorithms converges to the set of $\vec{\Phi}$-equilibria. Moulin and Vial [25] also define a general class of game-theoretic equilibria, ranging from pure strategy Nash equilibria to coarse correlated equilibria. To our knowledge, their generalizations have not been widely applicable in practice. Similarly, our generalized notions of equilibria may not be of significant practical value – at present, we know of no other interesting classes of $\vec{\Phi}$-equilibria besides Φ_{EXT}- and Φ_{INT}-equilibria. Still, we believe it is of theoretical interest to observe that no-external-regret and no-internal-regret can be viewed along the same continuum, and moreover, that they correspond to game-theoretic equilibria along an analogous continuum.

In future work, we hope to extend our results to handle more general action sets (e.g., convex and compact, following Stoltz and Lugosi [28]) and more general kinds of action transformations (e.g., history-dependent, following Lehrer [23], and time-varying, following Herbster and Warmuth [21]). Nevertheless, our present interest in finite action sets and finite sets of action transformations is not limiting in our ongoing simulation work, where we are investigating the short- and medium-term dynamics of no-Φ-regret learning algorithms.

7.9 Proof of Lemma 7.4

Lemma 7.4. *Let (X,d_X) be a compact metric space and let (Y,d_Y) be a metric space. Let $\{x_t\}$ be an X-valued sequence, and let S be a nonempty, closed subset of Y. If $f : X \to Y$ is continuous and if $f^{-1}(S)$ is nonempty, then $d_X(x_t, f^{-1}(S)) \to 0$ as $t \to \infty$ if and only if $d_Y(f(x_t),S) \to 0$ as $t \to \infty$.*

Proof. We write $d = d_X$ and $d = d_Y$, since the appropriate choice of distance metric is always clear from the context. To prove the forward implication, assume $d(x_t, f^{-1}(S))) \to 0$ as $t \to \infty$. Choose t_0 *s.t.* for all $t \geq t_0$, $d(x_t, f^{-1}(S)) < \frac{\delta}{2}$. Observe that for all x_t and for all $\gamma > 0$, there exists $q_t^{(\gamma)} \in f^{-1}(S)$ *s.t.* $d(x_t, q_t^{(\gamma)}) < d(x_t, f^{-1}(S)) + \gamma$. Now, since $d(x_t, q_t^{(\frac{\delta}{2})}) < \frac{\delta}{2} + \frac{\delta}{2} = \delta$, by the continuity of f, $d(f(x_t), f(q_t^{(\frac{\delta}{2})})) < \varepsilon$, for all $\varepsilon > 0$. Therefore, $d(f(x_t),S) < \varepsilon$, since $f(q_t^{(\frac{\delta}{2})}) \in S$.

To prove the reverse implication, assume $d(f(x_t),S) \to 0$ as $t \to \infty$. We must show that for all $\varepsilon > 0$, there exists a t_0 *s.t.* for all $t \geq t_0$, $d(x_t, f^{-1}(S)) < \varepsilon$. Define $T = \{x \in X \mid d(x, f^{-1}(S)) \geq \varepsilon\}$. If $T = \varnothing$, the claim holds. Otherwise, observe that T can be expressed as the complement of the union of open balls, so that T is closed and thus compact. Define $g : X \to \mathbb{R}$ as $g(x) = d(f(x),S)$. By assumption S is closed;

hence, $g(x) > 0$, for all x. Because T is compact, g achieves some minimum value, say $L > 0$, on T. Choose t_0 *s.t.* $d(f(x_t), S) < L$ for all $t \geq t_0$. Thus, for all $t \geq t_0$, $g(x_t) < L \Rightarrow x_t \notin T \Rightarrow d(x_t, f^{-1}(S)) < \varepsilon$.

Acknowledgements We gratefully acknowledge Dean Foster for sparking our interest in this topic, and for ongoing discussions that helped to clarify many of the technical points in this paper. We also thank Dave Seaman for providing a proof of the harder direction of Lemma 7.4 and anonymous reviewers for their constructive criticism. The content of this article is based on Greenwald and Jafari [15], which in turn is based on Jafari's Master's thesis [22]. This research was supported by NSF Career Grant #IIS-0133689 and NSF IGERT Grant #9870676.

References

1. R. Aumann. Subjectivity and correlation in randomized strategies. *Journal of Mathematical Economics,* 1:67–96, 1974.
2. D. Blackwell. An analog of the minimax theorem for vector payoffs. *Pacific Journal of Mathematics,* 6:1–8, 1956.
3. A. Blum and Y. Mansour. From external to internal regret. In *Proceedings of the 2005 Computational Learning Theory Conferences,* pages 621–636. Bertinoro. June 27–30, 2005, Lecture Notes in Computer Science 3559. Springer, Berlin, 2005.
4. G. Brown. Iterative solutions of games by fictitious play. In T. Koopmans, editor, *Activity Analysis of Production and Allocation.* Wiley, New York, NY, 1951.
5. A. Cahn. General procedures leading to correlated equilibria. *International Journal of Game Theory,* 33(1):21–40, December 2004.
6. N. Cesa-Bianchi and G. Lugosi. Potential-based algorithms in on-line prediction and game theory. *Machine Learning,* 51(3):239–261, 2003.
7. N. Cesa-Bianchi and G. Lugosi. *Prediction, Learning, and Games.* Cambridge University Press, New York, NY, 2006.
8. A. Cournot. *Recherches sur les Principes Mathematics de la Theorie de la Richesse.* Hachette, Paris, 1838.
9. F. Forges. Correlated equilibrium in two-person zero-sum games. *Econometrica,* 58(2):515, March 1990.
10. D. Foster and R. Vohra. A randomization rule for selecting forecasts. *Operations Research,* 41(4):704–709, 1993.
11. D. Foster and R. Vohra. Calibrated learning and correlated equilibrium. *Games and Economic Behavior,* 21:40–55, 1997.
12. D. Foster and R. Vohra. Regret in the on-line decision problem. *Games and Economic Behavior,* 29:7–35, 1999.
13. Y. Freund and R.E. Schapire. A decision-theoretic generalization of on-line learning and an application to boosting. In *Computational Learning Theory: Proceedings of the Second European Conference,* pages 23–37. Springer, New York, NY, 1995.
14. D. Fudenberg and D.K. Levine. Consistency and cautious fictitious play. *Journal of Economic Dynamics and Control,* 19:1065–1090, 1995.
15. A. Greenwald and A. Jafari. A general class of no-regret algorithms and gametheoretic equilibria. In *Proceedings of the 2003 Computational Learning Theory Conference,* pages 1–11. Washington, DC, August 2003.
16. A. Greenwald, A. Jafari, and C. Marks. *Blackwell's Approachability Theorem: A Generalization in a Special Case.* Tech. Rep. CS-06–01. Brown University, Department of Computer Science, Providence, RI, January 2006.

17. A. Greenwald, Z. Li, and C. Marks. *Bounds for Regret-Matching Algorithms.* Tech. Rep. CS-06–10. Brown University, Department of Computer Science, Providence, RI, June 2006.
18. J. Hannan. Approximation to Bayes risk in repeated plays. In M. Dresher, A. Tucker, and P. Wolfe, editors, *Contributions to the Theory of Games*, volume 3, pages 97–139. Princeton University Press, Princeton, NJ, 1957.
19. S. Hart and A. Mas-Colell. A simple adaptive procedure leading to correlated equilibrium. *Econometrica,* 68:1127–1150, 2000.
20. S. Hart and A. Mas-Colell. A general class of adaptive strategies. *Journal of Economic Theory*, 98(1):26–54, 2001.
21. M. Herbster and M.K. Warmuth. Tracking the best expert. *Machine Learning,* 32(2):151–178, 1998.
22. A. Jafari. On the notion of regret in infinitely repeated games. Master's Thesis, Brown University, Providence, RI, May 2003.
23. E. Lehrer. A wide range no-regret theorem. *Games and Economic Behavior*, 42(1):101–115, 2003.
24. N. Littlestone and M.K. Warmuth. The weighted majority algorithm. *Information and Computation*, 108:212–261, 1994.
25. H. Moulin and J.P. Vial. Strategically zero-sum games. *International Journal of Game Theory*, 7:201–221, 1978.
26. J. Nash. Equilibrium points in n-person games. *Proceedings of the National Academy of Sciences*, 36(1):48–49, 1950.
27. J. Robinson. An iterative method of solving a game. *Annals of Mathematics*, 54:298–301, 1951.
28. G. Stoltz and G. Lugosi. Learning correlated equilibria in games with compact sets of strategies. *Games and Economic Behavior*, 59:187–208, 2007.
29. P. Young. *Strategic Learning and Its Limits*. Oxford University Press, Oxford, 2004.

Chapter 8
Axioms of Distinction in Social Software

Vincent F. Hendricks

> 'Over a ten year period starting in the mid 90's I became convinced that all these topics – game theory, economic design, voting theory – belonged to a common area which I called Social Software.' – Rohit Parikh, [21]: p. 252

Around the turn of the millenium Rohit Parikh launched a new important program on the borderline of computer science and epistemology. "Social software" has aquired two meanings recently: One in which social software denotes various web-based software programs of a "social nature" from *Facebook* to *Wikipedia* and a meaning, now canonical, according to an entry exactly in *Wikipedia*, in which social software

> ... studies the procedures of society whether elections, conferences etc. as analogous to computer programs to be analyzed by similar logical and mathematical tools.

The formal tools of the trade are described in Parikh's seminal papers [19, 20] and subsequenly spelled out as "(1) logic of knowledge, (2) logic of games, and (3) game theory and economic design." [18]: p. 442.

These tools are also common to a related new trade called *formal epistemology* which also by a recent entry in *Wikipedia* is characterized as

> ... a subdiscipline of epistemology that utilizes formal methods from logic, probability theory and computability theory to elucidate traditional epistemic problems.

Social software and formal epistemology are both interdisciplinary approaches to the study of agency and agent interaction but are not quite to be confused with yet a new trend in philosophy recently *social epistemology* [6]. While social software and formal epistemology share the same toolbox, social epistemology pays stronger homage to the standard philosophical methodology of conceptual analysis and intuition pumps.

Vincent F. Hendricks
Department of Philosophy, University of Copenhagen, DK-2300 Copenhagen S, Denmark and Department of Philosophy, Columbia University, New York, NY 10027, USA,
e-mail: vincent@hum.ku.dk

J. van Benthem et al. (eds.), *Games, Norms and Reasons*, Synthese Library 353,
DOI 10.1007/978-94-007-0714-6_8,

Although sounding derogatory, conceptual analysis is not necessarily a bane as long as it is regimented by formal structure. In social software, the formal structure is provided by the aforementiond tools. While modelling agency and interaction certain methodological distinctions may have to be explicitly observed exactly in order to make conceptual sense of the use of logic of knowledge, logic of games, game theory and economic design in social software.

No better occassion to review but a few of these methdological distinctions and their conceptual impact in light of the 70th birthday of the founding architect of social software in the important sense – Rohit Parikh.

8.1 Perspectives, Agents and Axioms

Contemporary epistemology often draws a distinction between descriptive and normative theories of knowledge. There is a similar distinction in moral philosophy between descriptive and normative ethics. The former attempts to describe actual moral behavior while the latter sets the standards for correct moral conduct.

Similarly, descriptive epistemologics account for actual epistemic practice while normative ones are to prescribe rules of inquiry in terms of mechanisms for avoiding error and gaining truth, truth-conducive justification criteria, learning and winning strategies, procedures for revising beliefs etc. The distinction is sometimes blurred by the fact that while describing actual epistemic practice one may have to define various notions like, say, knowledge itself, justification and reliability inviting normative aspects.

Both descriptive and normative epistemologies usually subscribe to the common premiss that epistemic practice of agents by and large is "rational". What separates the two stances is whether epistemology is simply to describe this very practice or try to optimize it. Irrational epistemic behavior would be to follow some practice which a priori may be demonstrated to be en route to error (when this very practice is an available course of conduct to the agent in the environment) [7, 14]. It is not necessarily irrational on the other hand not to follow some prescription if the natural epistemic milieu sets the standards for what the agent is able to do and this prescription is not among them. The local epistemic circumstances may for one reason or the other bar the agent in question from choosing the best means for an end. Constraints could even be such that they reward "irrational" behavior. Calling such epistemic situations irrational would undermine the common premiss which the two approaches subscribe to. Not only may the environment limit the agent's behavior, other agents may as well. This is for instance illustrated by game theory's distinction between cooperative and non-cooperative games.

Sometimes agents would be able to have more knowledge than they actually have if they were not tied up in their local epistemic theater. Then they could freely pursue the optimal means for obtaining some desirable result whether truth, epistemic strength or winning in some other sense. One may then rightfully ask why epistemologists sometimes are in the business of means-ends prescriptions that no local agent is able to meet. There are two with each other related answers to this question:

- As epistemologists we are not only in the business of ascribing ourselves knowledge, but equally much in the business of ascribing knowledge to other agents. Lewis has pointed out that there is a significant difference between one agent ascribing himself knowledge in his local epistemic situation, and us ascribing him knowledge given the situation we are in. The two situations do not always coincide. First and third persons do not share the same real world in many contexts. There are rules to follow under knowledge attribution to oneself and others to know what we think we know.
- Rather principled information about what it would take to solve the epistemic task at hand, than no information at all. Epistemology is about whether knowledge is possible and about what agents can and cannot know insofar knowledge is possible. The problem is that it is not always clear from within whether something is knowable or not. One recurs to a perspective from without for a principled answer which may then spill over into the local circumstances.

According to Lewis [16], the agent may actually know more than we are able to ascribe to him. On his account this is due to the fact that the attribution of knowledge is highly sensitive to which world is considered actual and for whom in a given context. An agent in his environment is more likely to be aware of what the relevant possibilities are given the world considered actual by him than the knowledge ascriber standing by him or even outside. Lewis refers to these two stances as a *first* versus a *third* person perspective on inquiry.

Observe that an agent is free to be prescribe recommendations for himself to follow as long as the means suggested are available to him where he is. Outsiders may also freely prescribe recommendations for the agent inside as long as they are available to the agent in question. If the outsider decides to prescribe a course of conduct to solve an epistemic problem for the agent in the environment unavailable to the agent then the situation changes. Emphasized is then what it would take to solve the epistemic problem regardless of whether the agent is capable of actually performing the action(s) it would take .

Normative/descriptive, first person versus third person perspectives are not mutually exclusive distinctions. The distinction between descriptive and normative theories of knowledge together with a modified version of Lewis' first and third person perspective dichotomy are subsumed in the following two formulations:

First person perspective – A perspective on scientific inquiry is **first person** if it considers what an agent can solve, can do or defend considering the available means for an end given the epistemic environment he is sunk into

Third person perspective – A perspective on scientific inquiry is **third person** if it considers what an agent could solve, could do or defend considering the best means for an end independently of the epistemic environment he is sunk into

In criticizing some epistemological position, whether mainstream or formal, without noticing that the criticism is based on a third person perspective and the position advocated is first person may again turn out to be criticizing an apple for not being an orange [9]. The dichotomy has a significant bearing on the general epistemological and conceptual plausibility given the formal tools utilized by social software in modelling agent and agent interaction.

8.2 Setting Up the Matrix

What is being studied from the first and the third person perspective respectively using the logics of knowledge (and games) in social software are

- an agent or multiple agents in concert, and
- various epistemic axioms and systems characterizing the knowledge of one or more agents.

The task now is to tell a plausible epistemological story about the axioms valid while modelling, say, one agent, from a first person perspective and so forth for the remaining possibilities in the matrix[1]

	First person	Third person
One agent	x	x
Multiple agents	x	x

(8.1)

for $x \in \{\mathsf{T},\mathsf{K},\mathsf{D},4,5\}$

Here is an example of an epistemological story told about modelling one agent from a first person perspective all the way to **S4**: Hintikka stipulated that the axioms or principles of epistemic logic are conditions descriptive of a special kind of general (strong) *rationality* for a single agent and on a first person perspective [12]. The statements which may be proved false by application of the epistemic axioms are not inconsistent meaning that their truth is logically impossible. They are rather rationally "indefensible". Indefensibility is fleshed out as the agent's epistemic laziness, sloppiness or perhaps cognitive incapacity whenever to realize the implications of what he in fact knows. Defensibility then means not falling victim of "epistemic neglience" as Chisholm calls it. The notion of indefensibility gives away the status of the epistemic axioms and logics. Some epistemic statement for which its negation is indefensible is called "self-sustaining". The notion of self-sustenance actually corresponds to the concept of validity. Corresponding to a self-sustaining statement is a logically valid statement. But this will again be a statement which is rationally indefensible to deny. So in conclusion, epistemic axioms are descriptions of rationality.

[1] Only these 5 canonical axioms are considered here, others may be discussed as well – see [9].

There is evidence to the effect that Hintikka early on was influenced by the autoepistemology of Malcolm [17] and took, at least in part, their autoepistemology to provide a philosophical motivation for epistemic logic. There is an interesting twist to this motivation which is not readily read out of autoepistemology. Epistemic axioms may be interpreted as principles describing a certain strong rationality. The agent does not have to be aware of this rationality, let alone able to immediately compute it from the first person perspective as Hintikka argues when it comes to axiom K:

> In order to see this, suppose that a man says to you, 'I know that *p* but I don't know whether *q*' and suppose that *p* can be shown to entail logically *q* by means of some argument which he would be willing to accept. Then you can point out to him that what he says he does not know is already implicit in what he claims he knows. If your argument is valid, it is irrational for our man to persist in saying that he does not know whether *q* is the case. [12], p. 31.

The autoepistemological inspiration is vindicated while Hintikka argues for the plausibility of 4 as a governing axiom of his logic of knowledge as he refers to Malcolm:

> This is especially interesting in view of the fact that Malcolm himself uses his strong sense of knowing to explain in what sense it might be true that whenever one knows, one knows that one knows. In this respect, too, Malcolm's strong sense behaves like mine. [13, p. 154].

Besides the requirement of closure and the validity of the 4, axiom T is also valid to which Malcolm would object. A logic of autoepistemology is philosophically congruent with Hintikka's suggestion for a **S4** epistemic logic describing strong rationality from a first person point of view for a singular agent.

The key debate of whether epistemic axioms are plausibly describing agenthood or not seems much to depend on whether one subscribes to a first or a third person perspective on inquiry. Given an autoepistemological inspiration epistemic axioms describe a first person knowledge operator as Hintikka suggested. If epistemic axioms are describing *implicit knowledge* as Fagin et al. suggest [5], then what is being modelled is what follows from actual knowledge independently of agent computations. Agents can on this third person perspective not be held actually responsible for failing to exercise some reflective disposition. Closure principles may be problematic from the point of view of the agent, not necessarily from point of view of the ones studying the agent third person. Logical omniscience as a consequence of the epistemic axioms is a problem from a first person perspective but not necessarily from a third person perspective.

We are not going to fill in all the cells of the matrix with epistemological stories, just discuss one additional axiom which is tricky for knowledge, games and economic design, agents and any point of view.

8.3 Negative Introspection of Knowledge

One of the most celebrated motivations for the plausibility of Axiom 5, or the axiom of negative introspection, is a closed-world assumption in data-base applications [5]: An agent examining his own knowledge base will be led to conclude that whatever is not in the knowledge base he does not know and hence he will know that he does not. This is a first person motivation, but in the same breath, the argument for dodging logical omniscience is based on a third-person operative as seen above. So there is some meandering back and forth while arguing for the axiomatic plausibility of agency.

Axiom 5 may seem unrealistically strong for a singular agent in his environment, unless his environment is defined solipsistically, e.g. the closed world assumption. Solipsism is not necessarily a human or a real agent condition but a philosophical thesis; a thesis making idealized sense standing outside looking at the agent in his, admittedly, rather limited epistemic environment. Being a stone-hearted solipsist on a first person basis is hard to maintain coherently as W.H. Thorpe once reported:

> Bertrand Russell was giving a lesson on solipsism to a lay audience, and a woman got up and said she was delighted to hear Bertrand Russell say he was a solipsist; she was one too, and she wished there were more of us.

A reason for adopting the first person perspective and pay homage to axiom 5 for singular agents is that these assumptions provide some nice technical advantages / properties especially with respect to information partition models. There is now also a philosophical basis for doing things in this idealized way – epistemic solipsism and no false beliefs, e.g. infallibilism. Both of these philosophical theses have little to do with logic but plenty to do with the preconditions for studying knowledge from any point of view.

8.4 Negative Introspection in Games

It is more sticky to argue that Axiom 5 is reasonable to assume in multi-agent setups. But when game theorists for instance model non-cooperative extensive games of perfect information an **S5** logic of knowledge is used to establish the backward induction equilibrium [3].

For game theory the untenability of **S5** in multi-agent systems is quite severe. The problem concerns the knowledge of action as Stalnaker has pointed out: It should be possible for a player Ξ to know what a player Θ is going to do. For instance it should be rendered possible in case Θ only has one rational choice, and Ξ knows Θ to be rational, that Ξ can predict what Θ is going to do. This should not imply however that it is impossible for Θ to act differently as he has the capacity to act irrationally. In order to make sense of this situation what is needed is a counterfactually possible world such that

(a) Θ acts irrationally, but
(b) is incompatible with what Ξ knows.

Now Ξ's prior beliefs in that counterfactual world must be the same as they are in the actual world for Θ could not influence Ξ's priors beliefs by making a contrary choice (by definition of the game, Ξ and Θ act independently). Then it has to be the case in the counterfactual world, that Ξ believes he knows something (e.g. that Θ is irrational) which he in fact does not know. This is incompatible with **S5**.

Additionally, acting in the presence of other agents requires the information to be explicitly available to the agents first person, but it may only be implicitly at the agents' disposal if the over-all model is of implicit knowledge. It is not much help to have the knowledge explicitly available on the third person level if you have to make an informed move on the first person level featuring other agents trying to beat you as you are trying to beat them.

8.5 Negative Introspection in Economics

This discussion of the untenability of **S5** is not in any way linked to a view of inappropriateness of modelling a third person notion of knowledge via the axiom of veridicality T. One may reasonably argue like Stalnaker and Aumann that knowledge requires truth referring to a notion of third-person knowledge. The unintuitive results obtained by Aumann and others indicate that there is something wrong in the information model used by economists, which assumes that agents engaged in economic interactions actually have common knowledge rather than common belief. Thus one can infer that the impossibility of trade can be concluded from assuming that the agents engaged in economic interaction have more powers than they actually have. Once one endows agents with a more realistic epistemic model, it is possible to agree to disagree and trade is made plausible again.

Collins' explanation of what is wrong in Aumann's models is quite plausible. If agents have common belief rather than common knowledge then they cannot share a common prior, a crucial probabilistic assumption in Aumann's seminal paper "Agreeing to Disagree" [2]. An intuitive explanation is provided by Collins:

> Finally, an opponent might challenge my claim that it is belief rather than knowledge that ought to be central to *interactive epistemology*. My response to this is simply to point out that agents, even rational agents, can and do get things wrong. This is not a controversial claim, just the commonplace observation that rational agents sometimes have false beliefs. The reason for this is not hard to find. It is because the input on which we update is sometimes misleading and sometimes downright false. To demand that everything an agent fully believes be true is not to state a requirement of rationality but rather to demand that the agent be invariably lucky in the course of her experience. Being completely rational is one thing; always being lucky is another. [4].

Nothing is here said about what it would actually mean to have knowledge in economic exchanges. Perhaps to be always lucky aside from rational. This entails that the notion of knowledge does require truth in order for it to be intelligible. Collins points out that agents get things wrong all the time, even while being completely rational. Aumann's theorem demonstrate how alien to our everyday endeavors the notion of knowledge is. The notion of rationality can at most require that the agent only holds beliefs that are full beliefs, i.e., beliefs which the agent takes as true from his first person point of view.

Disagreement alone does not suffice for altering anyone's view. Each agent will therefore have some type of acceptance rule that will indicate to him whether it is rational or not to incorporate information. Sometimes the agent might lend an ear to an incompatible point of view for the sake of the argument and this might end up in implementing a change of view. When a network of agents is modeled from the outside, endowing these agents with third-person knowledge (as is customarily done in economic models) seems inappropriate. Be that as it may, if the agents are rational one should assume that their theories are in autoepistemic equilibrium, and this leads to the assumption that the first person views are each one **S5** **[1]**. These two things are perfectly compatible, which does not imply that certain type of inputs (the ones that are marked positive by your preferred theory of acceptance) might require perturbing the current autoepistemic equilibrium via additions or revisions. The philosophical results questioning the use of **S5** in interactive epistemology, question the fact that economic agents can be modeled *by the game theorist* as having third person knowledge. These results do not necessarily have a bearing for what one might or must assume about the first-person point of view of each agent.

8.6 Axioms of Distinction

In a recent interview, Parikh takes stock of the contemporary situation in epistemology:

> Currently there is sort of a divide among two communities interested in knowledge. One is the community of epistemologists for whom the 1963 paper by Gettier has created a fruitful discussion. The other community is the community of formal epistemologists who trace their roots back to modal logic and to Hintikka's work on knowledge and belief. There is relatively little overlap between the two communities, but properties of the formal notion of knowledge, e.g., positive and negative introspection, have caused much discussion. I would suggest that knowledge statements are not actually propositions. Thus unlike "There was a break-in at Watergate," "Nixon knew that there was a break-in at Watergate," is not a proposition but sits inside some social software. It may thus be a mistake to look for a fact of the matter. The real issue is if we want to hold Nixon responsible. [21]: 145.

Parikh is right in emphasizing that too much ink perhaps has been spilled over formal properties of knowledge rather than on how knowledge is acquired. Surely, much epistemic and moral success has to do with procedures and processes for getting it right rather than static epistemic definitions and axioms of charaterization.

The two endavours are not mutually exclusively. Axioms are used to a procedural end in agency and agent interaction because much of the (game theoretical) dynamics is dependent on the epistemic powers of the agents given axiomatically. "The real issue is if we want to hold Nixon responsible", or put differently, the real issue is whether there is a reliable (possibly effective) procedure, or strategy, for converging to the correct epistemic result (axiom or verdict) or winning the game against nature or other agents. Formal learning theory or "computational epistemology" as Kelly recently redubbed the paradigm is extensively concerned with the former, social software with the latter. By the end of the day, the two approaches are very similar in their procedural outlook on inquiry and success.

A successful epistemic story told is dependent on

- what the agent(s) can do in the environment,
- given the
  - the axioms describing their epistemic powers, and
  - the procedures of knowledge acquisition,
- pace the first person / third person point of view.

And so to finish off with a question based on the matrix above:

> **Are there axioms of distinction exclusively separating the first-person perspective from the third-person perspective?**

Such an axiomatic separation would hopefully supplement the great epistemological range and plausibility of the tools employed in Rohit Parikh's seminal and important program, *social software*.

References

1. H. Arló-Costa. Qualitative and probabilistic models of full belief. S. Buss, P. Hajék, and R. Pudlák, editors, *Proceedings of Logic Colloquim'98, Lecture Notes on Logic*, volume 13, ASL, Prague, 1998.
2. R.J. Aumann. Agreeing to disagree. *Annals of Statistics,* 4:1236–1239, 1976.
3. C. Bicchieri. *Rationality and Coordination.* Cambridge University Press, New York, NY, 1993.
4. J. Collins. *How We Can Agree to Disagree*, Columbia University, version of July 2, 1997, http://collins.philo.columbia.edu
5. R. Fagin. J.Y. Halpern, Y. Moses, and M.Y. Vardi. *Reasoning About Knowledge.* MIT Press, Cambridge, 1995.
6. A.I. Goldman. *Knowledge in a Social World.* Oxford University Press, New York, NY, 1999.
7. V.F. Hendricks. *The Convergence of Scientific Knowledge: A View from the Limit.* Trends in Logic: Studia Logica Library Series. Springer, Dordrecht, 2001.
8. V.F. Hendricks. Active agents. *Journal of Logic, Language, and Information*, J. van Benthem and R. van Rooy, editors, volume 12, Autumn, 4:469–495, 2003.
9. V.F. Hendricks. *Mainstream and Formal Epistemology.* Cambridge University Press, New York, NY, 2006.
10. V.F. Hendricks and J. Symons, editors. *Formal Philosophy.* Automatic Press/VIP, London & New York, NY, 2005.
11. V.F. Hendricks and P.G. Hansen, editors. *Game Theory: 5 Questions.* Automatic Press/VIP, London & New York, NY, 2007.
12. J. Hintikka. *Knowledge and Belief: An Introduction to the Logic of the Two Notions.* Cornell University Press, Cornell, 1962.

13. J. Hintikka. Knowing that one knows revisited. *Synthese*, 21:141–162, 1970.
14. K. Kelly. *The Logic of Reliable Inquiry*. Oxford University Press, New York, NY, 1996.
15. W. Lenzen. Recent work in epistemic logic. *Acta Philosophica Fennica*, 30:1–219, 1978.
16. D. Lewis. Elusive knowledge. *The Australian Journal of Philosophy*, 74:549–567, 1996.
17. N. Malcolm. Knowledge and belief. In M.F. Goodman and R.A. Snyder, editors, *Mind LXI, 242. Reprinted in Contemporary Readings in Epistemology*, pages 272–279. Prentice Hall, Englewood Cliffs, NJ (1993), 1952.
18. E. Pacuit and R. Parikh. Social interaction, knowledge and social software. In. D. Goldin, S. Smolka, and P. Wegner, editors, *Interactive Computation: The New Paradigm*, pages 441–461. Springer, Dordrecht, 2007.
19. R. Parikh. Language as social software. In J. Floyd and S. Shieh, editors, *Future Pasts: The Analytic Tradition in the Twentieth Century Philosophy*, pages 339–350. Oxford University Press, New York, NY, 2001.
20. R. Parikh. Social software. *Synthese*, 132:187–211, 2002.
21. R. Parikh. Interview: 5 Questions on Formal Philosophy, 2005. In [10].
22. R. Parikh. Interview: 5 Questions on Game Theory, 2007. In [11].
23. R. Stalnaker. Knowledge, belief and counterfactual reasoning in games. *Economics and Philosophy*, 12:133–163, 1996.

Chapter 9
Publication/Citation: A Proof-Theoretic Approach to Mathematical Knowledge Management*

Dexter Kozen and Ganesh Ramanarayanan

9.1 Introduction

There are many real-life examples of formal systems that support certain constructions or proofs, but that do not provide direct support for remembering them so that they can be recalled and reused the future. This task is usually left to some metasystem that is typically provided as an afterthought. For example, programming language design usually focuses on the programming language itself; the mechanism for accumulating useful code in libraries is considered more of a systems issue and is generally treated as a separate design task. Mathematics deals with the construction of proofs, but not with their publication and citation; that is the domain of the journals.

Automated deduction systems such as NuPrl [4, 6] and Mizar [16] have language support for accumulating results in libraries for later reference. However, the mechanisms for providing this support are typically not considered interesting enough to formalize in the underlying logic, although it is possible in principle to do so.

We regard publication/citation as an instance of *common subexpression elimination* on proof terms. These operations permit proofs to be reused, perhaps specialized to a particular context, without having to reconstruct them in every application.

In this paper we attempt to develop this idea from a proof-theoretic perspective. We describe a simple complete proof system for universal Horn equational logic with three new proof rules, **publish**, **cite** and **forget**. The first two rules allow the inclusion of proved theorems in a library and later citation. The last allows removal of theorems from the library. These rules encapsulate all the bookkeeping that must be done to ensure correctness in their application.

Dexter Kozen
Department of Computer Science, Cornell University, Ithaca, NY 14853-7501, USA,
e-mail: kozen@cs.cornell.edu
Ganesh Ramanarayanan
Department of Computer Science, Cornell University, Ithaca, NY 14853-7501, USA,
e-mail: graman@cs.cornell.edu

* In Honor of Professor Rohit Jivanlal Parikh on the Occasion of his 70th Birthday

J. van Benthem et al. (eds.), *Games, Norms and Reasons*, Synthese Library 353,
DOI 10.1007/978-94-007-0714-6_9,

The **publish** rule has the effect of publishing the universal closure of a theorem to the library. The combinator corresponding to **publish** wraps the proof term to inhibit β-reduction upon later citation. The result of this operation is a *citation token* that is type-equivalent to the original proof, thus interchangeable with it, but that does not perform the specialization that is supplied with the citation.

The **cite** rule allows for the application of published theorems using the type-equivalent citation token. The theorem is applied under a specialization given by a variable substitution and provided at the time of the citation. However, because of the wrapping done by **publish**, the substitution is not actually performed on the proof term.

The **forget** rule is correspondingly more involved, since it must remove all citations to the forgotten theorem from the library. This is accomplished by unwrapping all occurrences of the citation token, allowing the deferred substitutions and β-reductions to take place during proof normalization. Effectively, this replaces each citation of the forgotten theorem with an inlined specialization of the original proof, where the specialization is the one supplied when the theorem was cited.

A major advantage of our approach is that it avoids namespace management issues, allowing us to focus on the pure structure of publication/citation. In real systems, when a lemma is added to a library, it is usually given a name, and all subsequent references are by this name. This introduces the possibility of name collisions. To address this issue, these systems typically introduce some form of scoping or module structure. However, our proof-theoretic treatment of publication/citation allows us to avoid the problem entirely.

In this paper, we develop this idea for constructive universal equational Horn logic. However, it is clear that the mechanisms are more general and could be adapted to richer theories. Significant progress along these lines has already been made [1–3].

9.2 A Classical Proof System

We first present a classical proof system for constructive universal equational Horn logic as a basis for comparison. Let Σ be a signature consisting of function symbols $\Sigma = \{f, g, \ldots\}$, each with a fixed finite arity. Let Σ_n denote the set of elements of Σ of arity n. Let X be a set of individual variables $X = \{x, y, \ldots\}$, and let $T_\Sigma(X)$ denote the set of individual terms over Σ and X. Formally, an *individual term* is either

- a variable $x \in X$, or
- an expression of the form $ft_1 \ldots t_n$, where $t_1, \ldots, t_n$ are individual terms and $f \in \Sigma_n$.

For example, the signature of groups consists of a binary operator $\cdot$, a unary operator $^{-1}$, and a constant 1.

A *formula* is either

- an equation $s = t$ between individual terms,

- an implication of the form $s = t \rightarrow \varphi$, where φ is a formula, or
- a quantified expression for the form $\forall x\ \varphi$, where x is an individual variable and φ is a formula.

We use $d, e, \ldots$ to denote equations and $\varphi, \psi, \ldots$ to denote formulas.

We are primarily interested in the universal Horn formulas, which are universally quantified formulas of the form

$$\forall x_1\ \ldots\ \forall x_n\ e_1 \rightarrow \cdots \rightarrow e_n \rightarrow e. \tag{9.1}$$

In the system presented in this section, the quantification is implicit.

Let S denote the set of all substitutions $\sigma : X \mapsto T_\Sigma(X)$. The notation $[x/t]$ denotes the substitution that simultaneously substitutes the term t for all occurrences of the variable x in a term or formula.

The following axioms E are the axioms of classical equational logic, implicitly universally quantified.

$$\begin{aligned}
&x = x \\
&x = y \rightarrow y = x \\
&x = y \rightarrow y = z \rightarrow x = z \\
&x_1 = y_1 \rightarrow \cdots \rightarrow x_n = y_n \rightarrow f x_1 \ldots x_n = f y_1 \ldots y_n, \quad f \in \Sigma_n.
\end{aligned}$$

Besides E, we will also allow an *application theory* Δ of universal Horn formulas to serve as additional, application-specific axioms. For example, for group theory, Δ would consist of the equational axioms for groups.

We now give the deduction rules. Let A be a set of equations, $d, e, \ldots$ equations, and φ a Horn formula.

$$\vdash \sigma(\varphi), \quad \varphi \in \Delta \cup E, \quad \sigma \in S$$

$$e \vdash e$$

$$\frac{A \vdash \varphi}{A, e \vdash \varphi}$$

$$\frac{A, e \vdash \varphi}{A \vdash e \rightarrow \varphi}$$

$$\frac{A \vdash e \rightarrow \varphi \quad A \vdash e}{A \vdash \varphi}$$

The following rule is derived:

$$\frac{A \vdash e}{A[x/t] \vdash e[x/t]}$$

provided x does not occur in t. This rule obviates the need for an explicit universal quantifier and corresponding introduction and elimination rules.

One can also give annotated versions of these rules in which formulas are annotated with explicit proof terms, which are terms of the simply typed λ-calculus. Let $P = \{p, q, r, \ldots\}$ be a set of *proof variables*. A *proof term* is either

- a variable $p \in P$,
- a constant ref, sym, trans, cong_f for $f \in \Sigma$, or axiom_φ for $\varphi \in \Delta$,
- an application $\pi\,\tau$, where π and τ are proof terms,
- an application πt, where π is proof term and t is an individual term,
- an abstraction $\lambda p.\tau$, where p is a proof variable and τ is a proof term,
- an abstraction $\lambda x.\tau$, where x is an individual variable and τ is a proof term, or
- an expression $\mathsf{pub}\,\tau$, where τ is a proof term.

The combinator pub is just a fixed constant. Proof terms are denoted $\pi, \rho, \tau, \ldots$.

As usual in constructive mathematics, according to the Curry–Howard isomorphism, we can view proofs as constructions, formulas as types, and the deduction system as a set of typing rules. An annotated formula takes the form of a type judgement $\tau : \varphi$, where τ is a proof term. The interpretation of these constructs is the same as in the simply-typed λ-calculus.

The annotated rules are as follows.

$$\vdash \mathsf{axiom}_\varphi\, \sigma : \sigma(\varphi), \quad \varphi \in \Delta \cup E, \quad \sigma \in S$$

$$p : e \vdash p : e$$

$$\frac{A \vdash \tau : \varphi}{A,\ p : e \vdash \tau : \varphi}$$

$$\frac{A,\ p : e \vdash \tau : \varphi}{A \vdash \lambda p.\tau : e \to \varphi}$$

$$\frac{A \vdash \pi : e \to \varphi \quad A \vdash \rho : e}{A \vdash \pi\rho : \varphi}$$

Not all proof terms as defined above are well typed. In particular, abstractions over an individual variable $\lambda x.\tau$ and the pub combinator will only become relevant in Section 9.3, when we reintroduce explicit universal quantification.

9.3 A New System

The new system we present in this section builds directly on the classical system presented in Section 9.2, adding in the notion of a *library* $\mathscr{L}$. This library minimally contains all of the axioms, and we introduce rules to add and remove new theorems to and from the library along with their proofs.

In contrast to the system of Section 9.2, the system of the present section will have explicit universal quantification for all axioms and theorems in the library.

This will allow arbitrary specialization via substitution of individual terms for the quantified variables upon citation.

As in the system of Section 9.2, our system has three main syntactic categories: *individual terms*, *formulas*, and *proof terms*. Also as in Section 9.2, we will start by presenting the unannotated version of our proof system, which does not have any proof terms.

As before, let $X = \{x, y, \ldots\}$ be a set of *individual variables*, $\Sigma = \{f, g, \ldots\}$ a first-order signature, and $T_\Sigma(X) = \{s, t, \ldots\}$ the set of *individual terms* over X and Σ. Let Δ be a set of universal Horn formulas over X and Σ that serve as the application theory.

The unannotated version of our proof system consists of the following axioms and rules. We restate the equational axioms E, this time with explicit universal quantification. We use $\overline{x}$ to denote a tuple $x_1, \ldots, x_n$.

$$\forall x\ x = x$$

$$\forall x\, \forall y\ x = y \to y = x$$

$$\forall x\, \forall y\, \forall z\ x = y \to y = z \to x = z$$

$$\forall \overline{x}\, \forall \overline{y}\ x_1 = y_1 \to \ldots \to x_n = y_n \to f x_1 \ldots x_n = f y_1 \ldots y_n, f \in \Sigma$$

In the descriptions below, the separator character ; is used to distinguish proof tasks from the contents of the library. Judgements are of the form $\mathscr{L}\,;\, A \vdash \varphi$, where $\mathscr{L}$ is the current library consisting of a list of universally quantified Horn formulas of the form (9.1), A is a list of unquantified equational premises, and φ is an unquantified Horn formula. Elements of $\mathscr{L}$ are separated by commas, as are elements of A.

$$\textbf{(ident)}\ \frac{\mathscr{L}\,;}{\mathscr{L}\,;\, e \vdash e}$$

$$\textbf{(assume)}\ \frac{\mathscr{L}\,;\, A \vdash \varphi}{\mathscr{L}\,;\, A, e \vdash \varphi}$$

$$\textbf{(discharge)}\ \frac{\mathscr{L}\,;\, A, e \vdash \varphi}{\mathscr{L}\,;\, A \vdash e \to \varphi}$$

$$\textbf{(mp)}\ \frac{\mathscr{L}\,;\, A \vdash e \to \varphi \quad \mathscr{L}\,;\, A \vdash e}{\mathscr{L}\,;\, A \vdash \varphi}$$

$$\textbf{(publish)}\ \frac{\mathscr{L} \qquad ;\ \vdash \varphi}{\mathscr{L}, \forall \overline{x}\, \varphi\ ;}$$

$$\textbf{(cite)}\ \frac{\mathscr{L}, \forall \overline{x}\, \varphi\ ;}{\mathscr{L}, \forall \overline{x}\, \varphi\ ;\ \vdash \varphi[\overline{x}/\overline{t}]}$$

$$\textbf{(forget)}\ \frac{\mathscr{L}, \forall \overline{x}\, \varphi\ ;}{\mathscr{L} \qquad ;}, \quad \forall \overline{x}\, \varphi \notin \Delta \cup E$$

where in the **publish** rule, $\overline{x} = x_1, \ldots, x_n$ are the free variables of φ.

The rules of the proof system build on the classical set of rules, with the addition of the three new rules **publish**, **cite** and **forget**. We do not allow the equational and application theory axioms to be removed from the library, thus it is always be the case that $\Delta \cup E \subseteq \mathscr{L}$. The rules **publish** and **cite** serve as introduction and elimination rules for the universal quantifier, respectively, but quantifiers appear only in published theorems (i.e., those in the library). The **forget** rule is simply an ordinary weakening rule in this version; however, once annotations are added, the effect of this rule on proof terms is much more involved.

Now we add annotations. For the equational axioms and the application theory,

$$\begin{aligned}
\mathsf{ref} &: \forall x\ x = x \\
\mathsf{sym} &: \forall x\, \forall y\ x = y \to y = x \\
\mathsf{trans} &: \forall x\, \forall y\, \forall z\ x = y \to y = z \to x = z \\
\mathsf{cong}_f &: \forall \overline{x}\, \forall \overline{y}\ x_1 = y_1 \to \ldots \to x_n = y_n \to f x_1 \ldots x_n = f y_1 \ldots y_n, \quad f \in \Sigma \\
\mathsf{axiom}_\varphi &: \varphi, \quad \varphi \in \Delta.
\end{aligned}$$

Thus each axiom of equational logic and the application theory ($\Delta \cup E$) is inhabited by a constant. These type judgements are always present in $\mathscr{L}$.

In addition to these, we have the following annotated rules:

$$(\textbf{ident})\ \frac{\mathscr{L}\,;}{\mathscr{L}\,; p : e \vdash p : e}$$

$$(\textbf{assume})\ \frac{\mathscr{L}\,; A \vdash \tau : \varphi}{\mathscr{L}\,; A, p : e \vdash \tau : \varphi}$$

$$(\textbf{discharge})\ \frac{\mathscr{L}\,; A, p : e \vdash \tau : \varphi}{\mathscr{L}\,; A \vdash \lambda p.\tau : e \to \varphi}$$

$$(\textbf{mp})\ \frac{\mathscr{L}\,; A \vdash \pi : e \to \varphi \qquad \mathscr{L}\,; A \vdash \rho : e}{\mathscr{L}\,; A \vdash \pi\rho : \varphi}$$

$$(\textbf{publish})\ \frac{\mathscr{L} \qquad ; \vdash \tau : \varphi}{\mathscr{L}, \mathsf{pub}\, \lambda\overline{x}.\tau : \forall \overline{x}\ \varphi\,;}$$

$$(\textbf{cite})\ \frac{\mathscr{L}, \pi : \forall \overline{x}\ \varphi\,;}{\mathscr{L}, \pi : \forall \overline{x}\ \varphi\,; \vdash \pi\overline{t} : \varphi[\overline{x}/\overline{t}]}$$

$$(\textbf{forget})\ \frac{\mathscr{L}, \mathsf{pub}\,\pi : \forall \overline{x}\ \varphi\,;}{\mathscr{L}[\mathsf{pub}\,\pi/\pi] \qquad ;}, \quad \forall \overline{x}\ \varphi \notin \Delta \cup E$$

Publication forms the universal closure of the formula and the corresponding λ-closure of the proof term before wrapping with pub. Thus published theorems, like axioms, are always closed universal formulas. The proof term is closed by binding all the free individual variables appearing in the body of the proof term in the same order as they were bound in the formula (the free individual variables in formulas and corresponding proof terms are always the same).

As in Section 9.2, the interpretation of annotated formulas is the same as in the simply-typed λ-calculus. However, our type system is somewhat more restrictive than the usual one. For example, the type system prevents a binding operator λx from occurring in the scope of a binding operator λp. This enforces the universal Horn form (9.1) for published theorems.

The constant pub is polymorphic of type $\varphi \to \varphi$. Its main purpose is to wrap proof terms to inhibit β-reduction without altering their type. An expression of the form $\mathsf{pub}\,\pi$ is called a *citation token*. Intuitively, we can think of a citation token as a short abbreviation (a name or a pointer) for the proof π in the library. Since the proof and its citation token are type-equivalent, we can use them interchangeably.

Ordinarily, when a universally quantified theorem is cited in a special case defined by a substitution $[\overline{x}/\overline{t}]$, the proof term would be specialized as well by applying it to the sequence of individual terms $t_1,\ldots,t_n$. Without the pub wrapper, proof normalization would cause those terms to be substituted for $x_1,\ldots,x_n$ in the body of the λ-expression as in ordinary β-reduction. The pub wrapper prevents this from happening, since the expression $(\mathsf{pub}\,\pi)\tau$ is in normal form.

An alternative approach might use new names and bindings for published proofs, but this would introduce namespace management issues that are largely orthogonal to the publish/cite structure and which our approach circumvents.

For an accurate complexity analysis on the size of proofs, one could define the size of proof terms inductively in some reasonable way, taking the size of citation tokens to be 1. This would reflect the fact that in practice, a proof term would only be represented once, and citations would reference the original proof by name or by pointer.

9.4 An Example

To illustrate the operation of this proof system, we will go through a simple example. Supposing we wanted to prove the theorem

$$\forall x \;\; x = fx \to x = f(fx).$$

We will provide a proof of this fact, along with the corresponding extraction of proof terms. For the first part, we will omit the library $\mathscr{L}$ for readability, but we will reintroduce it later when we show the operation of publication.

Defining

$$\begin{aligned}\pi_1 &= \mathsf{cong}_f\, x\,(fx)\\ \pi_2 &= \mathsf{trans}\, x\,(fx)\,(f(fx)),\end{aligned}$$

we have by citation of the transitivity axiom and the congruence axiom for f that

$$\vdash \pi_1 \; : \; x = fx \to fx = f(fx) \tag{9.2}$$

$$\vdash \pi_2 \; : \; x = fx \to fx = f(fx) \to x = f(fx). \tag{9.3}$$

From (9.3) and **(assume)**, we have

$$p : x = fx \vdash \pi_2 : x = fx \to fx = f(fx) \to x = f(fx). \tag{9.4}$$

Also, by **(ident)**,

$$p : x = fx \vdash p : x = fx. \tag{9.5}$$

Applying **(mp)** with premises (9.4) and (9.5) gives

$$p : x = fx \vdash \pi_2\, p : fx = f(fx) \to x = f(fx). \tag{9.6}$$

Similarly, from (9.2) and **(assume)**, we have

$$p : x = fx \vdash \pi_1 : x = fx \to fx = f(fx), \tag{9.7}$$

and applying **(mp)** with premises (9.5) and (9.7) gives

$$p : x = fx \vdash \pi_1\, p : fx = f(fx). \tag{9.8}$$

Now applying **(mp)** with premises (9.6) and (9.8), we obtain

$$p : x = fx \vdash \pi_2\, p(\pi_1\, p) : x = f(fx), \tag{9.9}$$

and we conclude from **(discharge)** that

$$\vdash \lambda p.\pi_2\, p\,(\pi_1\, p) : x = fx \to x = f(fx). \tag{9.10}$$

We can now publish the universal closure of (9.10) using the publication rule, which adds the annotated theorem

$$\mathsf{pub}\,(\lambda x.\lambda p.\pi_2\, p\,(\pi_1\, p)) \; : \; \forall x \;\; x = fx \to x = f(fx) \tag{9.11}$$

to the library.

Now we show how (9.11) can be cited in a special case by proving the theorem

$$\forall y \;\; gy = f(gy) \to gy = f(f(gy)). \tag{9.12}$$

This is a more specific version of (9.11) obtained by substituting gy for x. We start by citing (9.11) with the term gy using the rule **(cite)**, which gives

$$\mathsf{pub}\,(\lambda x.\lambda p.\pi_2\, p\,(\pi_1\, p))(gy) \; : \; gy = f(gy) \to gy = f(f(gy)). \tag{9.13}$$

Publishing this theorem using **(publish)** results in the annotated theorem

$$\mathsf{pub}\,(\lambda y.\mathsf{pub}\,(\lambda x.\lambda p.\pi_2\,p\,(\pi_1\,p))(gy)) : \forall y\ \ gy = f(gy) \to gy = f(f(gy)) \quad (9.14)$$

being added to the library.

Now suppose we wish to use the **(forget)** rule to forget the original theorem (9.11) that was cited in the proof of (9.12). This removes (9.11) from the library and strips the pub combinator from all citations. The theorem (9.14) becomes

$$\mathsf{pub}\,(\lambda y.(\lambda x.\lambda p.\pi_2\,p\,(\pi_1\,p))(gy)) : \forall y\ \ gy = f(gy) \to gy = f(f(gy)) \quad (9.15)$$

The theorem itself remains in the library unchanged, but its proof is no longer in normal form, since the inner pub combinator has been stripped. Normalizing the proof, we obtain

$$\mathsf{pub}\,(\lambda y.\lambda p.\pi_2[x/gy]\,p\,(\pi_1[x/gy]\,p)) : \forall y\ \ gy = f(gy) \to gy = f(f(gy)),$$

where

$$\begin{aligned} \pi_1[x/gy] &= \mathsf{cong}_f\,(gy)\,(f(gy)) \\ \pi_2[x/gy] &= \mathsf{trans}\,(gy)\,(f(gy))\,(f(f(gy))). \end{aligned}$$

The proof now has the specialization of the proof of (9.11) "inlined" into it. This is equivalent to what we would have obtained had we set out to prove (9.12) directly, without appealing to the more general theorem (9.11) first.

9.5 Related and Future Work

Proof generalization, proof reuse, and mathematical knowledge management are active areas of research. Much of the work on proof generalization and reuse is oriented toward heuristic methods for discovering simple modifications of old proofs that apply in new but similar situations. Schairer et al. [14, 15] have suggest replaying tactics to develop proofs by analogy. Hutter [8, 9] has given a system of proof annotations and rules for manipulating them. The annotations are used to include planning information in proofs to help guide the proof search. Kolbe and Walther [10, 11] study the process of proof generalization by abstracting existing proofs to form *proof shells*. Their approach involves replacing occurrences of function symbols by second-order variables. Felty and Howe [7] also suggest a system of proof reuse using higher-order unification and metavariables to achieve abstraction. Melis and Whittle [12] study proof by analogy, focussing on the process of adapting existing proofs to new theorems with a similar structure. Piroi and Buchberger [5, 13] present a graphical environment for editing mathematics and managing a mathematical knowledge library. Allen et al. [4] also propose a structure for a formal digital library and discuss the problem of naming conflicts.

It would be interesting to explore the possibility of identifying similar proofs and finding common generalizations in a more proof-theoretic context such as the publication/citation mechanism presented in this paper. It would also be useful to extend the system to handle full first-order and higher-order logics.

One area of recent progress is the proof-theoretic representation of tactics [1, 2]. Another recent advance is the enhancement of the proof-theoretic apparatus to better capture natural dependencies among theorems, lemmas, and corollaries in the library and the locality of definitions. Most libraries are flat, which does not adequately capture the richness of mathematical knowledge. Recent progress in this direction has been made by Aboul-Hosn and Andersen [3], who present a hierarchical representation along with natural proof rules for restructuring.

Acknowledgements We are indebted to Kamal Aboul-Hosn, Samson Abramsky, Terese Damhøj Andersen, and Anil Nerode for valuable ideas and comments. This work was supported in part by NSF grant CCF-0635028. Any views and conclusions expressed herein are those of the authors and do not necessarily represent the official policies or endorsements of the National Science Foundation or the United States government.

References

1. K. Aboul-Hosn. A proof-theoretic approach to tactics. In J.M. Borwein and W.M. Farmer, editors, *Proceedings of the 5th International Conference on Mathematical Knowledge Management (MKM'06)*, volume 4108 of *Lecture Notes in Computer Science*, pages 54–66. Springer, Berlin, Germany, August 2006.
2. K. Aboul-Hosn. *A Proof-Theoretic Approach to Mathematical Knowledge Management*. PhD thesis, Cornell University, January 2007.
3. K. Aboul-Hosn and T.D. Andersen. A proof-theoretic approach to hierarchical math library organization. In *Proceedings of the 4th International Conference on Mathematical Knowledge Management (MKM'05)*, pages 1–16. Springer, Berlin, Germany, October 2005.
4. S. Allen, M. Bickford, R. Constable, R. Eaton, C. Kreitz, and L. Lorigo. *FDL: A prototype formal digital library*, 2002. http://www.nuprl.org/FDLproject/02cucs-fdl.html
5. B. Buchberger. Mathematical knowledge management using theorema. In B. Buchberger, O. Caprotti, editors, *1st International Conference on Mathematical Knowledge Management (MKM 2001)*, RISC-Linz, A-4232 Schloss Hagenberg, Austria, September 24–26, 2001.
6. R.L. Constable, S.F. Allen, H.M. Bromley, W.R. Cleaveland, J.F. Cremer, R.W. Harper, D.J. Howe, T.B. Knoblock, N.P. Mendler, P. Panangaden, J.T. Sasaki, and S.F. Smith. *Implementing Mathematics with the Nuprl Development System*. Prentice-Hall, Englewood Cliffs, NJ, 1986.
7. A. Felty and D. Howe. Generalization and reuse of tactic proofs. In *Proceedings of the 5th International Conference on Logic Programming and Automated Reasoning (LPAR94)*. Springer, Berlin, Germany, 1994.
8. D. Hutter. Structuring deduction by using abstractions. In T. Ellman, editor, *Proceedings of the International Symposion on Abstraction, Reformulation, and Approximation (SARA98)*, pages 72–78. Pacific Grove, CA, 1998.
9. D. Hutter. Annotated reasoning. *Annals of Mathematics and Artificial Intelligence, Special Issue on Strategies in Automated Deduction*, 29:183–222, 2000.
10. T. Kolbe and C. Walther. Reusing proofs. In *Proceedings of the 11th European Conference on Artificial Intelligence (ECAI-11)*, pages 80–84. Seattle, WA, May 21–23, 1994.
11. T. Kolbe and C. Walther. Proof management and retrieval. In *Proceedings of the Workshop on Formal Approaches to the Reuse of Plans, Proofs and Programs (IJCAI-14)*. Montreal, CA, 1995.

12. E. Melis and J. Whittle. Analogy in inductive theorem proving. *Journal of Automated Reasoning*, 22(2):117–147, 1999.
13. F. Piroi and B. Buchberger. An environment for building mathematical knowledge libraries. In A. Trybulec, A. Asperti, G. Bancerek, editor, *Proceedings of the 3rd International Conference Mathematical Knowledge Management (MKM 2004)*, volume 3119 of *Lecture Notes in Computer Science*. Springer, Berlin, Germany, September 2004.
14. A. Schairer. *A Technique for Reusing Proofs in Software Verification*. PhD thesis, FB 14 (Informatik) der Universität des Saarlandes und Institut A für Mechanik der Universität Stuttgart, Stuttgart, AT, 1998.
15. A. Schairer, S. Autexier, and D. Hutter. A pragmatic approach to reuse in tactical theorem proving. *Electronic Notes in Theoretical Computer Science*, 58(2), 2001.
16. A. Trybulec. The Mizar system. http://mizar.uwb.edu.pl/system/

Chapter 10
Generalizing Parikh's Theorem

Johann A. Makowsky*

10.1 Generalizing Parikh's Theorem

R. Parikh's celebrated theorem, first proved in [37], counts the number of occurrences of letters in words of a context-free languages L over an alphabet of k letters. For a given word w, the numbers of these occurrences is denoted by a vector $n(w) \in \mathbb{N}^k$, and the theorem states

Theorem 10.1 ([37]). *For a context-free language L, the set* $Par(L) = \{n(w) \in \mathbb{N}^k : w \in L\}$ *is semi-linear.*

A set $X \subseteq \mathbb{N}^s$ is *linear in* $\mathbb{N}^s$ iff there is vector $\overline{a} \in \mathbb{N}^s$ and a matrix $M \in \mathbb{N}^{s \times r}$ such that $X = A_{\overline{a},\overline{M}} = \{\overline{b} \in \mathbb{N}^s : \text{ there is } \overline{u} \in \mathbb{N}^r \text{ with } \overline{b} = \overline{a} + M \cdot \overline{u}\}$. Singletons are linear sets with $M = 0$. If $M \neq 0$ the series is *nontrivial*. $X \subseteq \mathbb{N}^s$ is *semi-linear in* $\mathbb{N}^s$ iff X is a finite union of linear sets $A_i \subseteq \mathbb{N}^s$. For $s = 1$ the semi-linear sets are exactly the ultimately periodic sets. The terminology is from [37], and has since become standard terminology in formal language theory. Several alternative proofs of Parikh's Theorem have appeared since. D. Pilling [38] put it into a more algebraic form, and more recently, L. Aceto et al. [3] showed that it depends only on a few equational properties of least pre-fixed points. B. Courcelle [14] has generalized Theorem 10.1 further to certain graph grammars, and relational structures which are definable in Monadic Second Order Logic (MSOL) in labeled trees, counting not only letter occurrences but the size of MSOL-definable subsets. In [23] a similar

Johann A. Makowsky
Department of Computer Science, Technion–Israel Institute of Technology, Haifa, Israel,
e-mail: janos@cs.technion.ac.il

* Partially supported by the Israel Science Foundation for the project "Model Theoretic Interpretations of Counting Functions" (2007–2010) and the Grant for Promotion of Research by the Technion–Israel Institute of Technology.

For R. Parikh,
at the occasion of
his 70th birthday

J. van Benthem et al. (eds.), *Games, Norms and Reasons*, Synthese Library 353,
DOI 10.1007/978-94-007-0714-6_10, © Springer Science+Business Media B.V. 2011

generalization was proven, inspired by a theorem due to Gurevich and Shelah [30] on spectra of MSOL-sentences with one unary function symbol and a finite number of unary relation symbols. The results of [23] are formulated in a model theoretic framework rather than using the language of graph grammars. But it turned out that some of their result could have been obtained also using the techniques of [14].

In this paper we explain discuss these generalizations of Parikh's Theorem without detailed proofs, but with the emphasis on the concepts involved and on applications. Like in the well known characterization of regular languages using Monadic Second Order Logic (MSOL) we also use MSOL, and an extension thereof, CMSOL, which allows for modular counting quantifiers. However, the languages in Parikh's Theorem are replaced by *arbitrary finite relational structures* which are of *bounded width*. The most general notion of width we shall use is *patch-width*, which was first introduced in [23]. It generalizes both tree-width and clique-width. The detailed discussion of these notions of width is given in Section 10.4. Finally, like in Courcelle's generalization of Parikh's Theorem, rather than counting occurrences of letters, we count cardinalities of CMSOL-definable unary relations. We shall first explain this for the case where these relations are given by unary predicates, where we speak of many-sorted spectra. The applications consist mostly in proving that certain classes of graphs and relational structures have unbounded tree-width, clique-width or patch-width. These are given in Section 10.5.

We assume the reader is familiar with the basics of Monadic Second Order Logic MSOL, cf. [12, 20]. For convenience, we collect the basics in Section 10.2. Otherwise this paper is rather self-contained.

10.2 Monadic Second Order Logic and Its Extension

First Order Logic FOL restricts quantification to elements of the structure, Monadic Second Order Logic MSOL also allows for quantification over subsets, but not over binary relations,, or relations of arity $r \geq 2$. The logic MSOL can be extended by modular counting quantifiers $C_{k,m}$, where $C_{k,m}x\ \varphi(x)$ is interpreted as "there are, modulo m, exactly k elements satisfying $\varphi(x)$". We denote the extension of MSOL obtained by adding, for all $k, m \in \mathbb{N}$ the quantifiers $C_{k,m}$, by CMSOL. Over structures which have a linear order the quantifiers $C_{k,m}x\ \varphi(x)$ can be eliminated without loss of expressive power.

Typical graph theoretic concepts expressible in FOL are the presence or absence (up to isomorphism) of a fixed (induced) subgraph H, and fixed lower or upper bounds on the degree of the vertices (hence also r-regularity). Typical graph theoretic concepts expressible in MSOL but not in FOL are connectivity, k-connectivity, reachability, k-colorability (of the vertices), and the presence or absence of a fixed (topological) minor. The latter includes planarity, and more generally, graphs of a fixed genus g. Typical graph theoretic concepts expressible in CMSOL but not in MSOL are the existence of an Eulerian circuit (path), the size of a connected component being a multiple of k, and the number of connected components is a multiple of k. All the non-definability statements above can be proved using Ehrenfeucht-Fraïssé games. The definability statements are straightforward.

10.3 Spectra of Sentences of Monadic Second Order Logic

Let φ be sentence of Monadic Second Order Logic MSOL over a vocabulary τ. The *spectrum* $\mathrm{SPEC}(\varphi)$ is the set of finite cardinalities $n \in \mathbb{N}$ of finite model of φ. We note that for unary languages, Parikh's Theorem looks at the spectrum of context-free languages. H. Scholz introduced spectra in [41], where he asked to characterize spectra. Spectra of first order sentences have been studied ever since. For a history and survey of the study of spectra, cf. [18].

10.3.1 Spectra of Sentences with One Unary Function Symbol

Let $S \subseteq \mathbb{N}$ be an ultimately periodic set. It is not difficult to construct even a first order sentence φ such the $\mathrm{SPEC}(\varphi) = S$. Surprisingly, ultimately periodic sets are precisely the spectra of sentences with one unary function symbol, and a finite set of unary relation symbols [16, 30].

Theorem 10.2 (Durand et al. [16], Gurevich and Shelah [30]). *Let φ be a sentence of* $\mathrm{MSOL}(\tau)$ *where τ consists of*

- *finitely many unary relation symbols,*
- *one unary function and equality only.*

Then $\mathrm{SPEC}(\varphi)$ *is ultimately periodic,*

10.3.2 From One Unary Function to Bounded Tree-Width

The finite structures which have only one unary function consist of disjoint unions of components of the form given in Figure 10.1. They look like directed forests where the roots are replaced by a directed cycle. The unary predicates are just colors attached to the nodes. The similarity of labeled graphs to labeled trees can be measured by the notion of *tree-width*, and in fact, these structures have tree width at most 2. The necessary background on tree-width will be given in Section 10.4.1. Inspired by Theorem 10.2, E. Fischer and J.A. Makowsky [23] generalized Theorem 10.2 by replacing the restriction on the vocabulary by a purely model theoretic condition involving the width of a relational structure.

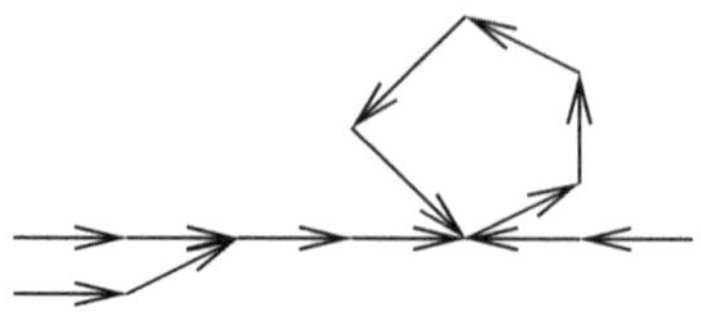

Fig. 10.1 Models of one unary function

Theorem 10.3 (E. Fischer and J.A. Makowsky [23]).
Let φ be an MSOL *sentence and $k \in \mathbb{N}$. Assume that all the models of φ have tree-width at most k. Then* SPEC(φ) *is ultimately periodic.*

To prove Theorem 10.3, and its generalizations below, the authors use three tools:

(a) A generalization of the Feferman-Vaught Theorem for k-sums of labeled graphs to the logic CMSOL, due to B. Courcelle, [11], and further refined by J.A. Makowsky in [33].
(b) A reduction of the problem to spectra of labeled trees by a technique first used by B. Courcelle in [14] in his study of graph grammars.
(c) An adaptation of the Pumping Lemma for labeled trees, cf. [29].

The proof of Theorem 10.3 is quite general. However, the details are rather technical. Its proof can be adapted to stronger logics, in particular to CMSOL introduced in Section 10.2. Second, one observes that very little of the definition of tree-width is used in the proof. The techniques used in the proof of Theorem 10.3 can be adapted to other notions of width of relational structures, such as *clique-width*, which was introduced first in [7] and studied more systematically in [10], and to *patch-width*, introduced in [23]. These notions will be discussed in detail in Section 10.4. Before we get to the various definitions of width we refer to any of these notions just as the *width* of a structure.

10.3.3 Many-Sorted Spectra

We want to generalize Theorem 10.1 to spectra. Rather than counting occurrences of letters, we look at many-sorted structures and the sizes of the different sorts, which we call many-sorted spectra.

For our purposes, a *k-sorted τ-structure* is a structure of the form

$$\mathfrak{A} = \langle A, P_1, \ldots, P_k, R_1, \ldots R_m \rangle$$

where A is a finite set, $P_i \subseteq A$, $\bigcup_{i=1}^{k} P_i = A$ and for all $i, j \leq k, i \neq j$ we have that $P_i \cap P_j = \varnothing$. The sets P_i are the sorts of $\mathfrak{A}$. Furthermore, the relations $R_\ell, \ell \leq m$ are typed. τ now consists of the relation symbols $\mathbf{P}_i, \mathbf{R}_\ell$, which are interpreted in $\mathfrak{A}$ in the obvious way. For a finite k-sorted τ-structure $\mathfrak{A}$ we denote by $\text{MCARD}(\mathfrak{A})$ the k-tuple $(n_1, \ldots, n_k)$ of the cardinalities n_i of P_i. The many-sorted spectrum $\text{MSPEC}(\varphi)$ of a k-sorted τ-sentence $\varphi \in \text{CMSOL}(\tau)$ is the set

$$\text{MSPEC}(\varphi) = \{\text{MCARD}(\mathfrak{A}) \in \mathbb{N}^k : \mathfrak{A} \models \varphi\}.$$

These definitions can be extended to MSOL augmented by the modular counting quantifiers $C_{k,m}$, where $C_{k,m}x\ \varphi(x)$ is interpreted as "there are, modulo m, exactly k elements satisfying $\varphi(x)$". We denote the extension of MSOL obtained by adding, for all $k, m \in \mathbb{N}$ the quantifiers $C_{k,m}$, by CMSOL.

In [23] the following theorem is proved:

Theorem 10.4 (E. Fischer and J.A. Makowsky [23]).
Let φ be a many-sorted $\text{CMSOL}(\tau)$-sentence, such that all of its models have width at most k. Then the many-sorted spectrum $\text{MSPEC}(\varphi)$ is a semi-linear set.

Recall that width stands here for any of the notions, tree-width, clique-width, patch-width to be discussed next.

10.4 Structures of Bounded Width

10.4.1 Tree-Width

In the eighties the notion of tree-width of a graph became a central focus of research in graph theory through the monumental work of Robertson and Seymour on graph minor closed classes of graphs, and its algorithmic consequences [40]. The literature is very rich, but good references and orientation may be found in [4, 5, 17]. Tree-width is a parameter that measures to what extent a graph is similar to a tree. Additional unary predicates do not affect the tree-width. Tree-width of directed graphs is defined as the tree-width of the underlying undirected graph.[1]

Definition 10.1 (Tree-width). A k-tree decomposition of a graph $G = (V, E)$ is a pair $(\{X_i \mid i \in I\}, T = (I, F))$ with $\{X_i \mid i \in I\}$ a family of subsets of V, one for each node of T, and T a tree such that

(a) $\bigcup_{i \in I} X_i = V$.
(b) for all edges $(v, w) \in E$ there exists an $i \in I$ with $v \in X_i$ and $w \in X_i$.
(c) for all $i, j, k \in I$: if j is on the path from i to k in T, then $X_i \cap X_k \subseteq X_j$ in other words, the subset $\{t \mid v \in X_t\}$ is connected for all v.
(d) for all $i \in I$, $|X_i| \leq k + 1$.

A graph G is of *tree-width at most k* if there exists a k-tree decomposition of G. A class of graphs K is a $TW(k)$-class iff all its members have tree width at most k.

[1] In [32] a different definition is given, which attempts to capture the specific situation of directed graphs. But the original definition is the one which is used when dealing with hyper-graphs and general relational structures.

Given a graph G and $k \in \mathbb{N}$ there are polynomial time, even linear time, algorithms, which determine whether G has tree-width k, and if the answer is yes, produce a tree decomposition, cf. [5]. However, if k is part of the input, the problem is **NP**-complete [2]

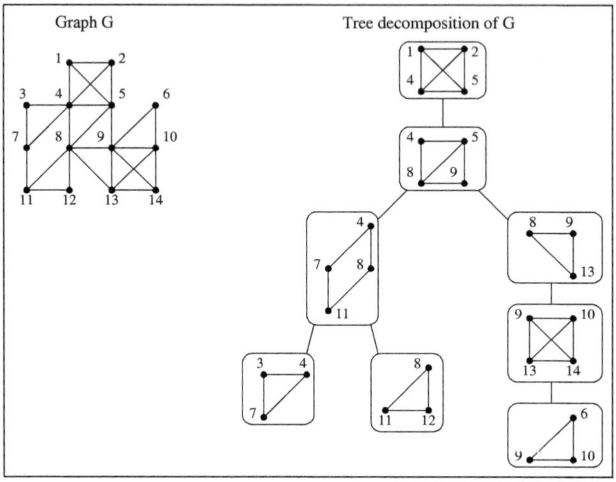

Fig. 10.2 A graph and one of its tree-decompositions

Trees have tree-width 1. The clique K_n has tree-width $n-1$. Furthermore, for fixed k, the class of finite graphs of tree-width at most k denoted by $TW(k)$, is MSOL-definable.

Example 10.1. The following graph classes are of tree-width at most k:

(a) Planar graphs of radius r with $k = 3r$.
(b) Chordal graphs with maximal clique of size c with $k = c - 1$.
(c) Interval graphs with maximal clique of size c with $k = c - 1$.

Example 10.2. The following graph classes have unbounded tree-width and are all MSOL-definable.

(a) All planar graphs and the class of all planar grids $G_{m,n}$.
Note that if $n \leq n_0$ for some fixed $n_0 \in \mathbb{N}$, then the tree-width of the grids $G_{m,n}, n \leq n_0$, is bounded by $2n_0$.
(b) The regular graphs of degree $r, r \geq 3$ have unbounded tree-width.

Tree-width for labeled graphs can be generalized to arbitrary relational structures in a straightforward way. Clause (ii) in the above definition is replaced by

(ii-rel) For each r-ary relation R, if $\bar{v} \in R$, there exists an $i \in I$ with $\bar{v} \in X_i^r$.

This was first used in [26].

10.4.2 Clique-Width

A k-colored τ-structures is a $\tau_k = \tau \cup \{P_1, \ldots, P_k\}$-structure where $P_i, i \leq k$ are unary predicate symbols the interpretation of which are disjoint (but can be empty).

Definition 10.2. Let $\mathfrak{A}$ be a k-colored τ-structure.

(a) (Adding hyper-edges) Let $R \in \tau$ be an r-ary relation symbol. $\eta_{R,P_{j_1},\ldots,P_{j_r}}(\mathfrak{A})$ denotes the k-colored τ structure $\mathfrak{B}$ with the same universe as $\mathfrak{A}$, and for each $S \in \tau_k$, $S \neq R$ the interpretation is also unchanged. Only for R we put

$$R^B = R^A \cup \{\bar{a} \in A^r : a_i \in P^A_{ji}\}.$$

We call the operation η *hyper edge creation*, or simply *edge creation* in the case of directed graphs. In the case of undirected graphs we denote by $\eta_{P_{j_1},P_{j_2}}$ the operation of adding the corresponding undirected edges.

(b) (Recoloring) $\rho_{i,j}(\mathfrak{A})$ denotes the k-colored τ structure $\mathfrak{B}$ with the same universe as $\mathfrak{A}$, and all the relations unchanged but for P^A_i and P^A_j. We put

$$P^B_i = \emptyset \text{ and } P^B_j = P^A_j \cup P^A_i.$$

We call this operation *recoloring*.

(c) (modification via quantifier free translation) More generally, for $S \in \tau_k$ of arity r and $B(x_1, \ldots, x_r)$ a quantifier free τ_k-formula, $\delta_{S,B}(\mathfrak{A})$ denotes the k-colored τ structure $\mathfrak{B}$ with the same universe as $\mathfrak{A}$, and for each $S' \in \tau_k$, $S' \neq S$ the interpretation is also unchanged. Only for S we put

$$S^B = \{\bar{a} \in A^r : \bar{a} \in B^A\}.$$

where B^A denotes the interpretation of B in $\mathfrak{A}$.

Note that the operations of type ρ and η are special cases of the operation of type δ.

Definition 10.3 (Clique-Width [10, 33]).

(a) Here $\tau = \{R_E\}$ consist of the symbol for the edge relation. Given a graph $G = (V, E)$, the *clique-width of G* (cwd(G)) is the minimal number of colors required to obtain the given graph as an $\{R_E\}$-reduct from a k-colored graph constructed inductively from colored singletons and closure under the following operations:

a. disjoint union ($\sqcup$)
b. recoloring ($\rho_{i \to j}$)
c. edge creation (η_{E,P_i,P_j})

(b) For τ containing more than one binary relation symbol, we replace the edge creation by the corresponding hyper edge creation $\eta_{R,P_{j_1},\ldots,P_{j_r}}$ for each $R \in \tau$.

(c) A class of τ-structures is a $CW(k)$-class if all its members have clique-width at most k.

If τ contains a unary predicate symbol U, the interpretation of U is not affected by the operations recoloring or edge creation. Only the disjoint union affects it.

A description of a graph or a structure using these operations is called a *clique-width parse term* (or *parse term*, if no confusion arises). Every structure of size n has clique-width at most n. The simplest class of graphs of unbounded tree-width but of clique-width at most 2 are the cliques. Given a graph G and $k \in \mathbb{N}$, determining whether G has clique-width k is in **NP**. A polynomial time algorithm was presented for $k \leq 3$ in [8]. It was shown in [24, 25] that for fixed $k \geq 4$ the problem is **NP**-complete. The recognition problem for clique-width of relational structures has not been studied so far even for $k = 2$. The relationship between tree-width and clique-width was studied in [1, 10, 27].

Theorem 10.5 (Courcelle and Olariu [10]). *Let K be a $TW(k)$-class of graphs. Then K is a $CW(m)$-class of graphs with $m \leq 2^{k+1} + 1$.*

Theorem 10.6 (Adler and Adler [1]). *For every non-negative integer n there is a structure $\mathfrak{A}$ with only one ternary relation symbol such that $\mathfrak{A} \in TW(2)$ and $\mathfrak{A} \notin CW(n)$.*

The following examples are from [28, 36].

Example 10.3 (Classes of clique-width at most k).

(a) The cographs with $k = 2$.
(b) The distance-hereditary graphs with $k = 3$.
(c) The cycles C_n with $k = 4$.
(d) The complement graphs $\overline{C_n}$ of the cycles C_n with $k = 4$.

The cycles C_n have tree-width at most 2, but the other examples have unbounded tree-width.

Example 10.4 (Classes of unbounded clique-width).

(a) The class of all finite graphs.
(b) The class of unit interval graphs.
(c) The class of permutation graphs.
(d) The regular graphs of degree 4 have unbounded clique-width.
(e) The class grids *Grid*, consisting of the graphs $Grid_{n\times n}$.

For more non-trivial examples, cf. [28, 36]. In contrast to $TW(k)$, we do not know whether the class of all $CW(k)$-graphs is MSOL-definable.

To find more examples it is useful to note, cf. [34]:

Proposition 10.1. *If a graph is of clique-width at most k and G' is an induced subgraph of G, then the clique-width of G' is at most k.*

In [23] the following is shown:

Theorem 10.7 (E. Fischer and J.A. Makowsky [23]).
Let $\varphi \in \mathrm{CMSOL}(\tau)$ be such that all its finite models have clique-width at most k. Then there are $m_0, n_0 \in \mathbb{N}$ such that if φ has a model of size $n \geq n_0$ then φ has also a model of size $n + m_0$.

From this we get immediately a further generalization of Theorem 10.3.

Corollary 10.1. *Let $\varphi \in \mathrm{CMSOL}(\tau)$ be such that all its finite models have clique-width at most k. Then $spec(\varphi)$ is ultimately periodic.*

10.4.3 Patch-Width

Here is a further generalization of clique-width for which our theorem still works. The choice of operation is discussed in detail in [9].

Definition 10.4. Given a τ-structure $\mathfrak{A}$, the *patch-width of G* (pwd(G)) is the minimal number of colors required to obtain $\mathfrak{G}$ as a $\{\tau\}$-reduct from a k-colored τ-structure inductively from fixed finite number of τ_k-structures and closure under the following operations:

(a) disjoint union ($\sqcup$),
(b) recoloring ($\rho_{i \to j}$) and
(c) modifications ($\delta_{S,B}$).

A class of τ-structures is a $PW_\tau(k)$-class if all its members have patch-width at most k.

A description of a τ-structure using these operations is called a *patch term.*

Example 10.5.

(a) In [10] it is shown that if a graph G has clique-width at most k then its complement graph $\overline{G}$ has clique-width at most $2k$. However, its patch-width is also k as $\overline{G}$ can be obtained from G by $\delta_{E,\neg E}$.
(b) The clique K_n has clique-width 2. However if we consider graphs as structures on a two-sorted universe (respectively for vertices and edges), then K_n has clique-width $c(n)$ and patch-width $p(n)$ where $c(n)$ and $p(n)$ are functions which tend to infinity. This will easily follow from Theorem 10.4. For the clique-width of K_n as a two-sorted-structure this was already shown in [39].

In [9] it is shown that a class of graphs of patch-width at most k is of clique-width at most $f(k)$ for some function f. It is shown in [22] that this is not true for relational structures in general.

In the definition of patch-width we allowed only unary predicates as auxiliary predicates (colors). We could also allow r-ary predicates and speak of r-ary patch-width. The theorems where bounded patch-width is required are also true for this more general case. The relative strength of clique-width and the various forms of patch-width are discussed in [22].

In [23] the following is shown:

Theorem 10.8 (E. Fischer and J.A. Makowsky [23]).
Let $\varphi \in \mathrm{CMSOL}(\tau)$ be such that all its finite models have patch-width at most k. Then there are $m_0, n_0 \in \mathbb{N}$ such that if φ has a model of size $n \geq n_0$ then φ has also a model of size $n + m_0$.

From this we get yet another generalization of Theorem 10.3.

Corollary 10.2. *Let $\varphi \in \mathrm{CMSOL}(\tau)$ be such that all its finite models have patch-width at most k. Then $spec(\varphi)$ is ultimately periodic.*

More recent work on spectra and patch-width may be found in [19, 42].

10.5 Applications of Theorem 10.8

In this paper we give several applications of Theorem 10.8 showing that certain classes of graphs have unbounded clique-width or patch-width.

10.5.1 Classes of Unbounded Patch-Width

Theorem 10.8 gives a new method to show that certain classes K of graphs have unbounded tree-width, clique-width or patch-width.

To see this we look at the class *Grid* of all grids $Grid_{n \times n}$. They are known to have unbounded tree-width, cf. [17], and in fact, every minor closed class of graphs of unbounded tree-width contains these grids. They were shown to have unbounded clique-width in [10, 28]. However, for patch-width these arguments do not work. On the other hand *Grid* is MSOL-definable, and its spectrum consists of the numbers n^2, so by Theorems 10.7 and 10.8, the unboundedness follows directly.

In particular, as this is also true for every $K' \supseteq K$, the class of all graphs is of unbounded patch-width.

Without Theorem 10.8, there was only a conditional proof of unbounded patch-width available. It depends on the assumption that the polynomial hierarchy Σ^{P} does not collapse to **NP**. The argument then proceeds as follows:

(a) Checking patch-width at most k of a structure $\mathfrak{A}$, for k fixed, is in **NP**. Given a structure $\mathfrak{A}$, one just has to guess a patch-term of size polynomial in the size of $\mathfrak{A}$.
(b) Using the results of [33] one gets that checking a $\mathrm{CMSOL}(\tau)$-property φ on the class $PW_\tau(k)$ is in **NP**, whereas, by [35], there are Σ^{P}_n-hard problems definable in MSOL for every level Σ^{P}_n of the polynomial hierarchy.
(c) Hence, if the polynomial hierarchy does not collapse to **NP**, the class of all τ-structures is of unbounded patch-width, provided τ is large enough.

10.5.2 The Patch-Width of Incidence Graphs

We look at two presentation a graph G. The first consists of a set of vertices $V(G)$ and a binary symmetric relation $E(G)$. The second is the incidence graph $I(G)$, which consists of two sorts, $V(G)$ and $E(G)$ and an incidence relation $R(G) \subseteq V(G) \times E(G)$ with $(u,e) \in R(G)$ iff there is $v \in V(G)$ with $(u,v) = e$.

It is well known that if G has tree-width at most k, so has $I(G)$. This is not so for clique-width. The cliques K_n have clique-width 2, but the clique-width of $I(K_n)$ grows with n. This follows immediately from Theorem 10.4. The cliques are definable by an MSOL-sentence φ_{clique} whose spectrum MSPEC(φ_{clique}) consists of the pairs $(n, \frac{n(n-1)}{2})$, which is not semi-linear.

Similarly, the class of square grids $Grid_{n\times n}$ and the class of its incidence graphs $I(Grid_{n\times n})$ are known to be definable by MSOL-sentences $\varphi_{sqgrids}$ and $\psi_{sqgrids}$, respectively. But their spectra are of the form n^2 and $(n^2, 2n^2+2n)$, respectively, which are not semi-linear. Hence they are not of bounded patch-width.

Our main result in this section is a new proof of the following theorem, implicit in papers by B. Courcelle and J. Engelfriet [6, 15, 21].

Theorem 10.9. *Let K be a class of graphs and $I(K)$ be the class of its incidence graphs. Assume $I(K)$ is defined by an* CMSOL*-sentence ψ and has bounded clique-width. Then K and $I(K)$ have bounded tree-width.*

10.5.3 Proof of Theorem 10.9

The proof of Theorem 10.9 combines Theorem 10.4 with the fact, taken from [31]:

Proposition 10.2. *Graphs of clique-width at most k which do not have the complete bipartite graph $K_{n,n}$ as a subgraph, have tree-width at most $3k(n-1)-1$.*

The clique-width of $I(K)$ and K are related as follows:

Proposition 10.3. *If $I(K)$ has bounded clique-width, so has K.*

This is a consequence of results in [21], formulated for graph grammars.

Furthermore, when we expand a τ-structure $\mathfrak{A}$ by adding new unary predicates $P_1,\dots,P_m$, whether definable or not, the width (tree-width, clique-width, patch-width) does not change.

Proposition 10.4. *Let $\mathfrak{A}$ be a τ-structure of width k. Then $\langle \mathfrak{A}, P_1,\dots,P_m\rangle$ as a $\tau \cup \{P_1,\dots P_m\}$ has also width k.*

Now assume $I(K)$ is defined by an CMSOL-sentence ψ and has bounded clique-width. Then, by Proposition 10.3, K has bounded clique-width. Assume further, for contradiction, that K has unbounded tree-width. Then, by Proposition 10.2, K contains all the complete bipartite graphs $K_{n,n}$ is subgraphs of graphs in K, and hence all the graphs $I(K_{n,n})$ as subgraphs of graphs of $I(K)$. Let $K' \subseteq K$ be the subclass of K of graphs which do contain a copy of $K_{n,n}$. Clearly, $I(K')$ is CMSOL-definable. We now show that $I(K') \subseteq I(K)$ has unbounded clique-width. We expand each graph of $I(K')$ by three disjoint unary predicates $P_i : i = 1,2,3$, one for each vertex set of $I(K_{n,n})$ and one for the edge set of $I(K_{n,n})$. The many-sorted spectrum of $I(K')$ with respect to these predicates is of the from $(n, n, \binom{n}{2})$, which is not semi-linear. Using Theorem 10.8, we conclude that $I(K')$ and hence $I(K)$ has unbounded clique-width.

10.6 Conclusions and Open problems

We have seen how Parikh's Theorem can be extended to arbitrary relational structures, provided they are of bounded width. Classes of structures of bounded patch-width share all of the interesting algorithmic properties which were proven for classes of bounded clique-width. In particular, membership in CMSOL-definable classes of structures of bounded patch-width can be decided in polynomial time, whereas for classes of unbounded patch-width this can arbitrarily hard within the polynomial hierarchy. As a matter of fact all the theorems proven for clique-width in [33] are also valid for patch-width. But this is only interesting, if the structures carry more relations than graphs. For graphs, it was shown in [9] that any class K of graphs bounded patch-width is also of bounded clique-width.

The true usefulness of patch-width as a structural parameter of relational structures still has to be explored. A first step is done in [22], where it is shown that there are classes of relational structures with unbounded clique-width but of bounded patch-width.

Open Question 10.1. *What is the complexity of checking whether a τ-structure $\mathfrak{A}$ has patch-width at most k, for a fixed k? What is the complexity, if k is part of the input?*

We conclude with some directions of further research. Proposition 10.3 actually is a special case of a more general theorem dealing with MSOL-transductions of graphs. A *transduction T* is, roughly speaking, a (possibly partial) map which, in a uniform way, associates with a τ-structure a τ_1-structure where the new atomic relations are replaced by MSOL(τ)-definable relations. The transductions here are scalar, i.e., the universe may be duplicated by taking disjoint copies of it, but the arities of the relations may not be changed. For details, the reader may consult [13, 33]. In [21] it the following is shown:

Theorem 10.10. *Let K be a class of graphs of bounded clique-width, and let T be a* MSOL*-transduction of graphs into graphs. Then $T(K) = \{T(G) : G \in K\}$ is also of bounded clique-width.*

By Proposition 10.4 this holds also for unary expansions of graphs. Therefore we have the following corollary to Theorem 10.8:

Corollary 10.3. *Let K be a many-sorted class of graphs of bounded clique-width, and let T be a* MSOL*-transduction of graphs into graphs. Let φ be a* CMSOL*-sentence such that all its models are in $T(K)$. Then $mspec(\varphi)$ is semi-linear.*

This suggest the following questions:

Open Question 10.2. *Are the grids $Grid_{n,n}$ distinctive for unbounded patch-width in the following sense:*

> *(S-1): Let K be a class of many-sorted τ-structures of unbounded patch-width. Does there exist a* CMSOL*-transduction T of τ-structures into graphs, such for all $n \in \mathbb{N}$ the grid $G_{n,n} \in T(K)$?*

One could also ask for a weaker version of (S-1) where we only require in the conclusion that

> *(S-2): $(\ldots)$ for infinitely many $n \in \mathbb{N}$ the grid $G_{n,n} \in T(K)$?*

Open Question 10.3. *Can we generalize Theorem 10.10 to the following:*

> *(S-2): Let K be a class of many-sorted τ-structures of bounded patch-width, and let T be a* CMSOL*-transduction of τ-structures into τ_1-structures. Then the class of τ_1-structures $T(K)$ is also of bounded patch-width.*

Open Question 10.4. *Can we generalize Corollary 10.3 to*

> *(S-3): Let K be a many-sorted class of τ-structures bounded patch-width, and let T be a* CMSOL*-transduction of τ-structures into τ_1-structures. Let φ be a* CMSOL(τ_1)*-sentence such that all its models are in $T(K)$. Then $mspec(\varphi)$ is semi-linear.*

Open Question 10.5. *Is there a converse to Corollary 10.3? More precisely, is the following true?*

> *(S-4): Let K be a many-sorted class of τ-structures such that for every* CMSOL*-transduction T of τ-structures into τ_1-structures, and for every* CMSOL(τ_1)*-sentence φ such that all its models are in $T(K)$, $mspec(\varphi)$ is semi-linear. Then K is of bounded patch-width.*

Note that trivially, the class of all graphs has unbounded patch-width, but it spectrum consists of $\mathbb{N}$ which is semi-linear. So the closure under definable classes in $T(K)$ is essential here. The statements (S-i) ($i = 1,2,3,4,5$) are not independent. (S-3) together with Theorem 10.8 implies (S-4). Also (S-1) and (S-3) implies (S-5). Note that (S-2) and (S-3) are not enough to get (S-5), unless we know that $T(K)$ is CMSOL-definable.

Acknowledgements Some passages are taken verbatim from [23] and from [18]. I would like to thank my co-authors A. Durand, E. Fischer, M. More and N. Jones for kindly allowing me to do so. I would like to thank B. Courcelle for valuable comments and references, and to I. Adler for making [1] available to me before it was posted.

References

1. H. Adler and I. Adler. A note on clique-width and tree-width on structures. Arxiv preprint arXiv:0806.0103v2, [cs.LO], 2008.
2. S. Arnborg, D.G. Corneil, and A. Proskurowski. Complexity of finding embedding in a k–tree. *SIAM Journal of Algebraic Discrete Methods*, 8:277–284, 1987.
3. L. Aceto, Z. Ésik, and A. Ingólfsdóttir. A fully equational proof of Parikh's theorem. *Informatique theéorique et applications*, 36(2):129–153, 2002.
4. H. Bodlaender. A tourist guide through tree width. *Acta Cybernetica*, 11:1–23, 1993.
5. H. Bodlaender. Treewidth: Algorithmic techniques and results. In I. Privara and P. Ruzicka, editors, *Proceedings of the 22th International Symposium on the Mathematical Foundation of Computer Science, MFCS'97*, volume 1295 of *Lecture Notes in Computer Science*, pages 29–36. Springer, Berlin, 1997.
6. B. Courcelle and J. Engelfriet. A logical characterization of the sets of hypergraphs defined by hyperedge replacement grammars. *Mathematical Systems Theory*, 28:515–552, 1995.
7. B. Courcelle, J. Engelfriet, and G. Rozenberg. Handle-rewriting hypergraph grammars. *Journal of Computer System Science*, 46:218–270, 1993.
8. D.G. Corneil, M. Habib, J.-M. Lanlignel, B. Reed, and U. Rotics. Polynomial time recognition of clique-width ≤ 3 graphs. In *Proceedings of LATIN'2000*, volume 1776 of *Lecture Notes in Computer Science*, pages 126–134. Springer, Berlin, 2000.
9. B. Courcelle and J.A. Makowsky. Fusion on relational structures and the verification of monadic second order properties. *Mathematical Structures in Computer Science*, 12.2:203–235, 2002.
10. B. Courcelle and S. Olariu. Upper bounds to the clique–width of graphs. *Discrete Applied Mathematics*, 101:77–114, 2000.
11. B. Courcelle. The monadic second-order logic of graphs I: Recognizable sets of finite graphs. *Information and Computation*, 85:12–75, 1990.
12. B. Courcelle. Monadic second-order logic of graphs VII: Graphs as relational structures. *Theoretical Computer Science*, 101:3–33, 1992.
13. B. Courcelle. Monadic second order graph transductions: A survey. *Theoretical Computer Science*, 126:53–75, 1994.
14. B. Courcelle. Structural properties of context-free sets of graphs generated by vertex replacement. Information and computation, 116:275–293, 1995.
15. B. Courcelle. The monadic second order logic of graphs, XIV: Uniformly sparse graphs and edge set quantification. *Theoretical Computer Science*, 299(1):1–36, 2003.
16. A. Durand, R. Fagin, and B. Loescher. Spectra with only unary function symbols. In M. Nielsen and W. Thomas, editors, *CSL'97*, volume 1414 of *Lecture Notes in Computer Science*, pages 189–202. Springer, Berlin, 1997.
17. R. Diestel. *Graph Theory*. Graduate Texts in Mathematics. Springer, Berlin, 1996.
18. A. Durand, N. Jones, J.A. Makowsky, and M. More. 50 years of the spectrum problem: Survey and new results, Available on the Arxiv: arXiv:0907.5495v1, 2009.
19. M. Doron and S. Shelah. Bounded m-ary patch-width are equivalent for $m > 2$. Electronically available at arXiv:math/0607375v1, 2006.
20. H. Ebbinghaus and J. Flum. *Finite Model Theory*. Springer, Berlin, 1995.
21. J. Engelfriet and V. van Oostrom. Logical description of context-free graph-languages. *Journal of Computer and System Sciences*, 55:489–503, 1997.
22. E. Fischer and J.A. Makowsky. Patch-width, a generalization of clique-width for relational structures. under preparation, 2010.
23. E. Fischer and J.A. Makowsky. On spectra of sentences of monadic second order logic with counting. *Journal of Symbolic Logic*, 69(3):617–640, 2004.
24. M.R. Fellows, F.A. Rosamond, U. Rotics, and S. Szeider. Proving NP-hardness for clique width ii: Non-approximability of clique-width. Electronic Colloquium on Computational Complexity, 2005.

25. M.R. Fellows, F.A. Rosamond, U. Rotics, and S. Szeider. Clique-width minimization is np-hard. In *STOC '06: Proceedings of the Thirty-Eighth Annual ACM Symposium on Theory of Computing*, pages 354–362. ACM, Seattle, WA, USA, May 21–23, 2006.
26. T. Feder and M. Vardi. The computational structure of monotone monadic SNP and constraint satisfaction: A study through Datalog and group theory. *SIAM Journal on Computing*, 28:57–104, 1999.
27. A. Glikson and J.A. Makowsky. NCE graph grammars and clique-width. In H.L. Bodlaender, editor, *Proceedings of the 29th International Workshop on Graph-Theoretic Concepts in Computer Science (WG 2003), Elspeet (The Netherlands)*, volume 2880 of *Lecture Notes in Computer Science*, pages 237–248. Springer, Berlin, 2003.
28. M.C. Golumbic and U. Rotics. On the clique-width of some perfect graph classes. *Internation Journal of Foundations of Computer Science*, 11:423–443, 2000.
29. F. Gécseg and M. Steinby. Tree languages. In G. Rozenberg and A. Salomaa, editors, *Handbook of Formal Languages, Vol. 3 : Beyond Words*, pages 1–68. Springer, Berlin, 1997.
30. Y. Gurevich and S. Shelah. Spectra of monadic second-order formulas with one unary function. In *LICS'03*, IEEE, pages 291–300, 2003.
31. F. Gurski and E. Wanke. The tree-width of clique-width bounded graphs without $K_{n,n}$. *Lecture Notes in Computer Science*, 1928:196–205, 2000.
32. T. Johnson, N. Robertson, P. Seymour, and R. Thomas. Directed tree-width. *Journal of Combinatorial Theory, Serie B*, 81(1):138–154, 2001.
33. J.A. Makowsky. Algorithmic uses of the Feferman-Vaught theorem. *Annals of Pure and Applied Logic*, 126.1-3:159–213, 2004.
34. J.A. Makowsky and J.P. Mariño. Tree-width and the monadic quantifier hierarchy. *Theoretical Computer Science*, 303:157–170, 2003.
35. J.A. Makowsky and Y. Pnueli. Arity vs. alternation in second order logic. *Annals of Pure and Applied Logic*, 78(2):189–202, 1996.
36. J.A. Makowsky and U. Rotics. On the cliquewidth of graphs with few P_4's. *International Journal on Foundations of Computer Science*, 10:329–348, 1999.
37. R. Parikh. On context-free languages. *JACM*, 13:570–581, 1966.
38. D.L. Pilling. Commutative regular equations and Parikh's theorem. *Journal of London Mathematical Society*, 6:663–666, 1973.
39. U. Rotics. *Efficient Algorithms for Generally Intractable Graph Problems Restricted to Specific Classes of Graphs*. PhD thesis, Technion- Israel Institute of Technology, 1998.
40. N. Robertson and P. D. Seymour. Graph minors. ii. algorithmic aspects of tree-width. *Journal of Algorithms*, 7:309–322, 1986.
41. H. Scholz. Problem # 1: Ein ungelöstes Problem in der symbolischen Logik. *Journal of Symbolic Logic*, 17:160, 1952.
42. S. Shelah. Spectra of monadic second order sentences. Paper No. 817, Electronically available at arXiv:math/0405158, 2004.

Chapter 11
Syllogistic Logic with Complements

Larry Moss

11.1 Introduction

This paper presents a logic for statements of the form *All X are Y* and *Some X are Y*, where the X and Y are intended as (plural) nouns or other expressions whose natural denotation is as subsets of an underlying universe. Languages like this have been studied previously, and the novelty here is to add an *explicit complement operator* to the syntax. So we now can say, for example, *All* X' *are* Y, or *Some non-X are Y*. The point of the paper is to present a sound and complete proof system for the associated entailment relation. In its details, the work is rather different from previous work in the area (for example, [1, 3, 5–7] and references therein). Our particular system is new as far as we know (but see just below) In addition, the work here builds models using a representation theorem coming from quantum logic.

Closely related work. After this paper was written, Ian Pratt-Hartmann and I wrote [10], a paper which studies syllogistic logics with transitive verbs. Section 3 of that paper contains a result which is closely related to Theorem 2.5 of this paper. There are a few differences: the proof in this paper here makes direct connections with representation of orthoposets, an unexpected topic for a paper in logic for natural language. In the other direction, the work in [10] gives the additional result that the validity problem for the logic is in NLOGSPACE, and as a result the proof there takes more work.

The contents Sections 11.3 and 11.4 are new here.

After we present our basic systems, we discuss other papers which present less-closely-related work; see Section 11.1.3

Larry Moss
Department of Mathematics, Indiana University, Bloomington, IN 47405, USA,
e-mail: lsm@cs.indiana.edu

J. van Benthem et al. (eds.), *Games, Norms and Reasons*, Synthese Library 353,
DOI 10.1007/978-94-007-0714-6_11, © Springer Science+Business Media B.V. 2011

11.1.1 Syllogistic Logic with Complement

We start with the syntax and semantics of a language which we call $\mathcal{L}(all, some, ')$. Let $\mathcal{V}$ be an arbitrary set whose members will be called *variables*. We use $X, Y, \ldots,$ for variables. The idea is that they represent plural common nouns. We also assume that there is a *complementation operation* $' : \mathcal{V} \to \mathcal{V}$ on the variables such that $X'' = X$ for all X. This *involutive* property implies that complementation is a bijection on $\mathcal{V}$. In addition, to avoid some uninteresting technicalities, we shall always assume that $X \neq X'$. Then we consider sentences *All X are Y* and *Some X are Y*. Here X and Y are any variables, including the case when they are the same. We call this language $\mathcal{L}(all, some, ')$. We shall use letters like S to denote sentences.

Semantics. One starts with a set M and a subset $[\![X]\!] \subseteq M$ for each variable X, subject to the requirement that $[\![X']\!] = M \setminus [\![X]\!]$ for all X. This gives a *model* $\mathcal{M} = (M, [\![\ ]\!])$. We then define

$$\begin{array}{lll} \mathcal{M} \models \textit{All } X \textit{ are } Y & \text{iff} & [\![X]\!] \subseteq [\![Y]\!] \\ \mathcal{M} \models \textit{Some } X \textit{ are } Y & \text{iff} & [\![X]\!] \cap [\![Y]\!] \neq \varnothing \end{array}$$

We say $\mathcal{M}$ *satisfies* S iff $\mathcal{M} \models S$. We allow $[\![X]\!]$ to be empty, and in this case, recall that $\mathcal{M} \models$ *All X are Y* vacuously. (For that matter, we also allow a model to have an empty universe.) And if Γ is a set of sentences, then we write $\mathcal{M} \models \Gamma$ to mean that $\mathcal{M} \models S$ for all $S \in \Gamma$. $\Gamma \models S$ means that every model which satisfies all sentences in Γ also satisfies S.

Example 11.1. We claim that $\Gamma \models$ *All A are C*, where

$$\Gamma \quad = \quad \{\textit{All } B' \textit{ are } X, \textit{All } X \textit{ are } Y, \textit{All } Y \textit{ are } B, \textit{All } B \textit{ are } X, \textit{All } Y \textit{ are } C\}.$$

Here is an informal explanation. Since all B and all B' are X, everything whatsoever is an X. And since all X are Y, and all Y are B, we see that everything is a B. In particular, all A are B. But the last two premises and the fact that all X are Y also imply that all B are C. So all A are C.

No. In previous work, we took *No X are Y* as a basic sentence in the syntax. There is no need to do this here: we may regard *No X are Y* as a variant notation for *All X are Y'*. So the semantics would be

$$\mathcal{M} \models \textit{No } X \textit{ are } Y \quad \text{iff} \quad [\![X]\!] \cap [\![Y]\!] = \varnothing$$

In other words, if one wants to add *No* as a basic sentence forming-operation, on a par with *Some* and *All*, it would be easy to do so.

Proof trees. We have discussed the meager syntax of $\mathcal{L}(all, some, ')$ and its semantics. We next turn to the proof theory. A *proof tree over* Γ is a finite tree $\mathcal{T}$ whose nodes are labeled with sentences in our fragment, with the additional property that

$$\frac{}{\textit{All } X \textit{ are } X}\;\textit{Axiom} \qquad \frac{\textit{Some } X \textit{ are } Y}{\textit{Some } X \textit{ are } X}\;\textit{Some}_1 \qquad \frac{\textit{Some } X \textit{ are } Y}{\textit{Some } Y \textit{ are } X}\;\textit{Some}_2$$

$$\frac{\textit{All } X \textit{ are } Z \quad \textit{All } Z \textit{ are } Y}{\textit{All } X \textit{ are } Y}\;\textit{Barbara} \qquad \frac{\textit{All } Y \textit{ are } Z \quad \textit{Some } X \textit{ are } Y}{\textit{Some } X \textit{ are } Z}\;\textit{Darii}$$

$$\frac{\textit{All } Y \textit{ are } Y'}{\textit{All } Y \textit{ are } X}\;\textit{Zero} \qquad \frac{\textit{All } Y' \textit{ are } Y}{\textit{All } X \textit{ are } Y}\;\textit{One}$$

$$\frac{\textit{All } Y \textit{ are } X'}{\textit{All } X \textit{ are } Y'}\;\textit{Antitone} \qquad \frac{\textit{All } X \textit{ are } Y \quad \textit{Some } X \textit{ are } Y'}{S}\;\textit{Contrad}$$

Fig. 11.1 Syllogistic logic with complement

each node is either an element of Γ or comes from its parent(s) by an application of one of the rules for the fragment listed in Figure 11.1. $\Gamma \vdash S$ means that there is a proof tree $\mathscr{T}$ for over Γ whose root is labeled S.

We attached names to the rules in Figure 11.1 so that we can refer to them later. We usually do not display the names of rules in our proof trees except when to emphasize some point or other. The only purpose of the axioms *All X are X* is to derive these sentences from all sets; otherwise, the axioms are invisible. The names "Barbara" and "Darii" are traditional from Aristotelian syllogisms. But the (*Antitone*) rule is not part of traditional syllogistic reasoning. It is possible to drop ($\textit{Some}_2$) if one changes the conclusion of (*Darii*) to *Some Z are X*. But at one point it will be convenient to have ($\textit{Some}_2$), and so this guides the formulation. The rules (*Zero*) and (*One*) are concerned with what is often called vacuous universal quantification. That is, if $Y' \subseteq Y$, then Y is the whole universe and Y' is empty; so Y is a superset of every set and Y' a subset. It would also be possible to use binary rules instead; in the case of (*Zero*), for example, we would infer *All X are Z* from *All X are Y* and *All X are Y'*. The (*Contrad*) rule is *ex falso quodlibet*; it permits inference of any sentence S whatsoever from a contradiction. See also Section 11.1.2 for a different formulation, and Proposition 11.2 and Theorem 11.4 for their equivalence.

Example 11.2. Returning to Example 11.1, here is a proof tree showing $\Gamma \vdash$ *All A are C*:

$$\dfrac{\dfrac{\dfrac{\textit{All } B' \textit{ are } X \quad \dfrac{\textit{All } X \textit{ are } Y \quad \textit{All } Y \textit{ are } B}{\textit{All } X \textit{ are } B}}{\textit{All } B' \textit{ are } B}}{\textit{All } A \textit{ are } B} \qquad \dfrac{\textit{All } B \textit{ are } X \quad \dfrac{\textit{All } X \textit{ are } Y \quad \textit{All } Y \textit{ are } C}{\textit{All } X \textit{ are } C}}{\textit{All } B \textit{ are } C}}{\textit{All } A \textit{ are } C}$$

Example 11.3. Readers who desire an exercise might wish to show that

$$\{\textit{All } B \textit{ are } X, \textit{All } B' \textit{ are } X, \textit{All } Y \textit{ are } C, \textit{Some } A \textit{ are } C'\} \vdash \textit{Some } X \textit{ are } Y'.$$

A solution is displayed in the proof of Lemma 11.12.

Lemma 11.1. *The following are derivable:*

(a) Some X are $X' \vdash S$ *(a contradiction fact)*
(b) All X are Z, No Z are $Y \vdash$ No Y are X *(Celarent)*
(c) No X are $Y \vdash$ No Y are X *(E-conversion)*
(d) Some X are Y, *No* Y *are* $Z \vdash$ Some X are Z' *(Ferio)*
(e) All Y are Z, All Y are $Z' \vdash$ No Y are Y *(complement inconsistency)*

Proof. For the assertion on contradictions,

$$\dfrac{\dfrac{}{\textit{All X are X}}\ \textit{Axiom} \qquad \textit{Some X are X}'}{S}\ \textit{Contrad}$$

(*Celarent*) in this formulation is just a re-phrasing of (*Barbara*), using complements:

$$\dfrac{\textit{All X are Z} \quad \textit{All Z are Y}'}{\textit{All Y are Z}'}\ \textit{Barbara}$$

(*E-conversion*) is similarly related to (*Antitone*), and (*Ferio*) to (*Darii*). For complement inconsistency, use (*Antitone*) and (*Barbara*).

The logic is easily seen to be sound: if $\Gamma \vdash S$, then $\Gamma \models S$. The main contribution of this paper is the completeness of this system, and an extended discussion on the relation of the principle of *ex falso quodibet* in the form of (*Contrad*) with *reductio ad absurdum* in the form which we shall see shortly.

Some syntactic abbreviations. The language lacks boolean connectives, but it is convenient to use an informal notation for it. It is also worthwhile specifying an operation of *duals*.

$$\begin{array}{lcl|lcl} \neg(\textit{All X are Y}) & = & \textit{Some X are Y}' & (\textit{All X are Y})^d & = & \textit{All Y}' \textit{ are X}' \\ \neg(\textit{Some X are Y}) & = & \textit{All X are Y}' & (\textit{Some X are Y})^d & = & \textit{Some Y are X} \end{array}$$

Here are some uses of this notation. We say that Γ is *inconsistent* if for some S, $\Gamma \vdash S$ and $\Gamma \vdash \neg S$. The first part of Lemma 11.1 tells us that if $\Gamma \vdash$ *Some X is X'*, then Γ is inconsistent. Also, we have the following result:

Proposition 11.1. *If* $S \vdash T$, *then* $\neg T \vdash \neg S$.

This fact is not needed below, but we recommend thinking about it as a way of getting familiar with the rules.

11.1.2 The Indirect System: Reductio Ad Absurdum

Frequently the logic of syllogisms is set up as an *indirect* system, where one in effect takes *reductio ad absurdum* to be part of the system instead of (*Contrad*). We formulate a notion $\Gamma \vdash_{RAA} S$ of indirect proof in this section. It is easy to check that the (RAA) system is stronger than the one with (*Contrad*). It will turn out that the weaker system is complete, and then since the stronger one is sound, it is thus complete as well. In the final section of the paper, we even provide a proof-theoretic reduction.

We define $\Gamma \vdash_{RAA} S$ as follows:

(a) If $S \in \Gamma$ or S is *All X are X*, then $\Gamma \vdash_{RAA} S$
(b) For all rules in Figure 11.1 except the contradiction rule, if S_1 and S_2 are the premises of some instance of the rule, and T the conclusion, if $\Gamma \vdash_{RAA} S_1$ and $\Gamma \vdash_{RAA} S_2$, then also $\Gamma \vdash_{RAA} T$.
(c) If $\Gamma \cup \{S\} \vdash_{RAA} T$ and $\Gamma \cup \{S\} \vdash_{RAA} \neg T$, then $\Gamma \vdash_{RAA} \neg S$.

In effect, one is adding hypothetical reasoning in the manner of the sequent calculus.

Proposition 11.2. *If $\Gamma \vdash S$, then $\Gamma \vdash_{RAA} S$.*

Proof. By induction on the heights of proof trees for $\vdash$. The only interesting step is when $\Gamma \vdash S$ via application of the contradiction rule. So for some T, $\Gamma \vdash T$ and $\Gamma \vdash \neg T$. Using the induction hypothesis, $\Gamma \vdash_{RAA} T$ and $\Gamma \vdash_{RAA} \neg T$. Clearly we also have $\Gamma \cup \{\neg S\} \vdash_{RAA} T$ and $\Gamma \cup \{\neg S\} \vdash_{RAA} \neg T$. Hence $\Gamma \vdash_{RAA} S$.

It is natural to ask whether the converse holds. We show that it does in Section 11.4, using proof theory. Prior to this, we get the same result via a semantic argument using completeness.

11.1.3 Comparison with Previous Work

The proof system in this paper, the one presented by the rules in Figure 11.1, appears to be new. However, the indirect system of Section 11.1.2 appears to be close to the earlier work of Corcoran [3] and Martin [6]. These papers are mostly concerned with modern reconstruction of Aristotleian syllogisms, as is the pioneering work in this area, Łukasiewicz's book [5]. We are not so concerned with this project, but rather our interest lies in logical completeness results for fragments of natural language. The fragment in this paper is obviously quite small, but we believe that the techniques used in studying it may help with larger fragments. This is the main reason for this work.

We write $\mathscr{L}(all, some)$ for the fragment of $\mathscr{L}(all, some, ')$ without the complement. That is, one drops the complementation from the syntax, treating $\mathscr{V}$ as a set (simpliciter) rather than a set with an operation. One also drops the requirement that $[\![X']\!]$ be the complement of $[\![X]\!]$, since this now makes no sense. For the proof theory, use the rules on the top half of Figure 11.1. In [7] we checked the completeness of this fragment. We also considered the extension $\mathcal{L}(all, some, no)$ of $\mathcal{L}(all, some)$ with sentences *No X are Y*. For that, one needs the (*Contrad*) rule and also some additional axioms: the equivalence of *No X are Y* and *All X are Y'* cannot be stated without complements. Indeed, the language $\mathcal{L}(all, some, ')$ of this paper is more expressive than $\mathcal{L}(all, some, no)$ in the following precise sense: Consider the two models $\mathcal{M}$ and $\mathcal{N}$ shown below:

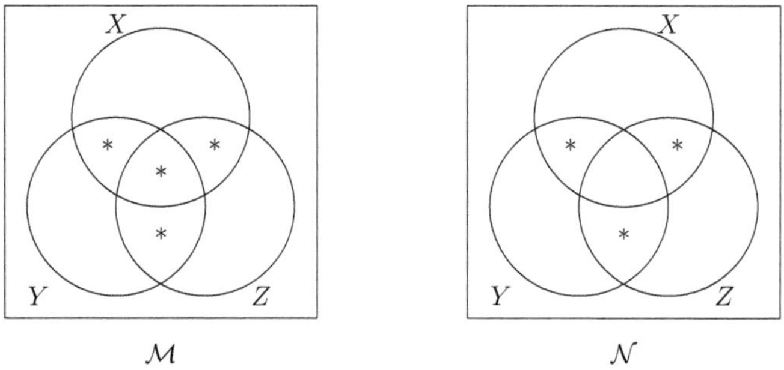

They satisfy the same sentences in $\mathcal{L}(all, some, no)$. (They also satisfy the same sentences of the form *Some A are B'*.) But let S be *Some X' are Y'* so that $\neg S$ is *All X' are Y*. $\mathcal{M} \models S$ but $\mathcal{N} \models \neg S$. We conclude from this example is that a logical system for the language with complements cannot simply be a translation into the smaller language.

I obtained the results in Section 11.2 below without knowing about the concept of an *orthoposet*. I later found the paper of Calude et al [2], and Theorem 11.1 was essentially stated there. [2] itself notes earlier work in the area, mentioning variations on Theorem 11.1 theorem in Zierler and Schlessinger [12] and also in Katrnoška [4]. We have included proofs in Section 11.2, mainly because we need to know in Theorem 11.1 that the map m preserves the order in *both* directions: the statement in [2] only has m preserving the order and being one-to-one. Still the proof is essentially the same as in [2].

11.2 Completeness via Representation of Orthoposets

An important step in our work is to develop an *algebraic semantics* for $\mathcal{L}(all, some, ')$. There are several definitions, and then a representation theorem. As with other uses of algebra in logic, the point is that the representation theorem is also a *model construction technique*.

An *orthoposet* is a tuple $(P, \leq, 0, ')$ such that

(a) $(P, \leq)$ is a partial order: $\leq$ is a reflexive, transitive, and antisymmetric relation on the set P.

(b) 0 is a minimum element: $0 \leq p$ for all $p \in P$.
(c) $x \mapsto x'$ is an antitone map in both directions: $x \leq y$ iff $y' \leq x'$.
(d) $x \mapsto x'$ is involutive: $x'' = x$.
(e) complement inconsistency: If $x \leq y$ and $x \leq y'$, then $x = 0$.

The notion of an orthoposet mainly appears in papers on quantum logic. (In fact, the stronger notion of an *orthomodular poset* appears to be more central there. However, I do not see any application of this notion to logics of the type considered in this paper.)

Example 11.4. For example, for all sets X we have an orthoposet $(\mathcal{P}(X), \subseteq, \varnothing, ')$, where $\subseteq$ is the inclusion relation, $\varnothing$ is the empty set, and $a' = X \setminus a$ for all subsets a of X.

Example 11.5. Let Γ be any set of sentences in $\mathcal{L}(all, some, ')$. Γ need not be consistent. We define a relation $\leq_\Gamma$ on the set $\mathcal{V}$ of variables of our logical system by

$$X \leq_\Gamma Y \quad \text{iff} \quad \Gamma \vdash \textit{All } X \textit{ are } Y.$$

We always drop the subscript Γ because it will be clear from the context which set Γ is used. We have an induced equivalence relation $\equiv$, and we take $\mathcal{V}_\Gamma$ to be the quotient $\mathcal{V}/\!\equiv$. It is a partial order under the induced relation. If there is some X such that $X \leq X'$, then for all Y we have $[X] \leq [Y]$ in $\mathcal{V}/\!\equiv$. In this case, set 0 to be $[X]$ for any such X. (If such X exists, its equivalence class is unique.) We finally define $[X]' = [X']$. If there is no X such that $X \leq X'$, we add fresh elements 0 and 1 to $\mathcal{V}/\!\equiv$. We then stipulate that $0' = 1$, and that for all $x \in \mathcal{V}_\Gamma$, $0 \leq x \leq 1$.

It is not hard to check that we have an orthoposet $\mathcal{V}_\Gamma = (\mathcal{V}_\Gamma, \leq, 0, ')$. The antitone property comes from the axiom with the same name, and the complement inconsistency is verified using the similarly-named part of Lemma 11.1.

A *morphism of orthoposets* is a map m preserving the order (if $x \leq y$, then $mx \leq my$), the complement $m(x') = (mx)'$, and minimum elements ($m0 = 0$). We say m is *strict* if the following extra condition holds: $x \leq y$ iff $mx \leq my$.

A *point* of a orthoposet $P = (P, \leq, 0, ')$ is a subset $S \subseteq P$ with the following properties:

(a) If $p \in S$ and $p \leq q$, then $q \in S$ (S is *up-closed*).
(b) For all p, either $p \in S$ or $p' \in S$ (S is *complete*), but not both (S is *consistent*).

Example 11.6. Let $X = \{1,2,3\}$, and let $\mathcal{P}(X)$ be the power set orthoposet from Example 11.4. Then S is a point, where

$$S \quad = \quad \{\{1,2\}, \{1,3\}, \{2,3\}, \{1,2,3\}\}.$$

(More generally, if X is any finite set, then the collection of subsets of X containing more than half of the elements of X is a point of $\mathcal{P}(X)$.) Also, it is easy to check that the points on this $\mathcal{P}(X)$ are exactly S as above and the three principal ultrafilters. S shows that a point of a boolean algebras need not be an ultrafilter or even a filter. Also, the lemma just below shows that for $\mathscr{P}(X)$, a collection of elements is included in a point iff every pair of elements has a non-empty intersection.

Lemma 11.2. *For a subset S_0 of an orthoposet $P = (P, \leq, ')$, the following are equivalent:*

(a) S_0 is a subset of a point S in P.
(b) For all $x, y \in S_0$, $x \nleq y'$.

Proof. Clearly (1) $\Longrightarrow$ (2). For the more important direction, use Zorn's Lemma to get a $\subseteq$-maximal superset S_1 of S_0 with the consistency property. Let $S = \{q : (\exists p \in S_1) q \geq p\}$. So S is up-closed. We check that consistency is not lost: suppose that $r, r' \in S$. Then there are $q_1, q_2 \in S_1$ such that $r \geq q_1$ and $r' \geq q_2$. But then $q_2' \geq r \geq q_1$. Since $q_1 \in S_1$, so too $q_2' \in S_1$. Thus we see that S_1 is not consistent, and this is a contradiction. To conclude, we only need to see that for all $r \in P$, either r or r' belongs to S. If $r \notin S$, then $r \notin S_1$. By maximality, there is $q \in S_1$ such that $q_1 \leq r'$. (For otherwise, $S_1 \cup \{r\}$ would be a consistent proper superset of S_1.) And as $r' \notin S$, there is $q_2 \in S_1$ such that $q_2 \leq r$. Then as above $q_1 \leq q_2'$, leading to the same contradiction.

We now present a representation theorem that implies the completeness of the logic. It is due to Calude et al. [2]. We also state an additional technical point.

Theorem 11.1 ([2]; see also [4, 12]). *Let $P = (P, \leq, ')$ be an orthoposet. There is a set* points(P) *and a strict morphism of orthoposets $m : P \to \mathscr{P}(\text{points}(P))$.*

Moreover, if $S \cup \{p\} \subseteq P$ has the following two properties, then $m(p) \setminus \bigcup_{q \in S} m(q)$ is non-empty:

(a) For all $q \in S$, $p \nleq q$.
(b) For all $q, r \in S$, $q \ngeq r'$.

Proof. Let *points*(P) be the collection of points of P. The map m is defined by $m(p) = \{S : p \in S\}$. The preservation of complement comes from the completeness and consistency requirement on points, and the preservation of order from the up-closedness. Clearly $m0 = \varnothing$. We must check that if $q \ngeq p$, then there is some point S such that $p \in S$ and $q \notin S$. For this, take $S = \{q\}$ in the "moreover" part. And for that, let $T = \{p\} \cup \{q' : q \in S\}$. Lemma 11.2 applies, and so there is some point $U \supseteq T$. Such U belongs to $m(p)$. But if $q \in S$, then $q' \in T \subseteq U$; so U does not belong to $m(q)$.

11.2.1 Completeness

The completeness theorem is based on algebraic machinery that we have just seen.

Lemma 11.3. *Let $\Gamma \subseteq \mathscr{L}(\text{all}, \text{some}, ')$. There is a model $\mathscr{M} = (M, [\![\ ]\!])$ such that*

(a) $\mathscr{M} \models \Gamma_{all}$.
(b) If T is a sentence in All *and $\mathscr{M} \models T$, then $\Gamma \vdash T$.*
(c) If Γ is consistent, then also $\mathscr{M} \models \Gamma_{some}$.

Proof. Let $\mathscr{V} = \mathscr{V}_\Gamma$ be the orthoposet from Example 11.5 for Γ. Let n be the natural map of $\mathscr{V}$ into $\mathscr{V}_\Gamma$, taking a variable X to its equivalence class $[X]$. If $X \leq Y$, then $[X] \leq [Y]$ by definition of the structure. In addition, n preserves the order in both directions. We also apply Theorem 11.1, to obtain a strict morphism of orthoposets m as shown below:

$$\mathscr{V} \xrightarrow{\quad n \quad} \mathscr{V}_\Gamma \xrightarrow{\quad m \quad} points(\mathscr{V}_\Gamma)$$

Let $M = points(\mathscr{V}_\Gamma)$, and let $[\![\ ]\!] : \mathscr{V} \to \mathscr{P}(M)$ be the composition $m \circ n$. We thus have a model $\mathscr{M} = (points(\mathscr{V}_\Gamma), [\![\ ]\!])$.

We check that $\mathscr{M} \models \Gamma$. Note that n and m are strict monotone functions. So the semantics has the property that the *All* sentences holding in $\mathscr{M}$ are exactly the consequences of Γ. We turn to a sentence in Γ_{some} such as *Some U are V*. Assuming the consistency of Γ, $U \not\leq V'$. Thus $[\![U]\!] \not\subseteq ([\![V]\!])'$. That is, $[\![U]\!] \cap [\![V]\!] \neq \varnothing$.

Unfortunately, the last step in this proof is not reversible, in the following precise sense. $U \not\leq V'$ does not imply that $\Gamma \vdash$ *Some U are V*. (For example, if Γ is the empty set we have $U \not\leq V'$, and indeed $\mathscr{M}(\Gamma) \models$ *Some U are V*. But Γ only derives valid sentences.

Lemma 11.4 (Pratt-Hartmann [9]). *Suppose that $\Gamma \models$* Some X are Y. *Then there is some existential sentence in Γ, say* Some A are B, *such that*

$$\Gamma_{all} \cup \{\text{Some } A \text{ are } B\} \models \text{Some } X \text{ are } Y.$$

Proof. If not, then for every $T \in \Gamma_{some}$, there is a model $\mathscr{M}_T \models \Gamma_{all} \cup \{T\}$ and $\mathscr{M} \models$ *All X are Y'*. Take the disjoint union of the models $\mathscr{M}_T$ to get a model of $\Gamma_{all} \cup \Gamma_{some} = \Gamma$ where S fails.

Theorem 11.2. *$\Gamma \vdash S$ iff $\Gamma \models S$.*

Proof. As always, the soundness half is trivial. Suppose that $\Gamma \models S$; we show that $\Gamma \vdash S$. We may assume that Γ is consistent.

If S is a sentence in *All*, consider $\mathscr{M}(\Gamma)$ from Lemma 11.3. It is a model of Γ, hence of S; and then by the property the second part of the lemma, $\Gamma \vdash S$.

For the rest of this proof, let S be *Some X are Y*. From Γ and S, we find A and B satisfying the conclusion of Lemma 11.4.

We again use Lemma 11.3 and consider the model $\mathscr{M} = \mathscr{M}(\mathscr{V}_{\Gamma_{all}})$ of points on $\mathscr{V}_{\Gamma_{all}}$. $\mathscr{M} \models \Gamma_{all}$.

Consider $\{[A],[B],[X']\}$. If this set were a subset of a point x, then consider $\{x\}$ as a one-point submodel of $\mathscr{M}$. In the submodel, $\Gamma_{all} \cup \{Some\ A\ are\ B\}$ would hold, and yet *Some X are Y* would fail since $[\![X]\!] = \varnothing$.

We use Lemma 11.2 to divide into cases:

(a) $A \leq A'$.
(b) $A \leq B'$.
(c) $A \leq X$.
(d) $B \leq B'$.
(e) $B \leq X$.
(f) $X' \leq X$.

(More precisely, the first case would be $[A] \leq [A']$. By strictness of the natural map, this means that $A \leq A'$; that is, $\Gamma_{all} \vdash All\ A\ are\ A'$.) In cases (1), (2), and (4), we easily see that Γ is inconsistent, contrary to the assumption at the outset. Case (6) implies that both (3) and (5) hold. Thus we may restrict attention to (3) and (5).

Next, consider $\{A, B, Y'\}$. The same analysis gives two other cases, independently: $A \leq Y$, and $B \leq Y$. Putting these together with the other two gives four pairs. The following are representative:

$A \leq X$ and $B \leq Y$: Using *Some A are B*, we see that $\Gamma \vdash$ *Some X are Y*.

$A \leq X$ and $A \leq Y$: We first derive *Some A are A*, and then again we see $\Gamma \vdash$ *Some X are Y*.

This completes the proof.

11.3 Going Further: Boolean Connectives Inside and Out

We continue our development a little further, mentioning a larger system whose completeness can be obtained by using our the results which we have seen.

We have in mind the language of *boolean compounds of* $\mathscr{L}(\text{all}, \text{some}, ')$ *sentences*. This language is just propositional logic built over $\mathscr{L}(all, some, ')$ as the set of atomic propositions. We call this larger language $\mathscr{L}(all, some, ', bc)$. For the semantics, we use the same kind of structures $\mathscr{M} = (M, [\![\ ]\!])$ as we have been doing in this paper. The semantics treats the boolean connectives classically. We have notions like $\mathscr{M} \models S$ and $\Gamma \models S$, defined in the usual ways.

The system is a Hilbert-style one, with axioms listed in Figure 11.2. The only rule is modus ponens.

We define $\vdash_{bc} S$ in the usual way, and then we say that $\Gamma \vdash_{bc} S$ if there are $T_1, \ldots, T_n$ from Γ such that $\vdash_{bc} (T_1 \wedge \cdots \wedge T_n) \rightarrow S$.

Rules 1–6 are essentially the system SYLL from [5]. Łukasiewicz and Słupecki proved a completeness and decidability result for SYLL, and different versions of this result may be found in Westerståhl [11] and in [7].

Proposition 11.3. $\vdash_{bc}$ All X are $Y \rightarrow$ All Y' are X'.

Lemma 11.5. *Let* $\Gamma \subseteq \mathscr{L}(\text{all}, \text{some}, ')$. *If* $\Gamma \vdash S$, *then* $\Gamma \vdash_{bc} S$.

a) All substitution instances of propositional tautologies.
b) $All\ X\ are\ X$
c) $(All\ X\ are\ Z) \wedge (All\ Z\ are\ Y) \rightarrow All\ X\ are\ Y$
d) $(All\ Y\ are\ Z) \wedge (Some\ X\ are\ Y) \rightarrow Some\ Z\ are\ X$
e) $Some\ X\ are\ Y \rightarrow Some\ X\ are\ X$
f) $\neg(Some\ X\ are\ X) \rightarrow All\ X\ are\ Y$
g) $Some\ X\ are\ Y' \leftrightarrow \neg(All\ X\ are\ Y)$

Fig. 11.2 Axioims for a system which adds sentential boolean connectives

Proof. By induction on the relation $\Gamma \vdash S$. (Recall that the rules are presented in Figure 11.1.) ($Some_2$) comes from axioms 2 and 4. For (*Antitone*), use the point about ($Some_2$) and also axiom 7 twice. For (*One*), use axiom 7 to see that $All\ X'\ are\ X \leftrightarrow \neg(Some\ X\ are\ X')$. This along with axiom 6 gives $All\ X'\ are\ X \leftrightarrow All\ X'\ are\ Y'$. Now we use our point about (*Antitone*) to see that $All\ X'\ are\ X \leftrightarrow All\ Y\ are\ X$. For (*Contrad*), use the Deduction Theorem and a propositional tautology.

Theorem 11.3. *The logical system above is sound and complete for* $\mathcal{L}(\text{all}, \text{some}, ', bc)$: $\Gamma \models S$ *iff* $\Gamma \vdash_{bc} S$.

Proof. The soundness being easy, here is a sketch of the completeness. We use compactness and disjunctive normal forms to reduce completeness to the verification that every consistent conjunction of $\mathcal{L}(all, some, ')$ sentences and their negations has a model. But $\mathcal{L}(all, some, ')$ essentially has a negation, so we need only consider consistent conjunctions. Now consistency here is in the propositional sense ($\vdash_{bc}$). But by Lemma 11.5, this implies consistency in the sense of our earlier logic ($\vdash$). And here we use Lemma 11.3.

A second proof We have another proof of this result, one which uses the completeness of SYLL and also the algebraic work from earlier in this paper.

The completeness of a system with classical negation reduces to the matter of showing that consistent sets Γ in $\mathcal{L}(all, some, ', bc)$ are satisfiable. Fix such a set Γ. We assume that Γ is maximal consistent. This implies first of all that Γ is closed under deduction in our logic, and so it contains all instances of the sentence in Proposition 11.3. It also implies that we need only build a model of $\Gamma \cap \mathcal{L}(all, some, ')$.

In our setting, we need a model in the style of this paper; in particular, we need the interpretation of complementary variables to be complementary sets. However, let us forget about this requirement for a moment, and pretend that U and U' are unrelated variables, except for what is dictated by the logic. That is, we consider Γ to be set of sentences in the boolean closure of syllogistic logic taken over a set of variables which is two copies of ours set $\mathcal{V}$. By completeness of that logic, $\Gamma \cap \mathcal{L}(all, some, ')$ has a model. Call it $\mathcal{M}$.

The problem is that since we forgot a key requirement, it probably will not hold in $\mathcal{M}$. So $[\![U]\!]_{\mathcal{M}}$ and $[\![U']\!]_{\mathcal{M}}$ need not be complementary sets: the best we can say is that these sets are disjoint by Axiom 7. In other words, $\mathcal{M}$ is *not* the kind of model which we use to define the semantic notions of the language, and thus we cannot use it is not directly of use in the completeness result. We must adjust the model in such a way as to (1) put each point in either $[\![U]\!]$ or $[\![U']\!]$, and at the same time (2) not changing the truth value of any sentence in the language $\mathscr{L}(all, some, ')$. The key is to use some of the algebraic work from earlier in the paper.

Consider the following orthoposet which we call $\mathscr{V}_{\mathscr{M}}$. Let $\mathscr{V}$ be the variables, let $\leq$ be defined by $U \leq V$ iff $[\![U]\!]_{\mathscr{M}} \subseteq [\![V]\!]_{\mathscr{M}}$. The points of $\mathscr{V}_{\mathscr{M}}$ are the equivalences classes of variables under the associated equivalence relation, and the order is the inherited one. Proposition 11.3 implies that if $U \leq V$, then also $V' \leq U'$. So we can define the complementation on $\mathscr{V}_{\mathscr{M}}$ by $[U]' = [U']$. Further, Proposition 11.3 also implies that the order is antitone. The complement inconsistency property comes from the fact that $[\![V]\!]$ and $[\![V']\!]$ are disjoint sets. (We may also need to add a 0 and 1 to $\mathscr{V}_{\mathscr{M}}$ if the structure so far has no minimum element. This is as in Example 11.5.)

For each $x \in M$, let

$$S_0(x) \quad = \quad \{[U] : x \in [\![U]\!]_{\mathscr{M}}\}.$$

This set is well-defined. We wish to apply Lemma 11.2 to each set $S_0(x)$. To be sure that the lemma applies, we must check that for $[U], [V] \in S_0(x)$, $[U] \not\leq [V']$. The point x itself belongs to $[\![U]\!]_{\mathscr{M}} \cap [\![V]\!]_{\mathscr{M}}$, and as $[\![V]\!] \cap [\![V']\!] = \varnothing$, we have $[\![U]\!] \not\subseteq [\![V']\!]$.

For each x, let $S(x)$ be a point of $\mathscr{V}_{\mathscr{M}}$ such that $S(x) \supseteq S_0(x)$. Define a model $\mathscr{N} = (N, [\![\ \ ]\!])$ by taking $N = M$ and setting

$$[\![U]\!]_{\mathscr{N}} \quad = \quad \{x \in M : U \in S(x)\}.$$

Since each $S(x)$ is a point, $\mathscr{N}$ is a bona fide structure for the language; that is, the semantic evaluation map preserves complements.

We claim that every sentence true in $\mathscr{M}$ is also true in $\mathscr{N}$. Let $\mathscr{M} \models$ *All U are V*. Thus $U \leq V$ in $\mathscr{V}_{\mathscr{M}}$. Then since points are closed upwards, $[\![U]\!]_{\mathscr{N}} \subseteq [\![V]\!]_{\mathscr{N}}$.

Finally, suppose that $\mathscr{M} \models$ *Some U are V*. Let $x \in [\![U]\!]_{\mathscr{M}} \cap [\![V]\!]_{\mathscr{M}}$. Then $\{U, V\} \subseteq S_0(x) \subseteq S(x)$, so $x \in [\![U]\!]_{\mathscr{N}} \cap [\![V]\!]_{\mathscr{N}}$.

We now know that $\mathscr{M}$ and $\mathscr{N}$ satisfy the same sentences. Since $\mathscr{N}$ is the kind of model we consider in the semantics, $\Gamma \cap \mathscr{L}(all, some, ')$ is satisfiable. This completes our second proof of Theorem 11.3.

It is possible to add boolean compounds of the NPs in this fragment. Once one does this, the axiomatization and completeness result become quite a bit simpler, since the system becomes a variant of boolean algebra.

11.4 *Ex Falso Quodlibet* Versus *Reductio ad Absurdum*

Our proof system employs the (*Contrad*) rule, also known as *ex falso quodlibet*. We also formulated the stronger system using *reductio ad absurdum* in Section 11.1.2. It follows trivially from the completeness of the (EFQ) system and the fact that it is stronger than the (RAA) system that the latter is complete. The argument is semantic. It might be of interest to those working on *proof-theoretic semantics* (see Ben Avi and Francez [1] and other papers by Nissim Francez) to see the explicit reduction.

Until the very end of this section, the only proof system used is for $\vdash_{RAA}$, and all trees shown are for that system.

11.4.1 Injective Proofs and Normal Forms

Let $\mathscr{T}$ be a proof tree over a set Γ, that is, a finite tree whose nodes are labeled with sentences according to the rules of our logic.

$\mathscr{T}$ is *injective* if different nodes have different labels.

$\mathscr{T}$ is *simple* if it is either an axiom alone, or if the leaves are labeled with sentences from Γ and the only rules used are (*Antitone*) at the leaves, and (*Barbara*).

Injectivity is a natural condition on proof trees. Unfortunately, the system does not have the property that all deductions have injective proofs. The simplest counterexample that I could find is the one in Example 11.2: there is no injective tree for the assertion shown there.

However, the system does admit normal forms which are basically two injective and simple trees put side-by-side. We say that $\mathscr{T}$ is a *normal form* if it is of one of the three forms shown in Figure 11.3. In the first form the root of $\mathscr{G}$ is labeled *All A are B*, and the root of $\mathscr{H}$ is labeled *All A are D*; $\mathscr{G}$ and $\mathscr{H}$ are the *principal subtrees*. Similar definitions apply to the other two forms. We require/permit that

(a) The principal subtrees must be injective and simple.
(b) One of the principal subtrees in each form might be missing. So any injective, simple tree is automatically a normal form.
(c) The label on the root of $\mathscr{T}$ does not occur as the label of any other node. In the first and third forms, we require that the label on the root of $\mathscr{G}$ not label anywhere on $\mathscr{H}$.

As a consequence of the second point, an injective and simple tree counts as normal, as does an injective tree that uses (*Zero*) or (*One*) at the root and is otherwise simple. A normal form tree need not be injective, because it might contain two principal subtrees which (though each is injective) have node labels in common.

The main advantage of working with simple trees is that the following results holds for them. Adding rules like (*Zero*) and (*One*) destroys Lemma 11.6, as easy counterexamples show.

Lemma 11.6. *Let* $\Gamma \cup \{\text{All } X \text{ are } Y\} \vdash_{RAA} \text{All } U \text{ are } V$ *via a simple proof tree* $\mathscr{T}$. *Then one of the following holds:*

(a) $\Gamma \vdash_{RAA} \text{All } U \text{ are } X$, *and* $\Gamma \vdash_{RAA} \text{All } Y \text{ are } V$.
(b) $\Gamma \vdash_{RAA} \text{All } U \text{ are } Y'$, *and* $\Gamma \vdash_{RAA} \text{All } X' \text{ are } V$.

$$\dfrac{\dfrac{\begin{array}{c}\mathcal{G}\\ \vdots\\ \textit{All } B' \textit{ are } B\end{array}}{\textit{All } A \textit{ are } B}\,\textit{One} \qquad \begin{array}{c}\mathcal{H}\\ \vdots\\ \textit{All } B \textit{ are } C\end{array}}{\textit{All } A \textit{ are } C} \qquad \dfrac{\dfrac{\begin{array}{c}\mathcal{G}_1\\ \vdots\\ \textit{All } B' \textit{ are } B\end{array}}{\textit{All } A \textit{ are } B}\,\textit{One} \qquad \dfrac{\begin{array}{c}\mathcal{G}_0\\ \vdots\\ \textit{All } B \textit{ are } B'\end{array}}{\textit{All } B \textit{ are } C}\,\textit{Zero}}{\textit{All } A \textit{ are } C}$$

$$\dfrac{\begin{array}{c}\mathcal{H}\\ \vdots\\ \textit{All } A \textit{ are } B\end{array} \qquad \dfrac{\begin{array}{c}\mathcal{G}\\ \vdots\\ \textit{All } B \textit{ are } B'\end{array}}{\textit{All } B \textit{ are } C}\,\textit{Zero}}{\textit{All } A \textit{ are } C}$$

Fig. 11.3 Normal forms. The principal subtrees must be simple and injective, but we allow degenerate cases where one of these is absent

Proof. By induction on $\mathcal{T}$. If $\mathcal{T}$ is a one-point tree, then $U = X$, and $Y = V$, and the statements in (1) are axioms. If $\mathcal{T}$ is one point followed by an application of the (*Antitone*) rule, we use (2). The induction step for (*Barbara*) is easy.

In the next two lemmas, we consider a graph $G = (\mathcal{V}, \rightarrow)$ related to Γ and to the overall logic. Here as before, $\mathcal{V}$ is the set of variables, and $U \rightarrow V$ iff Γ contains either *All U are V* or *All V are U*. Then to say that "$\Gamma \vdash_{RAA}$ *All U are V* via a proof tree which is simple" is exactly to say that Y is reachable from X in this graph, or that $X = Y$.

Lemma 11.7. *Let $\Gamma \vdash_{RAA}$* All U are V *via a proof tree which is simple.*

(a) There is a injective, simple proof tree for $\Gamma \vdash_{RAA}$ All U are V.
(b) Moreover, there is a tree in which (for all T) none of the nodes are labeled All T are U.

Proof. If $X = Y$, then a one-node tree is injective. Otherwise, take a path p of minimal length in G from X to Y. This path p contains no repeated edges, and if $A \rightarrow B$ is on p, then $B \rightarrow A$ is not on p. Moreover, X is not the target of any edge on p; this takes care of the second point in our lemma. There are several ways to turn p into an injective simple proof tree, using induction on the length of the path p.

Lemma 11.8. *Let V be such that $\Gamma \vdash_{RAA}$* All V are V' *and $\Gamma \vdash_{RAA}$* All V' are V *via simple proof trees. Then for all variables A and B, there is a normal form proof tree for $\Gamma \vdash_{RAA}$* All A are B.

Proof. Again we consider the graph $G = (\mathcal{V}, \rightarrow)$. The hypothesis tells us that there is a cycle $V \rightarrow \cdots \rightarrow V' \rightarrow \cdots \rightarrow V$. Consider the following proof tree:

$$\dfrac{\dfrac{\begin{array}{c}\vdots\\ \textit{All } V' \textit{ are } V\end{array}}{\textit{All } A \textit{ are } V}\,\textit{One} \qquad \dfrac{\begin{array}{c}\vdots\\ \textit{All } V \textit{ are } V'\end{array}}{\textit{All } V \textit{ are } B}\,\textit{Zero}}{\textit{All } A \textit{ are } B}$$

For the unshown parts, we use injective simple trees from the last lemma. We can also arrange that the tree overall be injective, by starting with a node V which minimizes the length of the cycle.

We remind the reader that until further notice, we use $U \leq V$ to mean that $\Gamma \vdash_{RAA}$ *All U are V*. Γ should be clear from context when we do so.

Lemma 11.9. *Let* $\Gamma \vdash_{RAA}$ All X are Y *via a proof tree without contradiction nodes. Then there is a normal form proof tree* $\mathcal{T}$ *for this deduction.*

Proof. One way to show this would be to use induction on proof trees. This is not how we proceed, but the details would not be substantially different. Our method is to modify a proof tree $\mathcal{T}$ without contradictions to obtain a normal tree. First, push all applications of (*Antitone*) to the leaves, out of the way. This is not hard, and we leave the details to the reader. Next, prune $\mathcal{T}$ so that all applications of (*Zero*) are on the right branch as we go up from the root, and all applications of (*One*) are on the left branch. This is accomplished by transformations such as

$$\dfrac{\mathit{All\ A\ are\ C} \qquad \dfrac{\dfrac{\mathit{All\ B'\ are\ B}}{\mathit{All\ C\ are\ B}}\,\mathit{One} \qquad \mathit{All\ B\ are\ D}}{\mathit{All\ C\ are\ D}}}{\mathit{All\ A\ are\ D}} \quad \Longrightarrow \quad \dfrac{\dfrac{\mathit{All\ B'\ are\ B}}{\mathit{All\ A\ are\ B}}\,\mathit{One} \qquad \mathit{All\ B\ are\ D}}{\mathit{All\ A\ are\ D}}$$

Once this is done, we now eliminate multiple applications of (*Zero*) on the right branch and (*One*) on the left. For example, we transform a tree whose root is as on the left to the tree whose root is as on the right.

$$\dfrac{\dfrac{\dfrac{\dfrac{\mathit{All\ B'\ are\ B}}{\mathit{All\ A\ are\ B}}\,\mathit{One} \qquad \mathit{All\ B\ are\ D}}{\mathit{All\ A\ are\ D}} \qquad \mathit{All\ D\ are\ A'}}{\mathit{All\ A\ are\ A'}}}{\mathit{All\ E\ are\ A'}}\,\mathit{One} \qquad \Bigg| \qquad \dfrac{\dfrac{\mathit{All\ B'\ are\ B}}{\mathit{All\ E\ are\ B}} \qquad \dfrac{\mathit{All\ B\ are\ D} \qquad \mathit{All\ D\ are\ A'}}{\mathit{All\ B\ are\ A'}}}{\mathit{All\ E\ are\ A'}}$$

We may then arrange all subtrees without (*Zero*) or (*One*) to be injective, using Lemma 11.7. Now we either have a tree as in Figure 11.3, or else we have a tree such as

$$\dfrac{\dfrac{\dfrac{\begin{matrix}\mathcal{G}\\ \vdots\\ \mathit{All\ B'\ are\ B}\end{matrix}}{\mathit{All\ A\ are\ B}}\,\mathit{One} \qquad \begin{matrix}\mathcal{H}\\ \vdots\\ \mathit{All\ B\ are\ C}\end{matrix}}{\mathit{All\ A\ are\ C}} \qquad \dfrac{\begin{matrix}\mathcal{I}\\ \vdots\\ \mathit{All\ C\ are\ C'}\end{matrix}}{\mathit{All\ C\ are\ D}}\,\mathit{Zero}}{\mathit{All\ A\ are\ D}}$$

This tree uses (*One*) and (*Zero*) and also a subsidiary derivation in the middle. In the notation of the figure, we would then have proofs from Γ of $B' \leq B$, $B \leq C$, and $C \leq C'$. These proofs only use (*Antitone*) and (*Barbara*). Then we put these together to get a similar proof of $C' \leq B' \leq B \leq C$. So at this point we may forget about our current tree altogether and use Lemma 11.8. This shows that we may assume that only two of $\mathcal{G}$, $\mathcal{H}$, and $\mathcal{I}$ are present.

Finally, we need to arrange the last requirement on normal trees. If this fails, then we replace $\mathscr{H}$ by a simple, normal proof tree of the same assertion with the "moreover" condition of Lemma 11.7. This automatically insures that *All A are B* does not occur in $\mathscr{H}$.

11.4.2 Proofs with and Without Contradiction

Lemma 11.10. *Let* $\Gamma \vdash_{RAA}$ Some C are D *without the contradiction rule. Then there is a proof tree* $\mathscr{T}$ *for this deduction such that* $\mathscr{T}$ *has exactly one use of* (Darii)*, and this is preceded and followed by zero or one applications of* (Some$_1$) *or* (Some$_2$)*.*

Proof. We eliminate successive applications of (*Darii*) at the root, using (*Barbara*):

$$\dfrac{\textit{All B are C} \qquad \dfrac{\textit{All A are B} \quad \textit{Some C are A}}{\textit{Some C are B}}}{\textit{Some C are D}} \quad \Longrightarrow \quad \dfrac{\dfrac{\textit{All A are B} \quad \textit{All B are C}}{\textit{All A are C}} \qquad \textit{Some C are A}}{\textit{Some C are D}}$$

The only other rules that involve *Some* sentences are (*Some*$_1$) and (*Some*$_2$). If more than one is used in either place, the series of uses may be collapsed to just one.

Recall Γ is *inconsistent* if for some S, $\Gamma \vdash S$ and $\Gamma \vdash \neg S$. (That is, this notion is defined using $\vdash$ and not $\vdash_{RAA}$.)

Lemma 11.11. *Let* Γ *be inconsistent. Then for some sentence* Some U are V *in* Γ*,* $\Gamma_{all} \vdash_{RAA}$ All U are V'.

Proof. There is a tree $\mathscr{T}$ for a derivation of an inconsistency that itself does not use the contradiction rule. By using Lemma 11.10, we have a tree such as

$$\dfrac{\begin{matrix}\mathscr{G}\\ \vdots \\ \textit{All U are W}\end{matrix} \qquad \dfrac{\begin{matrix}\mathscr{H}\\ \vdots \\ \textit{All V are W}'\end{matrix} \quad \textit{Some U are V}}{\textit{Some U are W}'}\,\textit{Darii}}{S}\,\textit{Contrad}$$

where $\mathscr{G}$ and $\mathscr{H}$ consist entirely of *All* sentences, so they are trees over Γ_{all}. This tree shows the general situation, except that *Some U are V* might be the result of one or two applications of *Some* rules from a *Some* sentence in Γ, and below *Some U are W*$'$ we might similarly find one or two steps leading to *Some W*$'$ *are U*, *Some W*$'$ *are W*$'$, or *Some U are U*. These extra possibilities are easy to handle. So we assume that *Some U are V* belongs to Γ. Using $\mathscr{G}$ and $\mathscr{H}$, we see that $\Gamma_{all} \vdash_{RAA}$ *All U are V*$'$. In the case of these "extra possibilities", we might need to add one or two steps afterwards.

Lemma 11.12. *If* $\Gamma \cup \{$All X are $Y\}$ *is inconsistent, then* $\Gamma \vdash_{RAA}$ Some X are Y'.

Proof. By what we have seen, there is a normal form proof tree $\mathcal{T}$ over $\Gamma_{all} \cup \{All\ X\ are\ Y\}$ to the negation of a sentence in Γ of the form *Some U are V*. We may assume that *All X are Y* labels a leaf; for if not, Γ is inconsistent. We argue by induction on the length n of a path from a leaf labeled *All X are Y* to the root of such a tree.

If the length is 0, then the leaf *All X are Y* is the root of $\mathcal{T}$. Therefore $X = U$ and $Y = V'$. So since Γ contains *Some U are V* = *Some X are* Y', we are done.

This is the base case of the induction. Assume our lemma for n (for all sets and all sentences as in the statement of our theorem, of course). There are two overall cases: (a) *All X are Y* labels one leaf of either $\mathcal{G}$ or $\mathcal{H}$ (but not both); (b) *All X are Y* labels one leaf of $\mathcal{G}$ and one leaf of $\mathcal{H}$. (Recall that $\mathcal{G}$ and $\mathcal{H}$ are injective, so *All X are Y* labels at most one leaf in each.)

Here is the argument in case (a). Suppose that *All X are Y* labels a leaf in $\mathcal{T}$ as in

$$\cfrac{\cfrac{All\ X\ are\ Y \quad All\ Y\ are\ Z}{All\ X\ are\ Z}}{\vdots}$$

so that the length of the path from *All X are Y* to the root is n. Drop *All X are Y* and *All Y are Z* from $\mathcal{T}$ to get a tree $\mathcal{T}'$. So *All X are Z* is a leaf of $\mathcal{T}'$, and the length of the path in $\mathcal{T}'$ from this leaf to the root is $n-1$, and $\mathcal{T}'$ is a normal form tree over $\Gamma_{all} \cup \{All\ X\ are\ Z\}$. By induction hypothesis, $\Gamma \vdash_{RAA}$ *Some X are* Z'. But *All Y are Z* belongs to Γ, as only one node of $\mathcal{T}$ is labeled with *All X are Y*. (We are using the hypothesis of case (a).) So $\Gamma \vdash_{RAA}$ *Some X are* Y'. The case when *All X are Y* labels a leaf of $\mathcal{T}$ participating in (*Barbara*) on the right is similar.

There are other subcases: *All X are Y* might label a leaf on $\mathcal{T}$ that participates in (*One*), (*Zero*), or the (*Antitone*) rule. We shall go into details on (*One*); (*Zero*) is basically the same, and (*Antitone*) is the easiest since it only occurs at the leaves. For (*One*), the tree $\mathcal{T}$ is

$$\cfrac{\cfrac{All\ B'\ are\ B}{All\ A\ are\ B}\,One \qquad \begin{array}{c}\mathcal{H}\\ \vdots \\ All\ B\ are\ C\end{array}}{All\ A\ are\ C}$$

So $X = B'$ and $Y = B$. By our third requirement on normal proof trees, $\mathcal{H}$ is a proof tree over Γ, and it does not contain *All A are B* anywhere. If we remove *All* B' *are B*, the tree is normal, but not because it matches the first type in Figure 11.3 (it does not match it, since it lacks an application of (*One*)); instead, it is a simple, injective tree. So by what we just did for (*Barbara*), we see that $\Gamma \vdash_{RAA}$ *Some A are* B'. Hence $\Gamma \vdash_{RAA}$ *Some* B' *are* B'; that is, $\Gamma \vdash_{RAA}$ *Some X are* Y'.

This concludes our work in case (a). Things are more interesting in case (b). There are three subcases, corresponding to which of the normal form trees in Figure 11.3 $\mathcal{T}$ exemplifies.

Subcase (i): $\mathcal{T}$ has an instance of (*One*) but not of (*Zero*) as on the top left in Figure 11.3. Its root *All A are C* is the negation of a sentence in Γ. Therefore, Γ

contains *Some A are C'*. Recall that $\mathscr{G}$ and $\mathscr{H}$ are simple. Now since both $\mathscr{G}$ and $\mathscr{H}$ have leaves labeled *All X are Y*, we have four possibilities:

(i.a) $B' \leq X, Y \leq B, B \leq X, Y \leq C.$
(i.b) $B' \leq X, Y \leq B, B \leq Y', X' \leq C.$
(i.c) $B' \leq Y', X' \leq B, B \leq X, Y \leq C.$
(i.d) $B' \leq Y', X' \leq B', B \leq Y', X' \leq C.$

Note that (i.a) and (i.c) are the same: the first two assertions in (i.a) are the duals of those in (i.c). Similar results apply to (i.b) and (i.d). So we only go into details on the (i.a) and (i.b). For (i.a), we have proofs from Γ of the three subtrees indicated by $\vdots$ below:

$$\dfrac{\dfrac{\dfrac{\dfrac{\vdots}{\textit{All } B \textit{ are } X}}{\textit{All } X' \textit{ are } B'} \quad \dfrac{\vdots}{\textit{All } B' \textit{ are } X}}{\dfrac{\textit{All } X' \textit{ are } X}{\textit{All } Y' \textit{ are } X}} \qquad \dfrac{\dfrac{\dfrac{\vdots}{\textit{All } Y \textit{ are } C}}{\textit{All } C' \textit{ are } Y'} \quad \dfrac{\textit{Some } A \textit{ are } C'}{\textit{Some } C' \textit{ are } C'}}{\dfrac{\textit{Some } C' \textit{ are } Y'}{\textit{Some } Y' \textit{ are } Y'}}}{\dfrac{\textit{Some } Y' \textit{ are } X}{\textit{Some } X \textit{ are } Y'}}$$

Thus, the tree above is a tree over Γ, with the desired conclusion.

(The displayed tree (without the three omitted subproofs) is the solution to the exercise mentioned in Example 11.3 in Section 11.1.1.)

In possibility (i.b), one gets $Y \leq B \leq Y'$, so informally, there are no Y. Also, $X \leq C'$, so $C' \leq X$. Since some A are C', some X are Y'.

For (ii), Γ contains *Some A are C'*. We again have four possibilities, similar to what we saw in (i) above. We'll only go into brief details on the first and third possibilities. The first would be when Γ derives $B' \leq X$, $Y \leq B$, $B \leq X$, and $Y \leq B'$. In this case, we have *All B' are B* and *All B are B'*. Hence we also get *All A are X* and *All C' are Y'* (see Lemma 11.8). Hence we have *Some X are Y'*. The third possibility finds $Y \leq B' \leq Y'$ and $X' \leq B \leq X$. So we get *All A are X* and *All C are Y'* again.

Subcase (iii) is entirely parallel to (i).

This concludes our proof.

Theorem 11.4. *If $\Gamma \vdash_{RAA} S$, then $\Gamma \vdash S$.*

Proof. By induction on the indirect proof relation $\vdash_{RAA}$. The key step is when $\Gamma \vdash_{RAA} S$ via proofs of $\Gamma \cup \{\neg S\} \vdash_{RAA} T$ and $\Gamma \cup \{\neg S\} \vdash_{RAA} \neg T$. By induction hypothesis, $\Gamma \cup \{\neg S\}$ is inconsistent in the direct logic. When S is a *Some* sentence, Lemma 11.12 tells us that $\Gamma \vdash S$. When S is an *All* sentence, we use Lemma 11.11.

11.5 Further Work in the Area

Readers of this paper might like to see [10] for more results on syllogistic systems. Of special note is that once the language adds *transitive verbs* and *relative clause subjects* (as in *every dog who sees a cat runs*), (RAA) is strictly stronger than (EFQ). And once one adds negation on verbs in addition to nouns, then it turns out to be impossible to obtain syllogistic proof systems.

Acknowledgement My thanks to Ian Pratt-Hartmann for many very useful conversations on this topic.

References

1. G.B. Avi and N. Francez. Proof-theoretic semantics for a syllogistic fragment. In P. Dekker, and M. Franke, editors, *Proceedings of the Fifteenth Amsterdam Colloquium*, http://www.illc.uva.nl/AC05/uploaded files/AC05Proceedings.pdf, ILLC/Department of Philosophy, University of Amsterdam, Amsterdam, 2005, 9–15.
2. C.S. Calude, P.H. Hertling, K. Svozil. Embedding quantum universes into classical ones. *Foundations of Physics*, 29(3): 349–379, 1999.
3. J. Corcoran. Completeness of an ancient logic. *Journal of Symbolic Logic,* 37: 696–702, 1972.
4. F. Katrnoška. On the representation of orthocomplemented posets. *Commentations Mathematiceas Universitatis Carolinae,* 23: 489–498, 1982.
5. J. Łukasiewicz. *Aristotle's Syllogistic from the Standpoint of Modern Formal Logic*. Clarendon Press, Oxford, 1951.
6. J.N. Martin. Aristotle's natural deduction reconsidered. *History and Philosophy of Logic*, 18(1): 1–15, 1997.
7. L.S. Moss. Completeness theorems for syllogistic fragments. In F. Hamm, and S. Kepser, editors, *Logics for Linguistic Structures* pages 143–173. Mouton de Gruyter, New York, NY, 2008.
8. I. Pratt-Hartmann. Fragments of language. *Journal of Logic, Language, and Information,* 13: 207–223, 2004.
9. I. Pratt-Hartmann. A natural deduction system for the syllogistic. Ms., University of Manchester, 2007.
10. I. Pratt-Hartmann and L.S. Moss. Logics for the relational syllogistic. *Review of Symbolic Logic*, 2(4): 647–683, 2009.
11. DagWesterståhl. Aristotelian syllogisms and generalized quantifiers. *Studia Logica* 48(4): 577–585, 1989.
12. N. Zierler and M. Schlessinger. Boolean embeddings of orthomodular sets and quantum logic. *Duke Mathematical Journal* 32: 251–262, 1965.

Chapter 12
From Unary to Binary Inductive Logic

Jeff B. Paris and Alena Vencovská

12.1 Introduction

Suppose you lived in a world of individuals $a_1, a_2, a_3, \ldots$ (which exhaust the universe) and a finite set of predicates $P(x), P_1(x), P_2(x), R(x,y), \ldots$ but no other constants or function symbols. You observe that $P(a_1)$ and $P(a_2)$ hold, and nothing else. In that case what probability, $w(P(a_3))$ say, in terms of willingness to bet, should you assign to $P(a_3)$ also holding?

As an analogy suppose for example you are standing by a road in Mumbai and the first two cars that go past are going from right to left (i.e. $P(a_1), P(a_2)$). In that case what probability should you assign to the next car also passing in this direction (i.e. $P(a_3)$)? Or of this being a one-way road (i.e. $\forall x P(x)$)? Of course in such a real world example one observes, and already knows, a great deal more than just the passing of the two cars. Nevertheless one can imagine a situation, for example where the "you" is an artificial agent and where $P(a_1) \wedge P(a_2)$ is genuinely all the available knowledge. It is this general *problem of induction*:

Given that you know only φ what probability should you assign to θ?

where θ and φ are sentences of our language, that we are interested in investigating. What we seek here is a *logical* approach based on explicit principles of rationality to constrain, if not specify exactly, what probability assignments are permissible.

Rudolf Carnap [1–3] is the person usually credited with such a logical approach to induction, since referred to as "Inductive Logic", though similar ideas and results were already present before that, in particular in the work of W.E. Johnson [14]. Nowadays we might equally refer to this area as Predicate Probabilistic Uncertain Reasoning.

Jeff B. Paris
School of Mathematics, University of Manchester, Manchester M13 9PL, UK,
e-mail: jeff.paris@manchester.ac.uk
Alena Vencovská
School of Mathematics, University of Manchester, Manchester M13 9PL, UK,
e-mail: alena.vencovska@manchester.ac.uk

J. van Benthem et al. (eds.), *Games, Norms and Reasons*, Synthese Library 353,
DOI 10.1007/978-94-007-0714-6_12,

The focus of Carnap's, and the other early researchers in the area, was very much on the case where the overlying languages had just unary predicates and, as we shall explain later, this endeavor enjoyed some notable successes at that time. Induction however is not entirely limited to the unary case. On occasions we seem to be applying induction to binary or even higher arity relations. For example if we know that Jane enjoys most people's company and that John's company is generally found to be most agreeable we may well hazard the guess that were Jane and John ever to meet Jane would indeed enjoy John's company.

It seems, for example from J.G. Kemeny's predictions in his 1963 paper [17], that the future intention of Carnap and his associates was indeed to take on languages with higher arity predicates etc. Unfortunately the discovery by N. Goodman [10, 11] of the GRUE paradox in the mid 1940s had by then so disheartened Carnap, and most other philosophers, that with a few notable exceptions this intended further development never happened.

The weakness in the Inductive Logic approach exposed by the GRUE paradox was that the assumption of being able to capture—in a real world application—all the available relevant knowledge by a single sentence within the context of a formal predicate language, is simply untenable. Since Carnap et al. were interested in developing an applicable system this was, as far as they were concerned, a death blow.

Nevertheless it seems to us that the programme is still worth developing, firstly because it would seem that in the simple worlds of artificial agents such a modelling is not entirely out of line with reality, secondly because, despite the wide gap, insights in Inductive Logic may on occasions offer illumination of the real world, and thirdly because even were the world of Inductive Logic impossibly far from the material world we inhabit it would nevertheless still seem a valid topic for academic investigation.

12.2 Notation and Background

Throughout we shall work with languages L for predicate logic with (just) the constant symbols $a_1, a_2, a_3, \ldots$ and *finitely* many predicate symbols, but without function symbols or equality. Our intended interpretation here is that these constants exhaust the universe. Let FL denote the formulae of L and SL the sentences of L (i.e. closed formulae).

Following Gaifman [9] we say a map $w : SL \longmapsto [0,1]$ is a *probability function* on the predicate language L if it satisfies that for all $\theta, \varphi, \exists x\, \psi(x) \in SL$:

(P1) If $\models \theta$ then $w(\theta) = 1$.
(P2) If $\models \neg(\theta \wedge \varphi)$ then $w(\theta \vee \varphi) = w(\theta) + w(\varphi)$.
(P3) $w(\exists x\, \psi(x)) = \lim_{m \to \infty} w(\bigvee_{i=1}^{m} \psi(a_i))$.

Given a probability function w on L we define, as usual, a corresponding two place *conditional probability function* on $SL \times SL$, denoted by $w(.|.)$, to be such that

$$w(\theta|\varphi) \cdot w(\varphi) = w(\theta \wedge \varphi).$$

With these definitions all the expected elementary properties concerning probability and conditional probability functions hold (see for example [27]), in particular if $w(\varphi) > 0$ then $w(\,.\,|\varphi)$ is itself a probability function.

Furthermore, by a theorem of Gaifman [9] any probability function defined on quantifier free sentences of the language L (i.e. satisfying (P1) and (P2) for such θ, φ) extends uniquely to a probability function on L. For this reason we can largely limit our considerations to probability functions defined just on quantifier free sentences. Furthermore, in that case $w(\theta)$ is determined by the values of w on the *state descriptions*, that is sentences of the form

$$\bigwedge_{j} \bigwedge_{\substack{b_1,\dots,b_{r_j} \in \\ \{a_{t_1},\dots,a_{t_m}\}}} Q_j^{\varepsilon_j}(b_1,\dots,b_{r_j}),$$

where the $\varepsilon_j \in \{0,1\}$, $Q^1 = Q$, $Q^0 = \neg Q$ and the r_j-ary predicates Q_j and the constants $a_{t_1},\dots,a_{t_m}$ include all those predicates and constants appearing in θ.

In particular if L_1 is a unary language, that is all the predicates appearing in L_1 are unary, then w is determined by its values on the sentences of the form

$$\bigwedge_{i=1}^{p} \alpha_{h_i}(a_i)$$

where $1 \leq p \in \mathbb{N}$ and the $\alpha_h(x)$ run through the *atoms* with respect to the set $P_1,\dots,P_n$ of unary predicates from L_1, that is the 2^n formulae of the form

$$\bigwedge_{j=1}^{n} P_j^{\varepsilon_j}(x).$$

Similarly if L_2 is a binary language then w is determined by its values on the sentences

$$\bigwedge_{i,j=1}^{p} \beta_{r_{ij}}(a_i,a_j)$$

where the $\beta_r(x,y)$ run through the *atoms* with respect to the set $R_1,\dots,R_m$ of binary predicates from L_2, that is the 2^m formulae of the form

$$\bigwedge_{j=1}^{m} R_j^{\varepsilon_j}(x,y).$$

In what follows n, m, L_1, L_2 will, as far as possible, be fixed in their roles as the number of unary/binary predicates in these unary/binary languages. We shall limit ourselves to these two languages because most of the results for binary carry over with suitable modifications to higher arities (see [19]) whereas in some aspects the unary case is rather special.

Returning now to the main *problem of induction*, within the subject it is widely assumed (though not necessarily without some misgivings) that the probability one should give to θ on the basis of knowledge φ is

$$\frac{w(\theta \wedge \varphi)}{w(\varphi)} \quad \text{i.e. } w(\theta | \varphi)$$

where $w(\theta \wedge \varphi), w(\varphi)$ are the probabilities one should give $\theta \wedge \varphi$ and φ respectively *on the basis of no knowledge at all.*[1] For this paper we shall continue in the tradition of making this assumption.

It is important to emphasize here that these probability values $w(\varphi)$ etc. are intended to be assigned subjectively on the basis of zero knowledge. Whilst that may seem to allow almost unlimited freedom we will be imposing rationality requirements on w, in the form of principles to be observed, which will considerably limit this freedom. In particular we explicitly assume (in case it was not obvious) that these assigned values are consistent with (P1-3), in other words that w is a probability function. In that sense then the whole problem of induction has now been reduced to:

What probability function w should one adopt in the absence of any knowledge whatsoever?

In later sections we shall suggest certain arguably rational principles which might bear on this question. For the present however one might think that there is a rather obvious answer to the above question. Namely if you know nothing then you have no reason to give any one state description for $a_1, a_2, \ldots, a_p$ any more (or less) probability then any other state description for $a_1, a_2, \ldots, a_p$. In the binary case this amounts to saying that your choice of probability function should be the *completely independent probability function* w_0 which gives each state description

$$\bigwedge_{i,j=1}^{p} \beta_{r_{ij}}(a_i, a_j)$$

probability 2^{-mp^2}. Attractive as this answer might appear it suffers the serious drawback of denying any induction. For example, in this case the answer to the question "what probability to assign to $R_1(a_{p+1}, a_{p+1})$ on the basis of having already observed $R_1(a_1, a_1), R_1(a_2, a_2), \ldots, R_1(a_p, a_p)$" would always be one half independently of how many (in this case p) previous supporting instances had been seen. In other words w_0 does not allow for any learning from experience. On the other hand it might not be so clear that one would want to ban w_0 outright as a possible choice. Already then we see that our efforts to solve our main problem are likely at best to only give us a restricted class of acceptable probability functions rather than one outright winner.

[1] We shall put aside the problem of what to do if $w(\varphi) = 0$, in what follows that difficulty will not arise. (For a thorough consideration of this problem see [5].)

12.3 Principles of Symmetry

As already indicated the main approach we shall take is to impose rationality requirements on the choice of w. Three such principles, based on the idea that there is no rational justification for w to break existing symmetries, come immediately to mind:

Constant Exchangeability, Ex: *$w(\theta)$ is invariant under permutations of the constant symbols of the language.*

For example, according to this principle we should have

$$w(R(a_1,a_3)\wedge\neg R(a_3,a_2)) = w(R(a_1,a_4)\wedge\neg R(a_4,a_3)).$$

Similarly

Predicate Exchangeability, Px: *$w(\theta)$ is invariant under permutations of the predicate symbols (of the same arity) in the language.*

and

Strong Negation, SN: *$w(\theta)$ does not change if we replace every occurrence of the predicate symbol Q in θ by $\neg Q$.*

Notice that for unary w if w satisfies Ex then its value on a state description,

$$w\left(\bigwedge_{i=1}^{p}\alpha_{h_i}(a_i)\right),$$

is actually a function $w(\langle h_1,h_2,\ldots,h_p\rangle)$ of the $\mathbf{h}$, or even more simply a function $w(\mathbf{r})$ of the vector $\langle r_1,r_2,\ldots,r_{2^n}\rangle$ where $r_j = |\{i\,|\,h_i = j\}|$. We shall occasionally use this notation in what follows.

In the binary case

$$w\left(\bigwedge_{i,j=1}^{p}\beta_{r_{ij}}(a_i,a_j)\right)$$

is a function $w(\mathbf{r})$ where $\mathbf{r}$ is the $p\times p$ matrix with ij'th element $r_{ij}\in\{1,2\ldots,2^m\}$. Again we will on occasions use this notation.

In the unary case de Finetti's Theorem [6–8] provides a powerful representation result for the probability functions satisfying Ex:

Theorem 12.1. *(de Finetti's Representation Theorem) If the probability function v on L_1 satisfies Ex then there is a countable additive measure μ on the Borel subsets of*

$$\mathbb{D}_{2^n} = \{\,\langle x_1,x_2,\ldots,x_{2^n}\rangle\in\mathbb{R}^{2^n}\,|\,x_i\geq 0, \sum_i x_i = 1\,\}$$

such that

$$v\left(\bigwedge_{i=1}^{p}\alpha_{h_i}(a_i)\right) = v(\mathbf{r}) = \int_{\mathbb{D}_{2^n}} x_1^{r_1}x_2^{r_2}\ldots x_{2^n}^{r_{2^n}}\,d\mu$$

where $r_j = |\{i\,|\,h_i = j\}|$. Conversely any probability function v on L_1 defined in this way satisfies Ex.

In other words v looks like a convex combination of Bernoulli multinomial trials giving probability $x_1^{r_1} x_2^{r_2} \dots x_{2^n}^{r_{2^n}}$ to $\bigwedge_{i=1}^p \alpha_{h_i}(a_i)$.

De Finetti's Theorem can also be expressed in another form, see [15, 16]:

Theorem 12.2. *(de Finetti's Representation Theorem, functional form) If the probability function v on L_1 satisfies Ex then there exist a measurable function $f : [0,1]^2 \to \{1,2,\dots,2^n\}$ and some independent uniformly distributed random variables ξ, ξ_i on [0,1], $i \in \mathbb{N}$ such that*

$$v\left(\bigwedge_{i=1}^{p} \alpha_{h_i}(a_i)\right)$$

is the probability that $f(\xi, \xi_i) = h_i$ for all $1 \leq i \leq p$. Conversely any probability function v on L_1 defined in this way satisfies Ex.

The early Inductive Logic community within Philosophy did not seem particularly aware of this seminal theorem (which in fact was available almost from the start of the area). Similarly this community has been largely unaware of subsequent results for binary probability functions satisfying Ex, firstly from P.H. Krauss [18] and later by D.N. Hoover [13], see also [15, 16], these latter generalizing also to higher arities. In the binary case, Hoover's Representation Theorem can be stated as follows:

Theorem 12.3. *(Hoover's Representation Theorem) If the probability function w on L_2 satisfies Ex then there exist a measurable function $f : [0;1]^4 \to \{1,2,\dots,2^m\}$ and some independent uniformly distributed random variables $\xi, \xi_i, \xi_{ij} (= \xi_{ji})$ on $[0,1]$, $i, j \in \mathbb{N}$ such that*

$$w\left(\bigwedge_{i,j=1}^{p} \beta_{r_{ij}}(a_i, a_j)\right)$$

is the probability that $f(\xi, \xi_i, \xi_j, \xi_{ij}) = r_{ij}$ for all $1 \leq i, j \leq p$. Conversely any probability function w on L_2 defined in this way satisfies Ex.

As we shall see this result provides a valuable source of binary probability functions.

The principles Ex, Px, SN are quite generally accepted in this subject and certainly we shall henceforth assume without further mention that our probability functions satisfy them. Indeed in the unary case Carnap, Johnson et al seemed happy to take symmetry even further and accept *Atom Exchangeability*. To explain this notice that in the unary case all there is to know about constants $a_1, a_2, \dots, a_p$ from the language is summed up by the state description

$$\bigwedge_{i=1}^{p} \alpha_{h_i}(a_i)$$

which $a_1, a_2, \dots, a_p$ satisfy. In this sense then a_j and a_i are *indistinguishable* iff $h_i = h_j$. Clearly indistinguishability is an equivalence relation. Define the *spectrum* of this state description to be the multiset $\{s_1, s_2, \dots, s_r\}$ of sizes of these equivalence classes.[2]

Atom Exchangeability Principle, Ax $w\left(\bigwedge_{i=1}^{p} \alpha_{h_i}(a_i)\right)$ *depends only on the spectrum of this state description.*

For example if we have (for $n = 2$)

$$\begin{array}{l} P_1(a_1),\ P_1(a_2),\ \neg P_1(a_3),\ P_1(a_4), \\ P_2(a_1),\ P_2(a_2),\ \neg P_2(a_3),\ \neg P_2(a_4), \end{array}$$

the spectrum of a_1, a_2, a_3, a_4 is $\{2, 1, 1\}$ and according to Ax this should be given the same probability as, for example,

$$\begin{array}{l} P_1(a_1),\ P_1(a_2),\ \ \neg P_1(a_3),\ P_1(a_4), \\ P_2(a_1),\ \neg P_2(a_2),\ \neg P_2(a_3),\ \neg P_2(a_4). \end{array}$$

where again the spectrum of a_1, a_2, a_3, a_4 is $\{2, 1, 1\}$.

Ax implies all the "Exchangeability Principles" mentioned earlier.

Despite some criticisms (see for example [3, 12, 20, 21, 23]) Johnson and Carnap appeared quite happy to accept Ax.

Turning to the binary analog of Ax we can observe that in this case, unlike the unary case, no finite set of sentences determines everything there is to know about a particular constant a_i. Nevertheless we could say that everything there is to know about $a_1, a_2 \dots, a_p$ in direct relation to each other is encapsulated in the state description

$$\bigwedge_{i,j=1}^{p} \beta_{r_{ij}}(a_i, a_j)$$

that they satisfy. Again, relative to this state description, for $1 \le i, j \le p$ we can define an equivalence relation of *indistinguishability* on $\{1, 2, \dots, p\}$ by

$$i \sim j \leftrightarrow r_{iq} = r_{jq} \text{ and } r_{qi} = r_{qj} \text{ for all } q = 1, 2, \dots, p.$$

Analogously to the unary case we define the *spectrum* of the state description to be the multiset $\{s_1, s_2 \dots, s_t\}$ where $s_1, s_2, \dots, s_t$ are the sizes of the non-empty equivalence classes with respect to $\sim$.[3] A binary version of Ax can now be stated as:

Principle of Spectrum Exchangeability, Sx

$$w\left(\bigwedge_{i,j=1}^{p} \beta_{h_{ij}}(a_i, a_j)\right)$$

is a function of just the spectrum of this state description.

---

[2] In [26] the spectrum was defined as the vector $\langle s_1, s_2, \dots, s_r \rangle$ where the s_i are the sizes of equivalence classes in non-increasing order. Clearly the two versions are equivalent.

[3] Again in [26] we adopted the equivalent formulation in terms the vector of the s_i in non-decreasing order.

Notice a key difference between the unary and binary here is that in the former if a_i, a_j are ever indistinguishable then they remain so no matter how many future $a_{p+1}, a_{p+2}, a_{p+3}, \ldots$ are subsequently observed. However in the binary case such later a_{p+g} can destroy earlier indistinguishabilities, for example by asserting $R(a_i, a_{p+1}) \wedge \neg R(a_j, a_{p+1})$.

Sx clearly generalizes to higher (and mixed) arities. We believe it to be a central principle in the study of higher arity induction and in what follows we shall almost entirely concentrate on probability functions satisfying it. Notice that since the value of w on a state description depends only on the spectrum $\overline{s}$ of that state description we can, with some useful economy, write $w(\overline{s})$ for that value.

It will be useful at this point to introduce a little notation. Given a state description

$$\bigwedge_{i,j=1}^{p} \beta_{r_{ij}}(a_i, a_j) \tag{12.1}$$

with spectrum $\overline{s} = \{s_1, s_2, \ldots, s_t\}$ we call $p = \sum_i s_i$ the *sum* of $\overline{s}$ and t the *length*, $|\overline{s}|$, of this spectrum. Let A^t_p be the set of all spectra with sum p and length t and given the state description (12.1) let $\mathcal{N}(\overline{s}, \overline{h})$ be the number of state descriptions for $a_1, a_2, \ldots, a_h$ where $h = \sum_i h_i$ which extend (12.1) and have spectrum $\overline{h}$. It is shown in [26] that this number depends only on the spectrum $\overline{s}$ and not on the particular state description (12.1).

12.4 Johnson's Sufficientness Principle

Not only did Johnson and Carnap seem reasonably happy to accept Ax but they in fact chose to embrace the following much stronger principle:

Johnson's Sufficientness Principle, JSP

$$w\left(\alpha(a_{p+1}) \mid \bigwedge_{i=1}^{p} \alpha_{h_i}(a_i)\right)$$

depends only on p and the number of a_i, $i = 1, 2, \ldots, p$ indistinguishable from a_{p+1} according to the state description

$$\alpha(a_{p+1}) \wedge \bigwedge_{i=1}^{p} \alpha_{h_i}(a_i).$$

This principle appears to depart from the idea of rational justification by symmetry to justification by *irrelevance*. The idea here is that given

$$\bigwedge_{i=1}^{p} \alpha_{h_i}(a_i)$$

whether or not a_{p+1} satisfies a particular atom α depends only on p and the number of previous a_i satisfying α whilst the distribution of the atoms satisfied by the remaining a_i is irrelevant.

Recognizing *irrelevance* seems to us much less reliable than recognizing symmetry, where of course one usually has some underlying mathematical notion of isomorphism to fall back on, see for example [12].

JSP implies Ax and (provided $n > 1$) cuts down the possible functions w to a one parameter family—Carnap's so called Continuum of Inductive Methods, see [2], making it a particularly appealing principle in practice. Its success at the unary level may make one hope for something similar higher up where the analogous version reads:

Johnson's Binary Sufficientness Postulate JBSP *For a natural number p and* $\mathbf{r} \in \{1,2,\ldots,2^m\}^{(p+1)\times(p+1)}$,

$$w\left(\bigwedge_{i,j=1}^{p+1} \beta_{r_{ij}}(a_i,a_j) \mid \bigwedge_{i,j=1}^{p} \beta_{r_{ij}}(a_i,a_j)\right) \tag{12.2}$$

depends only on p and on the number s of k, $1 \leq k \leq p$ *such that*

$$r_{p+1i} = r_{ki} \text{ and } r_{ip+1} = r_{ik} \text{ for all } 1 \leq i \leq p+1.$$

However one is in for a disappointment, the following theorem by the second author appears in [28]:

Theorem 12.4. *If w satisfies JBSP and Ex then w is either the completely independent probability function* w_0 *or is the probability function defined by*

$$w\left(\bigwedge_{i,j=1}^{p} \beta_{r_{ij}}(a_i,a_j)\right) = \begin{cases} 2^{-m} & \text{if all the } r_{ij} \text{ are equal} \\ 0 & \text{otherwise.} \end{cases}$$

As we have already pointed out the former of these two solutions is not really what we want. The second option given by this theorem (which corresponds to Carnap's c_0 in the unary case) is even worse, this probability function simply gives probability 1 to all the a_i being forever indistinguishable.

12.5 Representation Theorems for Functions Satisfying Sx

It turns out that there are two basic kinds of probability functions satisfying Sx, heterogeneous and homogeneous.

We say that w satisfying Sx is *t-heterogeneous* if $w(\overline{s}) = 0$ whenever the length of spectrum exceeds t and in addition

$$\lim_{k\to\infty} \sum_{\overline{k}\in\bigcup_{s<t} A_k^s} \mathscr{N}(\varnothing,\overline{k})w(\overline{k}) = 0. \tag{12.3}$$

In other words w is t-heterogeneous if in the limit all the probability is massed on the spectra of length exactly t. Since for spectrum $\overline{h}$ with $k \geq \sum h_i$,

$$w(\overline{h}) = \sum_{\overline{k} \in A_k} \mathscr{N}(\overline{h},\overline{k})w(\overline{k})$$

(where A_k is the set of all spectra with sum k) this means that

$$w(\overline{h}) = \lim_{k \to \infty} \sum_{\overline{k} \in A_k^t} \mathscr{N}(\overline{h},\overline{k})w(\overline{k}). \tag{12.4}$$

In contrast to heterogeneity we say that w satisfying Sx is *homogeneous* if for all t,

$$\lim_{k \to \infty} \sum_{\overline{k} \in A_k^t} \mathscr{N}(\varnothing,\overline{k})w(\overline{k}) = 0. \tag{12.5}$$

The following theorem, which justifies the assertion that there are two basic kinds of probability function satisfying Sx, is proved in [26].

Theorem 12.5. *Let w be a binary probability function on the binary language L_2 satisfying Sx. Then there are binary probability functions $w^{[t]}$ (satisfying Sx) and constants $\eta_t \geq 0$ for $t \in \mathbb{N}$ such that*

$$w = \sum_{i=0}^{\infty} \eta_i w^{[i]}, \qquad \sum_{i=0}^{\infty} \eta_i = 1,$$

$w^{[t]}$ is t-heterogeneous for $t > 0$ and $w^{[0]}$ is homogeneous. Furthermore the η_i are unique and so are the $w^{[i]}$ when $\eta_i \neq 0$.

Given this result the problem of representing probability functions satisfying Sx splits in finding such theorems for t-heterogeneous and homogeneous probability functions. For the remainder of this paper we shall confine ourselves to the t-heterogeneous case.

Two such representation theorems are given in [26] and another is to be found in [24]. The main aim of this section is to describe a further representation theorem which is in the style of a de Finetti theorem and seems potentially of some value in applications.

First recall that for a state description as in (12.1)

$$w\left(\bigwedge_{i,j=1}^{p} \beta_{r_{ij}}(a_i,a_j)\right)$$

is also somewhat more efficiently thought of as w of the matrix $\mathbf{r} \in \{1,2,\ldots,2^m\}^{p \times p}$ with ij'th entry r_{ij}. In the obvious sense then we shall talk about the spectrum, $\mathscr{S}(\mathbf{r})$, of this matrix, meaning of course the spectrum of the state description (12.1).

We now describe the construction of a family of particular binary probability functions $u^{\overline{e}}$ satisfying Sx. In the forthcoming theorem these will play the same role as the Bernoulli multinomial trials play in de Finetti's Theorem. Fix $t \geq 1$. For a matrix $\mathbf{q} \in \{1,2,\ldots,2^m\}^{q\times q}$ let $f(\mathbf{q})$ be the number of matrices $\mathbf{o} \in \{1,2,\ldots,2^m\}^{t\times t}$ with spectrum $\{1,1,\ldots,1\}$ $(= 1_t)$ which have $\mathbf{q}$ as their top left submatrix (i.e. $q_{ij} = o_{ij}$ for $1 \leq i,j \leq q$).

Let $e_i \in \mathbb{R}$, $e_i > 0$ for $i = 1,2,\ldots,t$ and $\sum_i e_i = 1$. For visualization purposes we shall think of these subscripts $i = 1,2,\ldots,t$ as different colours. Let $\overline{e}$ denote the multiset $\{e_1,e_2,\ldots,e_t\}$ and define a function $j^{\overline{e}}$ on colourings $\mathbf{c} = \langle c_1,c_2,\ldots,c_p\rangle \in \{1,2,\ldots,t\}^p$ of matrices $\mathbf{r} \in \{1,2,\ldots,2^m\}^{p\times p}$ inductively as follows.

Let $j(\varnothing,\varnothing) = 1$. Suppose we have defined the probability $j^{\overline{e}}(\mathbf{r},\mathbf{c})$ of getting, at stage p, a matrix $\mathbf{r} \in \{1,2,\ldots,2^m\}^{p\times p}$ and a colouring $\langle c_1,c_2,\ldots,c_p\rangle \in \{1,2,\ldots,t\}^p$. (It will turn out that if $c_i = c_j$ then i and j will be indistinguishable according to $\mathbf{r}$. However the converse may not hold.) Let $c_{i_1},c_{i_2},\ldots,c_{i_q}$ be the distinct colours in $\mathbf{c}$ in order of appearance and let $\mathbf{r}(\mathbf{c})$ be the $q\times q$ submatrix of $\mathbf{r}$ formed from its $i_1,i_2,\ldots,i_q$'th rows and columns.

Now pick colour c_{p+1} from $1,2,\ldots,t$ according to the probabilities $e_1,e_2,\ldots,e_t$ and let $\mathbf{c}^+$ is $\langle c_1,\ldots,,c_p,c_{p+1}\rangle$. If c_{p+1} is the same as an earlier colour, c_j say, extend $\mathbf{r}$ to a $(p+1)\times(p+1)$ matrix $\mathbf{r}^+$ by setting $r^+_{i(p+1)} = r_{ij}, r^+_{(p+1)i} = r_{ji}$ for $1 \leq i \leq p+1$. On the other hand if c_{p+1} is a previously unchosen colour then choose $\mathbf{r}^+$ (satisfying $r_{(p+1)i} = r_{(p+1)j}$ and $r_{i(p+1)} = r_{j(p+1)}$ whenever $c_i = c_j$) to extend $\mathbf{r}$ according to the probability $f(\mathbf{r}^+(\mathbf{c}^+))/f(\mathbf{r}(\mathbf{c}))$. Finally let $j^{\overline{e}}(\mathbf{r}^+,\mathbf{c}^+)$ be $j^{\overline{e}}(\mathbf{r},\mathbf{c})$ times the probability as described of then going from $\mathbf{r},\mathbf{c}$ to $\mathbf{r}^+,\mathbf{c}^+$.

Having defined these $j^{\overline{e}}(\mathbf{r},\mathbf{c})$ now set

$$u^{\overline{e}}(\mathbf{r}) = \sum_{\mathbf{c}} j^{\overline{e}}(\mathbf{r},\mathbf{c}).$$

Theorem 12.6. *The $u^{\overline{e}}$ are t-heterogeneous probability functions satisfying Sx. Furthermore if w is any t-heterogeneous probability function satisfying Sx then there is a countably additive measure μ on the Borel subsets of*

$$\mathbb{B}_t = \{\langle e_1,e_2,\ldots,e_t\rangle \in \mathbb{R}^t \mid e_i > 0, \sum_i e_i = 1\}$$

such that for $\theta \in SL$,

$$w(\theta) = \int_{\mathbb{B}_t} u^{\overline{e}}(\theta)\,d\mu.$$

Apart from its shedding light on the structure of t-heterogeneous probability functions this result is also of practical use in that to show that all t-heterogeneous probability functions satisfy some property it is enough to show that the property holds for all these relatively simple $u^{\overline{e}}$ and is preserved by convex combinations.

12.6 Instantial Relevance

For Carnap one of the successes of his programme was that the basic principle Ex actually implied a certain inequality, the *Principle of Instantial Relevance*, that Carnap felt should hold, even to the extent of having earlier suggested that it might be adopted as a basic principle in its own right, see [4]. In the unary case this principle can be stated as:

Principle of Instantial Relevance, PIR

$$w(\alpha(a_{p+2}) \,|\, \alpha(a_{p+1}) \wedge \theta(a_1,\dots,a_p)) \geq w(\alpha(a_{p+2}) \,|\, \theta(a_1,\dots,a_p)).$$

In essence then, the more times you have seen an atom[4] instantiated the higher your probability should be that the next individual observed will also satisfy it.

This is obviously a rather desirable principle,[5] one that we might well want to have at our disposal. For that reason the following theorem is particularly pleasing.

Theorem 12.7. *Ex implies PIR.*

The easiest proof of this result is a straightforward application of de Finetti's Theorem, expressed via that theorem it amounts simply to saying that the integral of a square is always non-negative.

In the binary case it is perhaps not obvious what the analog of PIR should be. One suggestion might be that we should have

$$w(R(a_{n+2},a_{n+2}) \,|\, R(a_{n+1},a_{n+1}) \wedge \psi(a_1,\dots,a_n)) \geq w(R(a_{n+2},a_{n+2}) \,|\, \psi(a_1,\dots,a_n)).$$

However this (and a number of similar attempts) follows from the *unary* version of PIR applied to the probability function v defined by

$$v\left(\bigwedge_{i=1}^{p} P^{\varepsilon}(a_i)\right) = w\left(\bigwedge_{i=1}^{p} R^{\varepsilon}(a_{n+i},a_{n+i}) \,|\, \psi(a_1,\dots,a_n)\right).$$

As far as a truly binary "PIR" is concerned one may feel that it was at least reasonable to expect that

$$w(R(a_2,a_2) \,|\, R(a_1,a_1) \wedge R(a_1,a_2) \wedge R(a_2,a_1)) \geq w(R(a_1,a_1)).$$

However for binary w satisfying just Ex this can fail as the following example shows.

Take $m = 1$ (so there is only one relation symbol $R(x,y)$) and define a binary probability function w satisfying Ex as follows. On a matrix $\mathbf{r} \in \{1,2\}^{p\times p}$ set $w(\mathbf{r})$ to be zero if it is *not* the case that all the entries off the diagonal are the same. On the other hand if they are all same, say they are all c where $c \in \{1,2\}$, set

[4] The principle applies equally to a formula $\varphi(x)$ in place of $\alpha(x)$.

[5] Nevertheless some of it's 'natural generalizations' have surprising consequences, see [25]

$$w(\mathbf{r}) = (1/2)\cdot\prod_{i=1}^{p}((c/2-1/4)(r_{ii}-1)+(5/4-c/2)(2-r_{ii})).$$

In other words off the diagonal the constant value is determined by a coin toss and on the diagonal it just looks like a Bernoulli process with probability $1/4$ of a 1 if there are 1s off the diagonal and probability 3/4 of a 1 if there are 2s off the diagonal. It is easy to see that simultaneously transposing the i'th and j'th columns and rows in $\mathbf{r}$ does not affect $w(\mathbf{r})$ so w satisfies Ex. However

$$w(R(a_2,a_2)\,|\,R(a_1,a_1)\wedge R(a_1,a_2)\wedge R(a_2,a_1)) = 1/4 < 1/2 = w(R(a_1,a_1)).$$

Returning again now to the unary case, if we strengthen the assumption Ex to Ax then we can obtain correspondingly stronger "instantial relevance" principles. To explain one such result, for two multisets $\{r_1,r_2,\ldots,r_s\}$ and $\{t_1,t_2,\ldots,t_q\}$ with $\sum_i r_i = \sum_j t_j$, $r_1 \geq r_2 \geq \ldots \geq r_s$ and $t_1 \geq t_2 \geq \ldots \geq t_q$ set

$$\{r_1,r_2,\ldots,r_s\} \trianglerighteq \{t_1,t_2,\ldots,t_q\}$$

if for all $1 \leq j \leq \max\{s,q\}$,

$$\sum_{i\leq j} r_i \geq \sum_{i\leq j} t_i,$$

where $r_i = 0$ for $s < i \leq \max\{s,q\}$ and $t_i = 0$ for $q < i \leq \max\{s,q\}$.

The following result for unary probability functions will appear in a forthcoming paper of the first author and P. Waterhouse.

Theorem 12.8. *Given state descriptions $\bigwedge_{i=1}^{p}\alpha_{h_i}(a_i)$, $\bigwedge_{i=1}^{p}\alpha_{g_i}(a_i)$ with spectra (as multisets) $\{r_1,r_2,\ldots,r_s\}$, $\{t_1,t_2,\ldots,t_q\}$ respectively,*

$$v\left(\bigwedge_{i=1}^{p}\alpha_{h_i}(a_i)\right) \geq v\left(\bigwedge_{i=1}^{p}\alpha_{g_i}(a_i)\right)$$

for all unary probability functions v on L_1 satisfying Ax iff

$$\{r_1,r_2,\ldots,r_s\} \trianglerighteq \{t_1,t_2,\ldots,t_q\}.$$

We would conjecture that an analogous result holds also in the binary (and higher arity) cases when the probability function satisfies Sx. Some few special cases along these lines have been proved, in particular that for binary w satisfying Sx, $w(\{2,1\}) \geq w(\{1,1,1\})$. In consequence of this special case we are at least able to decide some simple preferences, for example:

> *Given $\neg R(a_1,a_2), \neg R(a_2,a_1), R(a_3,a_2)$ can we assume that $R(a_2,a_3)$ is at least as probable as its negation?*

In this case should the connection

$$R(x,y) \leftrightarrow R(y,x)$$

suggested by $\neg R(a_1,a_2)$ and $\neg R(a_2,a_1)$ cause $R(a_3,a_2)$ to give predominant support to $R(a_2,a_3)$, or should the $\neg R(a_2,a_1)$ prove the deciding factor in giving preference to $\neg R(a_2,a_3)$? Or maybe neither of them can claim superiority (or at least equality) in all situations?

In fact assuming Sx we can answer this question in the affirmative. To see this notice that (for $m=1$) the sentence

$$\neg R(a_1,a_2) \wedge \neg R(a_2,a_1) \wedge R(a_3,a_2) \wedge R(a_2,a_3)$$

can be extended to a state description with spectrum $\{2,1\}$ in 4 ways (all other extensions have spectrum $\{1,1,1\}$) whilst

$$\neg R(a_1,a_2) \wedge \neg R(a_2,a_1) \wedge R(a_3,a_2) \wedge \neg R(a_2,a_3)$$

can only be extended to a state description with spectrum $\{1,2\}$ in 2 ways (again all others have spectrum $\{1,1,1\}$). Since for w satisfying Sx, $w(\{2,1\}) \geq w(\{1,1,1\})$ the required inequality follows.

12.7 Conclusion

In this note we have described some key concepts and results in unary inductive logic and discussed their generalizations to binary inductive logic. As well as a number of largely technical problems, for example extending these results to languages with infinitely many binary predicates, some quite basic problems concerning principles of "instantial relevance" remain largely open. Hopefully the representation theorems which have recently been developed will enable us to go some way towards answering them.

Dedication

We would like to dedicate this modest paper to Rohit Parikh on the occasion of his 70th birthday. His contribution to Mathematical Logic has been a source of inspiration and pleasure to us all.

References

1. R. Carnap. *Logical Foundations of Probability*. University of Chicago Press and Routledge & Kegan Paul Ltd., Chicago, IL, and London, 1950.
2. R. Carnap. *The Continuum of Inductive Methods*. University of Chicago Press, Chicago, IL, 1952.

3. R. Carnap. A basic system of inductive logic. In R.C. Jeffrey, editors, *Studies in Inductive Logic and Probability*, volume 2, pages 7–155. University of California Press, Berkeley, CA, 1980.
4. R. Carnap. Replies and systematic expositions. In P.A. Schlipp, editors, *The Philosophy of Rudolf Carnap*. La Salle, IL, Open Court, 1963.
5. G. Coletti and R. Scozzafava. *Probabilistic Logic in a Coherent Setting, Trends in Logic*, volume 15. Kluwer Academic Press, London, Dordrecht, 2002.
6. B. De Finetti. La prevision: ses lois logiques, ses sources subjetive. *Annales de l'Institut Henri Poincaré*, 7:1–68, 1937.
7. B. De Finetti. On the condition of partial exchangeability. In R.C. Jeffrey, *Studies in Inductive Logic and Probability*, volume 2. University of California Press, Berkley, CA and Los Angeles, CA, 1980.
8. B. de Finetti. *Theory of Probability*, volume 1, Wiley, New York, NY, 1974.
9. H. Gaifman. Concerning measures on first order calculi. *Israel Journal of Mathematics*, 2:1–18, 1964.
10. N. Goodman. A query on confirmation. *Journal of Philosophy*, 43:383–385, 1946.
11. N. Goodman. On infirmities in confirmation-theory. *Philosophy and Phenomenology Research*, 8:149–151, 1947.
12. M.J. Hill, J.B. Paris, and G.M. Wilmers. Some observations on induction in predicate probabilistic reasoning. *Journal of Philosophical Logic*, 31(1):43–75, 2002.
13. D.N. Hoover. *Relations on Probability Spaces and Arrays of Random Variables*. Preprint, Institute of Advanced Study, Princeton, NJ, 1979.
14. W.E. Johnson. Probability: The deductive and inductive problems. *Mind*, 41(164):409–423, 1932.
15. O. Kallenberg. *Probabilistic Symmetries and Invariance Principles*. Springer, New York, NY, ISBN-10: 0-387-25115-4, 2005.
16. O. Kallenberg. The Ottawa Workshop, http://www.mathstat.uottawa.ca/~givanoff/ wskallenberg.pdf
17. J.G. Kemeny. Carnap's theory of probability and induction. In ed. P.A. Schilpp, *The Philosophy of Rudolf Carnap*, pages 711–738. La Salle, IL, Open Court, 1963.
18. P.H. Krauss. Representation of symmetric probability models. *Journal of Symbolic Logic*, 34(2):183–193, 1969.
19. J. Landes. The Principle of spectrum exchangeability within inductive logic. Ph.D. dissertation, University of Manchester, Manchester, April 2009.
20. P. Maher. Probabilities for two properties. *Erkenntnis*, 52:63–91, 2000.
21. P. Maher. Probabilities for multiple properties: The models of Hesse, Carnap and Kemeny. *Erkenntnis*, 55:183–216, 2001.
22. F. Matúš. Block-factor fields of Bernoulli shifts. *Proceedings of Prague Stochastics'98*, volume 2. 383–389, 1998.
23. D. Miller. Popper's qualitative theory of versimilitude. *British Journal for the Philosophy of Science*, 25:166–177, 1974.
24. C.J. Nix. Probabilistic Induction in the Predicate Calculus Doctorial Thesis, Manchester University, Manchester, UK, 2005. See http://www.maths.man.ac.uk/~jeff/#students
25. C.J. Nix and J.B. Paris. A continuum of inductive methods arising from a generalized principle of instantial relevance. *Journal of Philosophical Logic*, 35(1):83–115, 2006.
26. C.J. Nix and J.B. Paris. A note on binary inductive logic. *Journal of Philosophical Logic*, 36(6):735–771, 2007.
27. J.B. Paris. *The Uncertain Reasoner's Companion*. Cambridge University Press, Cambridge, 1994.
28. A. Vencovská. Binary Induction and Carnap's Continuum. *Proceedings of the 7th Workshop on Uncertainty Processing (WUPES)*, Mikulov, 2006. See http://mtr.utia.cas.cz/wupes06/articles/data/vencovska.pdf

Chapter 13
Challenges for Decidable Epistemic Logics from Security Protocols

R. Ramanujam and S.P. Suresh

13.1 Summary

13.1.1 Knowledge and Communication

A central question in knowledge theory relates to how knowers update their knowledge on receipt of a communication. This is important, since the very purpose of communications is (typically) to create such an update of knowledge in the recipient. However, there is often a lack of concordance between the intended update and that which occurs, leading to interesting situations and much work for knowledge theorists

Communication protocols studied by computer scientists offer a restricted (but yet interesting) domain for the study of knowledge change. Protocol descriptions define (but also limit) how communications are to be interpreted, and in this sense potential knowledge change can be "calculated" and messages designed accordingly, as long as participants in the protocol game can be trusted to play by the rules.

The last caveat above is significant, and a good bit of the theory of distributed systems relates to what happens when some of the players do *not* strictly adhere to the protocol diktat. If a sender cannot be trusted, a recipient is uncertain how to interpret a received message. Computer scientists solve this problem in two ways:

- The **global** way: honest participants proceed as if all the world is honest, and protocol rules ensure that desired global coordination is achieved, as long as only a fraction (say, at most one third) are untrustworthy. This is the realm of *fault tolerant* distributed algorithms [13].
- The **local** way: honest participants run a pre-protocol using special facilities (and codes) to create a trusted subnetwork and decide on a protocol to be followed

R. Ramanujam
Chennai Mathematical Institute, Chennai, India, e-mail: `jam@imsc.res.in`
S.P. Suresh
Chennai Mathematical Institute, Chennai, India, e-mail: `spsuresh@cmi.ac.in`

J. van Benthem et al. (eds.), *Games, Norms and Reasons*, Synthese Library 353, DOI 10.1007/978-94-007-0714-6_13, © Springer Science+Business Media B.V. 2011 within it. Once this step is completed, they use the protocol agreed on. This is the approach used by *cryptographic protocols*.

One of the earliest applications of knowledge theory in distributed computing was in fault tolerance [8, 11]. BAN logics [3] initiated the latter study, that of epistemic logics for security analysis. This domain poses challenging questions for epistemic logics, because we are called upon to determine knowledge update in the presence of a great deal of uncertainty and distrust. Since this makes the update weak, any attempt to transfer knowledge must take such distrust into account as well. As remarked above, these are pre-protocols, so communications do not have much informational content; but their *form* is quite critical.

13.1.2 Cryptographic Protocols

Security protocols are specifications of communication patterns which are intended to let agents share secrets over a public network. They are required to perform correctly even in the presence of **malicious intruders** who listen to the message exchanges that happen over the network and also manipulate the system (by blocking or forging messages, for instance). Obvious correctness requirements include **secrecy**: an intruder cannot read the contents of a message intended for others; **authenticity**: if B receives a message that appears to be from agent A and intended for B, then A indeed sent the same message intended for B in the recent past.

Mechanisms for ensuring security typically use **encrypted communication**. However, even the use of the most perfect cryptographic tools does not always ensure the desired security goals. (See [1] for an illuminating account.) This situation arises primarily because of **logical flaws** in the design of protocols. It is widely acknowledged that security protocols are hard to analyze, bugs difficult to detect, and hence that it is desirable to look for automatic means by which **attacks** on protocols can be discovered. This means that formal models for reasoning about security protocols should be developed.

Here is a typical message exchange in a security protocol.

$$1.\ A \rightarrow B : \{n\}_{public(B)}$$
$$2.\ B \rightarrow A : \{n\}_{public(A)}$$

This notation represents the following intention on the part of the designer. A and B have been active in the network in the recent past. A wishes to talk to B, and ensure that she is talking to B. Both have strong faith in *public key encryption* and know each other's public keys. A generates a fresh nonce (a random, previously unused, unguessable number) n and sends it to B encrypted in his public key. When B receives the message, he can indeed decrypt and learn n. He returns it to A now encrypted with her public key.

Given perfect encryption, when A sends the message she *knows* that only somebody who *knows* the private key of B can *know* n. So, when she later receives n

encrypted in her own public key, she *knows* that someone who knows the private key of B has accessed n (and it has to be recently, since n was previously unused).

Note the repeated use of the word *know* in the above story. It is clear that one of the main goals of security protocols is the selective transfer of some kind of knowledge to certain select members of a possibly hostile, public network. The network is hostile in that some members might be saboteurs who *actively* try to manipulate the actions of the other members to try to learn new secrets. As we can see, cryptographic methods are used to aid honest participants.

13.1.3 Difficulties

Consider the implications for knowledge theory from this little story. We list below some immediate questions that arise, most of which can be easily answered, individually. The difficulty is in answering them all, in *one uniform* framework.

- What details of the encryption algorithm should be known to A and B for them to employ the reasoning above?
- If the algorithm employs random elements, should they be able to ascertain that they are indeed chosen randomly (in order for their understanding to be certified as knowledge)?
- What knowledge is needed on A's part to ensure that n is previously unused? Given that n comes from an unbounded set, does the representation used by A matter?
- Once A encrypts n with B's public key, she only has a bit string from which she cannot get any new information. (If she could, so could others.) Indeed, if she does this twice, she would get a different string each time. How can she be said to know that this is the term $\{n\}_B$?
- What B receives is a bit string. How does he know it is encrypted at all, let alone with his key? In effect, this means that he knows all possible messages encrypted with his key.
- Note that B has no way of telling when the n received was generated and by whom. What precisely does B know regarding the communication, that causes him to act?

A simple answer to most of these questions is the Dolev-Yao model [6], used extensively in formal studies of security protocols, which we describe in the next section. In this view, a receiver may expect to receive a term $\{t\}_k$ according to the protocol, but unless she also has k^{-1}, she cannot get t (using only this encrypted term). This is a form of database knowledge: an agent A has only that information explicitly stored in the agent's database. However, ascribing knowledge to agents involves inference as well, and it is this more general form of knowledge that is of interest in epistemic logics.

Note that we are using a variety of forms of knowledge, each of which has been extensively studied by knowledge theorists. We have *propositional knowledge*: for instance, that $\{\{x\}_k\}_{k^{-1}} = x$; *process knowledge*: that of efficacy of encryption; *algorithmic knowledge*: that of how to extract messages from codes; *implicit knowledge*: that of how B responds to a message; *explicit knowledge*: that of n in A's database; and so on. It is worth noting that the difference between *propositional knowledge* and *sentential knowledge* emphasized by Parikh [15] is also crucially relevant when we talk of knowing encoded text.

13.1.4 Decidability Issues

A central aim of formal methods in security theory is to find algorithmic solutions to the verification problem: do all runs of a given security protocol *Pr* satisfy a given security property φ ? When the answer is no, the counterexample is termed an *attack* on the protocol, and the need for automatic methods arises from the fact that finding attacks can be quite complicated, whereas the cost of any potential attack is very high.

Since the space of all runs of a security protocol typically constitute an infinite state system, even simple reachability properties are undecidable, and hence the project of automatic verification of security protocols is doomed to failure, unless some restrictions are imposed. Typically this is in the form of bounded verification, whereby we assume that the number of concurrent multisessions possible at any time is bounded, and syntactic conditions ensure that the term space of communications is also bounded.

When the security property φ involves epistemic modalities, there is a further complication. Suppose that we place external bounds so that all runs use only a fixed finite set of terms T. Then the semantics defines a finite state system, and the verification problem is decidable. However, with Hintikka-style semantics of knowledge, this also implies that T is common knowledge in the system, which goes against the very basics of security theory. In the example discussed above, we crucially used the fact that a nonce n was freshly generated and hence not known to others.

An alternative is that we check φ only over runs that use T, but let the knowledge modality range over all runs, modelling the fact that agents do not know T. However, this is easily seen to lead to undecidability as well.

An interesting complication arises due to the fact that we are talking of decidability of the verification problem and not that of the *satisfiability* problem for the logic. When the logic is sufficiently simple, deciding the satisfiability of knowledge properties of security protocols is not very different from that of other "interpreted" systems. However, when we are considering the runs of a given protocol, the situation is different: asserting the existence of an equivalent run requires a witness that is admissible according to the given protocol. In knowledge theory, this is typically achieved by forming a product with the given finite state system being checked. In the case of a security protocol, the system is infinite state, and hence a simple product does not suffice. As we will see below, we will let the knowledge formula guide us to an abstraction of the infinite state system to one that is finite state and then proceed to verify it in the latter.

13.1.5 This Paper

In this paper, we look for a minimal logic of knowledge that addresses some of the semantic issues discussed above, and for which the verification problem is decidable. The main idea is to limit expressiveness so that knowledge of data and encryption is described by an underlying inference system that operates at the level of terms used in messages, and propositional epistemic connectives are defined on top of this system.

The logic we study is a standard Hintikka style propositional logic of knowledge, where the atomic propositions have specific structure related to security protocols.

13.1.6 The Proposal

The syntax of the logic is as given below:

$$\mathcal{L} ::= A\ has\ x \mid sent(A,B,x) \mid received(A,B,x) \mid \neg\alpha \mid \alpha \vee \beta \mid \mathbf{G}\alpha \mid \mathbf{H}\alpha \mid \mathbf{K}_A\alpha$$

$\mathbf{G}\alpha$ asserts that α holds *always* in the future. Its dual $\mathbf{F}\alpha$ says that α holds *sometime* in the future. Similarly $\mathbf{H}\alpha$ and $\mathbf{P}\alpha$ refer to all the time points and some time point, respectively, in the past. $\mathbf{L}_A\alpha$ is the dual of $\mathbf{K}_A\alpha$, as usual.

The basic propositions require some explanation: the x that figures in the syntax stands for **secret nonces or keys** exchanged in an execution of a protocol. *A has x* asserts that A has the secret x in her database. $sent(A,B,x)$ specifies that a communication event happened with A sending a term containing x intended for B. $received(A,B,x)$ makes a similar assertion about A receiving a term purportedly from B.

The logic is admittedly rather weak. Note the absence of next-time or previous-time modalities, so we are not describing system transitions here. Moreover information accumulation is monotone, as we will see below. Critically, there are no encrypted terms in the syntax of formulas, and hence we cannot describe cryptographic protocols. All this is in the spirit of an abstract specification logic in which we only wish to specify security requirements, not describe mechanisms for implementing them.

One important reason for studying such a minimal logic is to address the semantic difficulties discussed earlier. The main idea is to limit the expressiveness of the logic as follows:

- The model uses an explicit primitive for talking about what is and what is not decipherable by agents, but this is not referred to in the logic.
- An indistinguishability relation is defined, which in effect makes pieces of messages that can only have originated with the intruder/adversary indistinguishable from those generated by honest principals. Knowledge is defined in terms of this relation.
- Thus knowledge quantifies over intruder capabilities, and describes properties invariant on an equivalence class of messages (compatible with a given protocol).
- Formulas, in themselves, describe only what principals know or do not know about who has access to which secret. Protocols use cryptographic mechanisms to achieve this, and the verification problem determines whether, under the perfect encryption assumption, the specifications are met.

Even in this limited specification language, we can easily specify many desirable properties of protocols. For instance, here is a simple version of secrecy, which says that in any run where A sends B (distinct from I) a secret m, I cannot get hold of m.

$$\mathbf{G}[sent(A,B,m) \supset \neg(I\ has\ m)]$$

An even stronger version states that A in fact knows this (and can therefore base her further actions on this knowledge, if she chooses to).

$$\mathbf{G}[sent(A,B,m) \supset \mathbf{K}_A\ \neg(I\ has\ m)]$$

Authentication is simply stated as:

$$\mathbf{G}[received(A,B,m) \supset \mathbf{P}sent(B,A,m)]$$

The main theorem of the paper asserts that the protocol verification problem of the logic is elementarily decidable, as long as we consider protocol runs with a fixed upper bound on the number of concurrent multi-sessions. Despite this bound, the inexact nature of agents' knowledge forces us to consider an infinite set of runs as possible, and hence the decision question is nontrivial.

13.1.7 BAN Logic

It is important to highlight the similarities and differences of our approach with that of BAN logic [3]. BAN logic is a highly abstract logic that is intended to be used for assertional reasoning about protocols, much in the style of Hoare logics. It is a propositional modal logic with formulas like *A believes* α, *A sees* α, *A sent* α, *fresh n*, *A controls* α etc. More critically, the logic comes with a deduction system with plausible rules like the following:

$$\frac{A\ sees\ \alpha \qquad A\ believes\ (fresh\ \alpha)}{A\ believes\ \alpha}$$

$$\frac{A\ believes\ (C\ controls\ \alpha) \qquad A\ believes\ (C\ sent\ \alpha)}{A\ believes\ \alpha}$$

There are many rules but the above two are crucial, as they pertain to *information transfer*. The key to reasoning in this logic is the process of *idealisation*, where message exchanges are themselves *propositionalized*. For example, a communication $A \rightarrow B : k$ may be rendered as

$$B\ sees\ (A\ sent\ (k\ is\ a\ good\ key\ for\ A\ and\ B))$$

Based on all this, non-trivial reasoning about the behaviour of protocols can be carried out assertionally. Many protocols have been proved correct using BAN logic, and flaws in many protocols have been detected using it as well. But this approach – the idealisation process, in particular – has met with a lot of criticism over the years ([14], for example), as there are many examples of protocols where there is a mismatch between the logic and the semantics of the protocol. The point is that the intruder behaviour is too rich to be captured by a simple set of rules.

Recent work by Cohen and Dam [4, 5] has made significant progress by providing a Kripke semantics for BAN logic that addresses the logical omniscience problem, and by exploring various completeness and expressibility issues.

An important feature of the logic discussed here is that the idealisation step is limited to assertions about which (atomic) secrets an agent has access to. This has implications for reasoning, as well as decidable verification, as we will see below.

13.2 Security Protocol Modelling

We briefly present our model for protocols in this section. A more detailed presentation can be found in [16]. Most of the elements of the model are standard in the literature on modelling security protocols. In particular, we use the Dolev-Yao adversary model [6].

Terms and Actions

We start with a (potentially infinite) set of **agents** Ag, which includes the **intruder** I, and the others, who are called **honest agents**. Fix a countable set of **fresh secrets** $\mathcal{N}$. (This includes *random, nonguessable nonces* as well as *temporary session keys*.) $\mathcal{K}$, the set of **potential keys**, is given by $\mathcal{N} \cup \{public(A), private(A) \mid A \in Ag\} \cup \{shared(A,B) \mid A, B \in Ag\}$. Here $public(A)$, $private(A)$, and $shared(A,B)$ denote the public key of A, private key of A, and (long-term) shared key between A and B. We assume an inverse $\overline{k}$ for each $k \in \mathcal{K}$ such that $\overline{\overline{k}} = k$.

$\mathcal{T}_0 \stackrel{\text{def}}{=} \mathcal{K} \cup Ag$ is the set of **basic terms**.

The set of **information terms** is defined to be

$$\mathcal{T} ::= m \mid (t_1, t_2) \mid \{t_1\}_k$$

where m ranges over $\mathcal{T}_0$, t_1 and t_2 range over $\mathcal{T}$, and k ranges over $\mathcal{K}$. Here (t_1, t_2) denotes the pair consisting of t_1 and t_2, and $\{t_1\}_k$ denotes the term t_1 encrypted using k. These are the terms used in the message exchanges (which will be presently introduced). We use $st(t)$ to denote the set of subterms of t.

We model communication between agents by *actions*. An action is either a *send action* of the form $A!B{:}t$ or a *receive action* of the form $A?B{:}t$. Here A and B are distinct agents, A is honest, and t is a term. For an action a of the form $A!B{:}t$ or $A?B{:}t$, we define $term(a)$ to be t. The agent B is (merely) the **intended receiver** in $A!B{:}t$ and the **purported sender** in $A?B{:}t$. Since the intruder is assumed to have access to the entire communication network at all times, every send action can be seen as an instantaneous receive by the intruder, and similarly, every receive action is an instantaneous send by the intruder.

Protocol Specifications

A protocol is given by the roles it contains, and a **role** is a finite sequence of actions. A **parametrized role** $\eta[m_1, \cdots, m_k]$ is a role in which the *basic terms* $m_1, \ldots, m_k$ are singled out as parameters. The idea is that an agent participating in the protocol can execute many *sessions* of a role in the course of a single run, by instantiating the parameters in many different ways. All the basic terms occurring in a parametrized role that are not counted among the parameters are the **constants** of the role. They do not change their meaning over different sessions of the role.

Suppose $\eta = a_1 \cdots a_k$ is a parametrized role. We say that a nonce n **originates** at $i (\leq k)$ in η if:

- n is a parameter of η,
- a_i is a send action, and
- $n \in st(term(a_i))$ and for all $j < i, n \notin st(term(a_j))$.

If a nonce n originates at i in a role it means that the agent sending the message a_i uses n for the first time in the role. This usually means that, in any session of that role, the agent playing the role has to generate a fresh, nonguessable random number and send it as a challenge. Subsequent receipt of the same number in the same session plays a part in convincing the agent that the original message reached the intended recipient.

A **protocol** is a finite set of parametrized roles $\{\eta_1, \ldots, \eta_n\}$. The set of constants of Pr, denoted $\mathbf{C}(Pr)$, consists of all constants of all roles of Pr. The semantics of a protocol is given by the set of all its runs. A run is got by instantiating each role of the protocol in an appropriate manner, and forming admissible interleavings of such instantiations. We present the relevant definitions below.

Substitutions and Events

A **substitution** σ is a map from $\mathcal{T}_0$ to $\mathcal{T}_0$ such that $\sigma(Ag) \subseteq Ag$, $\sigma(\mathcal{N}) \subseteq \mathcal{N}$, and $\sigma(I) = I$. For any $T \subseteq \mathcal{T}_0$, σ is said to be a T-substitution iff for all $x \in \mathcal{T}_0$, $\sigma(x) \in T$. A substitution σ is **suitable for a parametrized role** η if $\sigma(m) = m$ for all constants m of η. We say that σ is **suitable for a protocol** Pr if $\sigma(m) = m$ for all constants m of Pr.

A run of a protocol is any sequence of actions that can possibly be performed by the various agents taking part in the protocol. We model each run as a sequence of *event occurrences*, which are actions with some extra information about causality. (From now on, we will gloss over the difference between events and event occurrences.) An event of a protocol Pr is a triple (η, σ, lp) such that η is a role of Pr, σ is a substitution, and $1 \leq lp \leq |\eta|$. A T-event is one which involves a T-substitution. For an event $e = (\eta, \sigma, lp)$ with $\eta = a_1 \cdots a_\ell$, $act(e) \stackrel{\text{def}}{=} \sigma(a_{lp})$. If $e = (\eta, \sigma, lp)$ and $lp < |\eta|$ and $e' = (\eta, \sigma, lp + 1)$, then we say that e *locally precedes* e' and denote it by $e <_\ell e'$. We say that a nonce n is **uniquely originating** in a set of events E of a protocol Pr if there is at most one event (η, σ, lp) of E and at most one nonce m such that m originates at lp in η and $\sigma(m) = n$. (Note that the fact m originates in η implies that m is a parameter of η.)

Message Generation Rules

We intend a run of a protocol to be an admissible sequence of events. A very important ingredient of the admissibility criterion is the enabling of events given a particular information state. To treat this formally, we need to define how the agents (particularly the intruder) can build new messages from old. This is formalised by the notion of derivations.

A **sequent** is of the form $T \vdash t$ where $T \subseteq \mathcal{T}$ and $t \in \mathcal{T}$. A **derivation** or a **proof** π of $T \vdash t$ is a tree whose nodes are labelled by sequents and connected by one of the *analz*-rules or *synth*-rules in Figure 13.1; whose root is labelled $T \vdash t$; and whose leaves are labelled by instances of the *Ax* rule. We will use the notation $T \vdash t$ to denote both the sequent, and the fact that it is derivable. For a set of terms T, $\overline{T} \stackrel{\text{def}}{=} \{t \mid T \vdash t\}$ is the *closure* of T. Note that $\overline{T}$ is in general infinite even when T is finite, and hence the following proposition is useful. It is quite well known in the literature. A proof can be found in [18], for instance.

$$\frac{}{T \cup \{t\} \vdash t}\,Ax \qquad \frac{T \vdash t_1 \quad T \vdash t_2}{T \vdash (t_1, t_2)}\,pair$$

$$\frac{T \vdash (t_1, t_2)}{T \vdash t_i}\,split_i (i = 1, 2) \qquad \frac{T \vdash t \quad T \vdash k}{T \vdash \{t_1\}_k}\,encrypt$$

$$\frac{T \vdash \{t\}_k \quad T \vdash \overline{k}}{T \vdash t}\,decrypt$$

analz-rules *synth*-rules

Fig. 13.1 Message generation rules

Proposition 13.1. *Suppose that T is a finite set of terms and t is a term. Checking whether $T \vdash t$ is decidable.*

Information States, Updates, and Runs

An **information state** (or just *state*) is a tuple $(s_A)_{A \in Ag}$, where $s_A \subseteq \mathcal{T}$ for each $A \in Ag$. The **initial state** of *Pr*, denoted by *init*(*Pr*) is the tuple $(s_A)_{A \in Ag}$ such that for all $A \in Ag$,

$$s_A = \mathbf{C}(Pr) \cup Ag \cup \{private(A)\} \cup \{public(B), shared(A,B) \mid B \in Ag\}.$$

The notion of information state as simply a set of terms is rudimentary, but suffices for our purposes; it can be considered as a database of explicit but restricted knowledge. The notions of an action **enabled** at a state, and *update*(s,a), the **update** of a state s on an action a, are defined as follows:

- A send action a is enabled at s iff $term(a) \in \overline{s_A}$.
- A receive action a is enabled at s iff $term(a) \in \overline{s_I}$.
- $update(s, A!B{:}t) \stackrel{\text{def}}{=} s'$ where $s'_I = s_I \cup \{t\}$, and for all agents C other than I, $s'_C = s_C$.
- $update(s, A?B{:}t) \stackrel{\text{def}}{=} s'$ where $s'_B = s_B \cup \{t\}$ and for all agents C other than B, $s'_C = s_C$.

$update(s, \eta)$ for a state s and a sequence of actions η is defined in the obvious manner. Thus if s_A is a form of explicit knowledge, then $\overline{s_A}$ is a form of implicit knowledge, but one that is algorithmically constructible by the agent.

Given a protocol *Pr* and a sequence $\xi = e_1 \cdots e_k$ of events of *Pr*, *infstate*(ξ) is defined to be $update(init(Pr), act(e_1) \cdots act(e_k))$. Given a protocol *Pr*, a sequence $e_1 \cdots e_k$ of events of *Pr* is said to be a **run** of *Pr* iff the following conditions hold:

- for all $i, j \leq k$ such that $i \neq j$, $e_i \neq e_j$,

- for all $i \leq k$ and for all e such that $e \xrightarrow{+}_{\ell} e_i$, there exists $j < i$ such that $e_j = e$,
- for all $i \leq k$, $act(e_i)$ is enabled at $infstate(e_1 \cdots e_{i-1})$, and
- every nonce that is not a constant of Pr is uniquely originating in $\{e_1, \ldots, e_k\}$.

We say that ξ is a T-run of Pr, for any given $T \subseteq \mathcal{T}_0$, if for all $i \leq k$, $st(e_i) \cap \mathcal{T}_0 \subseteq T$. We say that ξ is a b-run of Pr, for any given $b \in \mathbb{N}$, if there are at most b nonces uniquely originating in ξ. We let $\mathcal{R}(Pr)$, $\mathcal{R}_T(Pr)$, and $\mathcal{R}_b(Pr)$ denote respectively the set of all runs, all T-runs, and all b-runs of Pr.

13.3 The Semantics of the Logic

We now formally present the syntax and semantics of our logic. From the model for security protocols presented earlier, it is clear that protocol descriptions mention abstract names for agents, nonces, and keys, but the runs use different instantiations for these abstract names. It is these concrete systems determined by protocols that we wish to specify properties of and verify. To be abstract, the specification logic should also mention only the abstract names, but the semantics must translate them into concrete names used in runs. A difficulty is that the denotation of an abstract name differs at different points in runs, so we have to find a way of resolving the different meanings. This is akin to the problem of **rigid designators** in first order modal logic.

We use the simple device of using logical variables. Their meaning is given by assignments, just as in first-order logic. But we do not allow quantification over variables in the logic itself. We let $\mathcal{X}$ denote the infinite set of logical variables. These variables are supposed to stand for nonces. We also need to use concrete agent names in the logic for specifications based on agent-based knowledge modalities to make sense. It bears emphasizing that variables do not have the same status as the abstract names in the protocol specification. These indeed refer to concrete nonces that occur in concrete executions of the protocol, but quite often we do not want to bother about which particular concrete nonce is being talked about.

Recall the syntax of the logic:

$$\mathcal{L} ::= A \; has \; x \mid sent(A,B,x) \mid received(A,B,x) \mid \neg\alpha \mid \alpha \vee \beta \mid \mathbf{G}\alpha \mid \mathbf{H}\alpha \mid \mathbf{K}_A \alpha$$

The semantics of the logic crucially hinges on an equivalence relation on runs, which is defined as follows. Intuitively, an agent cannot distinguish a term $\{t\}_k$ from any other bitstring in a state where she has no information about k.

We define the set $\mathcal{P}$ of patterns as follows (where $\square$ denotes an unknown pattern):

$$P, Q \in \mathcal{P} ::= m \in \mathcal{T}_0 \mid (P, Q) \mid \{P\}_k \mid \square$$

An *action pattern* is an action that uses a pattern instead of a term, and an *event pattern* is an event that uses an action pattern instead of an action.

We can now define the patterns derivable by an agent on seeing a term t in the context of a set of terms S. In a sense, this is the only *certain* knowledge that the agent can rely on at that state. A similar notion has been defined in [20] in the context of providing semantics for a BAN-like logic.

$$pattern(m,S) = \begin{cases} m & \text{if } m \in \mathcal{T}_0 \cap S \\ \square & \text{if } m \in \mathcal{T}_0 \setminus S \end{cases}$$

$$pattern((t_1,t_2),S) = (pattern(t_1,S), pattern(t_2,S))$$

$$pattern(\{t\}_k,S) = \begin{cases} \{pattern(t,S)\}_k & \text{if } \overline{k} \in \overline{S} \\ \square & \text{otherwise} \end{cases}$$

We extend the definition to $pattern(a,S)$ and $pattern(e,S)$ for an action a, event e and a set of terms S in the obvious manner. Note that $pattern(a,S)$ is an action pattern and $pattern(e,S)$ is an event pattern. For $\xi = e_1 \cdots e_n$, we define $pattern(\xi,S)$ to be the sequence $pattern(e_1,S) \cdots pattern(e_n,S)$.

Definition 13.1. An agent A's *view* of a run ξ, denoted $\xi \lceil A$, is defined as $pattern(\xi',S)$, where ξ' is the subsequence of all A-events of ξ, and $S = infstate(\xi)$. For two runs ξ and ξ' of *Pr* and an agent A, we define ξ and ξ' to be A-equivalent (in symbols $\xi \sim_A \xi'$) iff $\xi \lceil A = \xi' \lceil A$. For $\xi = e_1 \cdots e_n$, $\xi' = e'_1 \cdots e'_{n'}$, $i \leq n$, and $i' \leq n'$, we say that $(\xi,i) \sim_A (\xi',i')$ when $e_1 \cdots e_i \sim_A e'_1 \cdots e'_{i'}$.

The next issue in giving the semantics of $\mathcal{L}$ is how to handle the logical variables. This is standard. Along with the protocol we have an assignment $\mathbf{a} : X \to \mathcal{N}$. A T-assignment (for $T \subseteq \mathcal{N}$) is one which maps logical variables only to nonces in T.

The semantics of formulas in such logics of knowledge is typically given at *points*: $(\xi,i) \models_{\mathbf{a}} \alpha$ [10]. However, to emphasize the fact that knowledge semantics crucially depends on the set of runs being considered, we present it as $\mathcal{R}, (\xi,i) \models_{\mathbf{a}} \alpha$ below. As we vary the set $\mathcal{R}$, for instance to consider a subset, what is common knowledge to agents changes, and hence the semantics as well.

Fix a protocol *Pr* and an assignment **a**. For any subset $\mathcal{R}$ of $\mathcal{R}(Pr)$, we define the satisfaction relation $\mathcal{R}, (\xi,i) \models_{\mathbf{a}} \alpha$ as follows, where $\xi \in \mathcal{R}$, $i \leq |\xi|$, and α is a formula:

- $\mathcal{R}, (\xi,i) \models_{\mathbf{a}} A\ has\ x$ iff $\mathbf{a}(x) \in \overline{s_A}$ for $s = infstate((\xi,i))$.
- $\mathcal{R}, (\xi,i) \models_{\mathbf{a}} sent(A,B,x)$ iff $act(\xi(i)) = A!B{:}t$ for some t such that $\mathbf{a}(x)$ occurs as a non-key subterm of t.
- $\mathcal{R}, (\xi,i) \models_{\mathbf{a}} received(A,B,x)$ iff $act(\xi(i)) = A?B{:}t$ for some t such that $\mathbf{a}(x)$ occurs as a non-key subterm of t.
- $\mathcal{R}, (\xi,i) \models_{\mathbf{a}} \mathbf{G}\alpha$ iff for all i such that $i \leq i' \leq |\xi|$, it is the case that $\mathcal{R}, (\xi,i') \models_{\mathbf{a}} \alpha$.
- $\mathcal{R}, (\xi,i) \models_{\mathbf{a}} \mathbf{H}\alpha$ iff for all i' such that $1 \leq i' \leq i$, it is the case that $\mathcal{R}, (\xi,i') \models_{\mathbf{a}} \alpha$.
- $\mathcal{R}, (\xi,i) \models_{\mathbf{a}} \mathbf{K}_A\alpha$ iff for all $\xi' \in \mathcal{R}$ and $i' \leq |\xi'|$ such that $(\xi,i) \sim_A (\xi',i')$, it is the case that $\mathcal{R}, (\xi',i') \models_{\mathbf{a}} \alpha$.

For a protocol Pr and a formula α, we say that α is **valid** over Pr iff for all assignments **a**, and all runs (ξ, i) of Pr, $\mathcal{R}(Pr), (\xi, i) \models_{\mathbf{a}} \alpha$.

For a protocol Pr, formula α, and a fixed $T \subseteq \mathcal{T}_0$, we say that α is T**-valid** over Pr iff for all T-assignments **a**, and all T-runs (ξ, i) of Pr, $\mathcal{R}_T(Pr), (\xi, i) \models_{\mathbf{a}} \alpha$.

For a protocol Pr, formula α, and $b \geq |T|$, we say that α is b**-valid** over Pr iff for all assignments **a**, and all b-runs (ξ, i) of Pr, $\mathcal{R}_b(Pr), (\xi, i) \models_{\mathbf{a}} \alpha$.

Note that we have defined the semantics of formulas not over arbitrary sequences, but relative to the runs of a protocol. This is because we are not studying abstract notions of consistency in the logic but the more concrete questions of attacks on security protocols. If a formula is satisfiable as a sequence of information states which cannot be obtained as an admissible run of a protocol, such a property is deemed to be uninteresting. Since the logic itself has no access to encrypted terms and hence cannot describe protocols, it cannot constrain satisfiability to range over such admissible runs either.

13.4 Decidability

The technical problem we now consider is the *verification problem* for our logic. This asks for a given protocol Pr and a given formula α whether $Pr \models \alpha$. The first thing to note is that this problem is undecidable in general.

Theorem 13.1. *The verification problem for $\mathcal{L}$ is undecidable.*

Undecidability in the context of unboundedly long messages (but boundedly many nonces) was shown by [9] and for unboundedly many nonces (but bounded message length) by [7], by different techniques. Chapter 3 of [19] presents a uniform framework in which such undecidablity results can be seen.

We therefore look at restrictions of the problem and try to obtain decidability. A natural restriction is to confine our interest only to T-runs, for a fixed finite set $T \subseteq \mathcal{T}_0$. The following assertion shows that this is of no help.

Theorem 13.2. *For a given Pr and α, and for a fixed finite set $T \subseteq \mathcal{T}_0$, checking whether all T-runs of Pr satisfy α is undecidable.*

The proof is based on the same kind of Turing machine codings used in the proof of Theorem 13.1. The point is that even though we evaluate α only on T-runs of Pr, the knowledge modalities range over *all runs* of Pr, not just over all T-runs. A "simple" formula like $\mathbf{L}_A$(*I has m*) forces one to consider runs (including those in which A does not play a crucial role) in which I tries to obtain m through complicated means, and which might involve the honest agents generating lots of new nonces. The trouble is that the "simple" atomic formula *I has m* packs a lot of expressive power, and allows a trivial coding of secrecy.

In this context, it is reasonable to try to obtain decidability by restricting the semantics of protocols (since even very weak logics are undecidable). One approach would be to ask if α is T-valid over Pr, given Pr and α. But that would amount to common knowledge of T, and goes against the grain of security theory: after all, agents are assumed to have the ability to generate random nonces about which others know nothing!

We follow an approach that is standard in security analysis. We place an upper bound on the number of nonces that can be used in any run. This also implies a bound on the number of concurrent multi-sessions an agent can participate in. More formally, given $b > 0$, we ask if it is the case that α is b-valid over Pr.

Before we proceed with our specific decidability result, we make two preliminary observations.

Proposition 13.2. *Given a protocol Pr and a formula α, there exists a **finite** set of assignments $\mathcal{A}$ such that $Pr \models \alpha$ iff for all runs (ξ, i) of Pr and all assignments $\mathbf{a}$ from $\mathcal{A}$, $(\xi, i) \models_{\mathbf{a}} \alpha$.*

Proof. Fix an enumeration $x_0, x_1, \dots$ of $\mathcal{X}$ and an enumeration $n_0, n_1, \dots$ of $\mathcal{N}$. Let (without loss of generality) $T_0 = \{n_0, \dots, n_{K-1}\}$ be the *constants* of Pr and α (the constants of α are just the nonces occurring in α). Let $x_K, \dots, x_{L-1}$ be the variables occurring in α.

For every $\mathbf{a} : \mathcal{X} \to \mathcal{N}$, consider $\mathbf{a}^{-1} : \mathcal{N} \to \mathcal{N} \cup \{x_K, \dots, x_{L-1}\}$ defined as follows:

$$\mathbf{a}^{-1}(n) = \begin{cases} n & \text{if } n \in T_0 \text{ or } n \notin \mathbf{a}(\{x_K, \dots, x_{L-1}\}) \\ x_i & \text{if } i \text{ is the least such that } a(x_i) = n \end{cases}$$

Let $\widehat{\mathbf{a}} : \{x_K, \dots, x_{L-1}\} \to \{n_K, \dots, n_{L-1}\}$ be given by:

$$\widehat{\mathbf{a}}(x_i) = \begin{cases} n & \text{if } \mathbf{a}(x_i) = n \in T_0 \\ n_i & \text{if } \mathbf{a}(x_i) \notin T_0 \end{cases}$$

For every assignment $\mathbf{a}$, let $\widetilde{\mathbf{a}} : \mathcal{N} \to \mathcal{N}$ be given by:

$$\widetilde{\mathbf{a}}(n) = \begin{cases} \mathbf{a}^{-1}(n) & \text{if } \mathbf{a}^{-1}(n) \in \mathcal{N} \\ \widehat{\mathbf{a}}(\mathbf{a}^{-1}(n)) & \text{otherwise} \end{cases}$$

It is now straightforward to prove that for all runs (ξ, i) of Pr, all assignments $\mathbf{a}$, and all subformulas β of α:

$$(\xi, i) \models_{\mathbf{a}} \beta \text{ iff } (\widetilde{\mathbf{a}}(\xi), i) \models_{\widehat{\mathbf{a}}} \beta$$

and the proposition immediately follows.

On the strength of the above theorem, in what follows we only consider the problem of checking whether a particular set of runs of a protocol satisfy a formula under a fixed assignment (which we shall not mention explicitly).

For a protocol Pr and a set of agents Ag', call ξ an Ag'-run of Pr if only agents from Ag' occur in ξ. The following proposition is easy to prove.

Proposition 13.3. *Given a protocol Pr and a formula α, there is a finite set of agents Ag' such that $Pr \models \alpha$ iff $(\xi, 0) \models \alpha$ for all Ag'-runs ξ of Pr.*

Theorem 13.3. *Fix a bound b. The problem of checking for a given protocol Pr and a formula α of $\mathcal{L}$ whether α is b-valid over Pr is decidable in time $2^{(n \cdot b \cdot d)^{O(1)}}$, where n is the size of the protocol specification, and d is the modal depth of α.*

Fix a bound b for the rest of the section. Also fix a protocol Pr and a formula α. We want to check whether α is b-valid over Pr.

We let T be the set consisting of the constants of Pr, and the nonces occurring in α. For ease of notation, we refer to $\mathcal{R}_b(Pr)$ by $\mathcal{R}$ for the rest of the section. Assume that the modal depth of the formula is d. This means that it is enough to consider "chains" of runs of length at most d to determine the truth of the given formula on any run. Thus if we find an *finite* set of runs $\mathcal{R}'$ that is d-bisimilar to $\mathcal{R}$, it will turn out to be enough to check the truth of α over $\mathcal{R}'$, in order to verify α over $\mathcal{R}$.

Towards this, we define a set of *nonce patterns* $\{\Box_{i,j} \mid i \le d, j \le b\}$, and define for each $i \le d$ the set $\mathcal{P}_i = \{\Box_{i',j} \mid i' \le i, j \le b\}$. We denote by $\mathcal{R}_i (i \le d)$ the set of runs which only use nonces from T and patterns from $\mathcal{P}_i$ (in place of nonces). A run belonging to $\mathcal{R}_i$ is also referred to as an i-run.

For $i \le d$, a *zap function of rank i* is a *one-to-one, partial map* $\mu : \mathcal{N} \to T \cup \mathcal{P}_i$ such that for all $n \in T$, $\mu(n) = n$. We say that a zap function μ is suitable for a b-run ξ of the protocol if it is defined for all nonces occurring in ξ. For two b-runs ξ and ξ' of Pr, we say that $\xi \approx_i \xi'$ iff there exist zap functions μ, μ' of rank i suitable for ξ and ξ', respectively, such that $\mu(\xi) = \mu'(\xi')$. It is clear that $\approx_i$ is an equivalence relation for all $i \le d$.

The heart of the proof is the following lemma:

Lemma 13.1. *For all $i < d$, b-runs ξ_1 and ξ_2 such that $\xi_1 \approx_i \xi_2$, the following holds:*

for all agents A, all b-runs ξ_1', and all positions $\ell \le |\xi_1|$, $\ell' \le |\xi_1'|$ such that $(\xi_1, \ell) \sim_A (\xi_1', \ell')$, there exists a b-run ξ_2' such that $\xi_1' \approx_{i+1} \xi_2'$ and $(\xi_2, \ell) \sim_A (\xi_2', \ell')$.

Proof. Let μ_1, μ_2 be appropriate zap functions of rank i such that $\mu_1(\xi_1) = \mu_2(\xi_2)$. Let $\mathcal{N}'$ be the nonces occurring in $\xi_1 \lceil A$. Define ξ_2' to be $\tau(\xi_1')$ where τ is defined as follows (assuming a set of "fresh nonces" $n_1, \ldots, n_b$):

$$\tau(n) = \begin{cases} \text{undefined} & \text{if } n \text{ does not occur in } \xi_1' \\ \mu_2^{-1}(\mu_1(n)) & \text{if } n \in \mathcal{N}' \\ n_i & \text{if } n \text{ is the } i^{\text{th}} \text{ nonce occurring in } \xi_1' \text{ and not in } \mathcal{N}' \end{cases}$$

It is clear that we can find zap functions μ_1' and μ_2' of rank $i+1$ such that $\mu_1'(\xi_1') = \mu_2'(\xi_2')$. (They mimic μ_1 respectively on $\mathcal{N}'$ and map the other nonces (in order of occurrence) to $\Box_{i+1,0}, \Box_{i+1,1}, \ldots, \Box_{i+1,b}$).

Now we show that $(\xi_2, \ell) \sim_A (\xi_2', \ell')$. Note that $\xi_2 = \mu_2^{-1}(\mu_1(\xi_1))$ and $\xi_2' = (\mu_2')^{-1}(\mu_1'(\xi_1'))$. Also $\xi_1 \lceil A = \xi_1' \lceil A$. But notice that on $\mathcal{N}'$, μ_2' and μ_1' agree with μ_1, and therefore $\xi_2' \lceil A = \mu_1^{-1}(\mu_1(\xi_1' \lceil A)) = \xi_1' \lceil A$.

From this, it is a standard argument to prove the following lemma.

Lemma 13.2. *For any $i \leq d$, any formula β of modal depth $d-i$, and any two runs ξ and ξ' such that $\xi \approx_i \xi'$, and all $\ell \leq |\xi|$, $(\xi,\ell) \models \beta$ iff $(\xi,\ell) \models \beta$.*

Theorem 13.3 now follows from the fact that the verification problem reduces to verifying the truth of α over a finite set $\mathcal{R}'$ of runs of *Pr* (got from $\mathcal{R}_d$ by using a "fresh nonce" $n_{x,y}$ in place of each $\Box_{i,j} \in \mathcal{P}_d$). A naive decision procedure is to generate the set of all such runs and check whether the given formula is true. Proposition 13.1 is used crucially both to check the truth of atomic formulas of the form *A has x*, and to check admissibility conditions on actions sequences. We get the bound on $\mathcal{R}'$ as follows: each event of a run in $\mathcal{R}'$ is specified by an action a in the protocol specification, and nonces of the form $n_{i,j}$ that instantiate the ones in a. Thus the set of events we need to consider is of size $B = n \cdot (b \cdot d)^{O(1)}$. Now a run in $\mathcal{R}'$ is determined by a subset of the events and an ordering on them. Thus the number of runs is $(B+1)^B$. Thus we get the bound specified in the theorem.

13.5 Discussion

The central idea of the decision procedure above is to *lift* the decidability of the question $T \vdash t$ to that of the logic. This suggests that we can extend the result to protocols whose message terms over richer cryptographic primitives, as long as the question $T \vdash t$ remains decidable. We could include algebraic properties of the encryption operators, or extend the term algebra with primitives such as *blind signatures*, and yet achieve such decidability. In [2], we show this in the case of blind pairing.

The use of the inference system on terms suggests that knowledge of agents reduces to *provability* in the weaker system. However, there are many cryptographic contexts such as *zero knowledge proofs* where agents can verify that a term has a particular structure, without being able to construct it. Such a consideration takes us beyond Dolev-Yao models.

Halpern and Pucella [12] make an attempt to go beyond the Dolev Yao model, by also considering probabilistic notions and the intruder attempting to guess nonces and keys. They employ the idea of modelling adversary capabilities by restrictions on the algorithms used by adversaries; this is represented in our set-up by indexing the message derivation system: $T \vdash_A t$ describes the algorithm employed by A. This approach is explored in [17]. But our emphasis has been on decidability, and it will be interesting to explore decision questions in rich logics like the one in [12].

Another consideration, largely ignored in this paper, relates to how security protocols may be implemented in a manner consistent with knowledge specifications [21]. This line of work is important, but what we have attempted to argue here is that such considerations lead us not only to interesting security theory, but also new theories of knowledge.

References

1. R. Anderson and R.M. Needham. Programming Satan's computer. In *Computer Science Today*, volume 1000 of *Lecture Notes in Computer Science*, pages 426–441. Springer, Berlin, Germany, 1995.
2. A. Baskar, R. Ramanujam, and S.P. Suresh. Knowledge-based modelling of voting protocols. In D. Samet, editor, *Proceedings of the 11th Conference on Theoretical Aspects of Rationality and Knowledge*, pages 62–71. Brussels, Belgium, June 25–27, 2007.
3. M. Burrows, M. Abadi, and R.M. Needham. A logic of authentication. *ACM Transactions on Computer Systems*, 8(1):18–36, February 1990.
4. M. Cohen and M. Dam. A completeness result for BAN logic. In *2005 International Workshop on Methods for Modalities (M4M-05)*, pages 202–219. Berlin, Germany, December 1–2, 2005.
5. M. Cohen and M. Dam. A complete axiomatization of knowledge and cryptography. In *22nd IEEE Symposium on Logic in Computer Science (LICS 2007)*, pages 77–88. IEEE Computer Society, Wroclaw, Poland, 2007.
6. D. Dolev and A. Yao. On the security of public-key protocols. *IEEE Transactions on Information Theory*, 29:198–208, 1983.
7. N.A. Durgin, P.D. Lincoln, J.C. Mitchell, and A. Scedrov. The undecidability of bounded security protocols. In *Proceedings of the Workshop on Formal Methods and Security Protocols (FMSP'99)*. Trento, Italy, July 11–12, 1999.
8. C. Dwork and Y. Moses. Knowledge and common knowledge in a byzantine environment: Crash Failures. *Information and Computation*, 88(2):156–186, 1990.
9. S. Even and O. Goldreich. *On the Security of Multi-party Ping-pong Protocols*. Technical Report 285, Technion—Israel Institute of Technology, 1983.
10. R. Fagin, J.Y. Halpern, Y. Moses, and M.Y. Vardi. *Reasoning about Knowledge*. M.I.T. Press, Cambridge, MA, 1995.
11. J.Y. Halpern and Y. Moses. Knowledge and common knowledge in a distributed environment. *Journal of the ACM*, 3(3):549–587, 1990.
12. J.Y. Halpern and R. Pucella. Modeling adversaries in a logic for security protocol analysis. In *Formal Aspects of Security, First International Conference, FASec 2002*, volume 2629 of *Lecture Notes in Computer Science*, pages 115–132. Springer, Berlin, Germany, 2003.
13. N. Lynch. *Distributed Algorithms*. Morgan Kaufmann Publishers, San Francisco, CA, 1996.
14. D.M. Nessett. A critique of the Burrows, Abadi and Needham logic. *ACM Operating systems review*, 24(2):35–38, 1990.
15. R. Parikh. Logical omniscience and common knowledge: WHAT do we know and what do WE know? In *Proceedings of TARK 2005*, pages 62–77. Singapore, June 10–12, 2005.
16. R. Ramanujam and S.P. Suresh. Decidability of context-explicit security protocols. *Journal of Computer Security*, 13(1):135–165, 2005.
17. R. Ramanujam and S.P. Suresh. A (restricted) quantifier elimination for security protocols. *Theoretical Computer Science*, 367:228–256, 2006.
18. M. Rusinowitch and M. Turuani. Protocol insecurity with finite number of sessions and composed keys is NP-complete. *Theoretical Computer Science*, 299:451–475, 2003.
19. S.P. Suresh. *Foundations of Security Protocol Analysis*. PhD thesis, The Institute of Mathematical Sciences, Chennai, India, November 2003. Madras University. Available at `http://www.cmi.ac.in/~spsuresh`
20. P.F. Syverson and P.C. van Oorschot. On unifying some cryptographic protocol logics. In *Proceedings of the 13th IEEE Symposium on security and privacy*, pages 14–28. IEEE Press, Oakland, California, May 16–18, 1994.
21. R. van der Meyden and T. Wilke. Preservation of epistemic properties in security protocol implementations. In D. Samet, editor, *Proceedings of TARK '07*, pages 212–221. Brussels, Belgium, June 25–27, 2007.

MIX
Papier aus verantwortungsvollen Quellen
Paper from responsible sources
FSC® C105338

If you have any concerns about our products,
you can contact us on
ProductSafety@springernature.com

In case Publisher is established outside the EU,
the EU authorized representative is:
Springer Nature Customer Service Center GmbH
Europaplatz 3, 69115 Heidelberg, Germany

Printed by Libri Plureos GmbH
in Hamburg, Germany